W9-DJL-426

© Jerry Bauer

## *About the Authors*

LEWIS R. GOLDFRANK, M.D. (right), has been director of Emergency Medical Services at Bellevue Hospital since 1979 and is medical director of the New York City Poison Center.

EDWARD ZIEGLER (left) is a former editor at *Reader's Digest.*

# EMERGENCY DOCTOR

# EMERGENCY DOCTOR

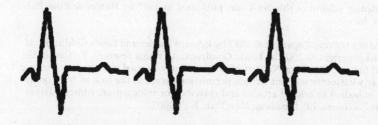

## Edward Ziegler

in Cooperation with

## Lewis R. Goldfrank, M.D.

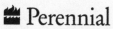 Perennial

*An Imprint of HarperCollinsPublishers*

# HARPER

A hardcover edition of this book was published in 1987 by Harper & Row, Publishers, Inc.

First Perennial edition published 2004.

*Designed by C. Linda Dingler*

Library of Congress Cataloging-in-Publication Data
Ziegler, Edward.
    Emergency doctor / Edward Ziegler, in cooperation with Lewis R. Goldfrank.— 1st Perennial ed.
      p. cm.
    Originally published: New York : Harper & Row, c1987.
    ISBN 0-06-059502-7
    1. Goldfrank, Lewis R., 1941– . 2. Emergency physicians—New York (State)—New York—Biography. 3. Emergency medicine—Case studies. 4. Bellevue Hospital. I. Goldfrank, Lewis R., 1941– . II. Title.
    [DNLM: 1. Goldfrank, Lewis R., 1941– . 2. Bellevue Hospital. 3. Emergency Medicine—Personal Narratives. 4. Emergency Service, Hospital—Personal Narratives. 5. Physicians—Personal Narratives. WZ 100 G6184Z 1987a]
RC86.7.G63Z54 2004
362.18'09747'1—dc22
                                 2004044347

17 18 RRD 20 19 18 17 16 15

This book is dedicated to my daughter

*Sally Warren Ziegler*

# Contents

Acknowledgments ix

Staff of Bellevue Hospital and Associated
Institutions in Order of Appearance xi

1. An Overturned Crane 1
2. "First, Do No Harm" 8
3. Learning to Juggle 16
4. A Morning's Rounds 28
5. Doing a Decent Thing 38
6. A Shock and a Dive 48
7. A Question of Poison 59
8. The Making of a Medical Student 71
9. Becoming an Emergency Doctor 80
10. The Carriage Trade 97
11. A Midday Arrival 108
12. An Alkaloid Plague 121
13. The Case of the Crazed Executives 131

# Contents

14.  Creepie Crawlies                                          142

15.  A Blonde and a Severed Leg                                153

16.  A Lesson and a Crash                                      164

17.  Human Warmth and a Drink of Gasoline                      176

18.  A Leap and an Inspection                                  187

19.  A Doctor on Call                                          197

20.  Notes on a Juggernaut                                     208

21.  About Children                                            220

22.  An Aneurysm and a Dangerous Diet                          228

23.  About AIDS                                                239

24.  A Peanut and a Mercury Injection                          255

25.  Moulage Day                                               265

26.  Knocking Back a Few                                       279

27.  Pokeweed, Headaches and Tribal Medicine                   293

28.  No Place to Go                                            308

29.  A Day and a Night to Remember                             322

30.  Toward Medical Victory                                    341

     Afterword                                                 353

# Acknowledgments

Special thanks to the staff and patients of Bellevue Hospital for the help they have rendered in making this account of daily life at that institution possible. A word of gratitude, also, for the memory of all those who, since 1736, have sustained a tradition that is unsurpassed in the history of American hospitals. Here everyone in need will be welcomed and cared for.

Grateful note should also be made of the efforts of Hugh Van Dusen of Harper & Row, who conceived of this book in the first place and has warmly encouraged its development from its inception.

A final word of thankful appreciation to my wife, Sally McIntosh Ziegler, for her unerring critical judgment and help in the creation of this work.

# Staff of Bellevue Hospital and Associated Institutions
in Order of Appearance

LEWIS R. GOLDFRANK, M.D., Director of Emergency Department (ED)

KATHLEEN A. DELANEY, M.D., Attending Physician, ED

NEAL E. FLOMENBAUM, M.D., Associate Director, ED

DAVID LIVINGSTON, M.D., Chief Resident, Department of Surgery

ROBERT A. HESSLER, Ph.D., M.D., Attending Physician, ED

NEAL A. LEWIN, M.D., Assistant Director, ED

BRENDA A. SMART, M.B., Ch.B. (Sheffield, England), Fellow, ED

ELIZABETH REYNOLDS, R.N., ED

SHARDA McGUIRE, R.N., B.S.N., M.S.N., Assistant Head Nurse, ED

HEDVA SHAMIR, M.D., Attending Physician, ED

LEON FIELDS, Nurse's Aide, ED

TERESA HAINES, B.S., M.A., FNP, Nurse Practitioner, ED

MARY DWYER, R.N., B.S.N., M.P.H., Head Nurse, Walk-in Clinic, ED

JOHN R. SISAS, New York Police Department, ED Liaison

ANA TORRES, Office Aide, ED

JOHN DUVAL, EMT-P, New York City Emergency Medical Services

MOETAHAR PADELLAN, EMT-P, New York City Emergency Medical Services

DANA GAGE, M.D., Attending Physician, ED; Medical Director, Bellevue Shelter Clinic

RICHARD WEISMAN, Pharm.D., Director, New York City Poison Control Center (NYCPCC), New York City Department of Health

ROBERTO BELLINI, M.D., Fellow in Clinical Toxicology, NYCPCC–Bellevue ED

MARY ANN HOWLAND, Pharm.D., Professor of Clinical Pharmacy, St. John's University College of Pharmacy; Consultant, NYCPCC–Bellevue ED

ELISABETH K. WEBER, R.N., Assistant Head Nurse, ED

# EMERGENCY DOCTOR

EMERGENCY DOCTOR

All the doctors, nurses, aides and other staff members mentioned in this work are identified by their actual names, and are described in typical daily activities. The patients' names and distinguishing characteristics, with the exception of Brigitte Gerney's—whose widely publicized accident is described in the opening chapters—have been changed.

Chronology, descriptions and actual circumstances have been rearranged in some instances. Dialogue is true to the spirit of all cases, although not always verbatim. All cases, however, are drawn directly from the life and work of Dr. Goldfrank and his staff and from daily events at Bellevue Hospital. These materials have been reviewed by Goldfrank and his colleagues to assure medical and scientific accuracy.

# 1 An Overturned Crane

It was almost noon on a bright spring day when the thirty-five-ton crane began to unload steel rods from a flat-bed truck idling at curbside. The site was that of a projected forty-two-story luxury condominium on Third Avenue between Sixty-third and Sixty-fourth streets in Manhattan.

As the extended arm of the machine began to lift a load of rods and swing them over toward the construction excavation, lunchtime pedestrians scurried uncomfortably along a sidewalk bordered by a plywood barrier, behind which was the pit from which the building would rise. Two women brushed past a third. This third woman was virtually in the shadow of the machine as its loaded arm began to counterbalance the base of the crane, causing it to tip up on two of its giant wheels.

For an instant, the machine wavered, angled oddly toward the excavation. The first two women sensed something amiss and rushed to get away. The third, Brigitte Gerney, was not so fortunate. She tried to retrace her steps toward Sixty-fourth Street.

"Get out of the way! We're going over!" a voice close to her shouted. It was the twenty-nine-year-old construction worker who was at the controls of the heavy machine. Below him, the sidewalk, which had been undermined by the excavation of the past few days, began to cave in. Then, amid the splintering of wood and the screech of bending metal, the crane turned over, trapping Mrs. Gerney suddenly underneath and coming to rest upside down at the edge of the abyss. Only a thin course of lightly framed plywood resting atop an

I-beam seemed to be keeping the crane from plunging with its victim some thirty feet into the construction pit below.

"It was like an earthquake," Mrs. Gerney later recalled. "I remember my bag flying out of my hands. I heard the noise of all the bones cracking in my legs."

For the forty-nine-year-old Manhattan mother of two, it was the beginning of a seemingly endless ordeal. Both of her legs were pinned at a point ten inches above the knees between an edge of the giant machine and the crumbled sidewalk. As the crane came to rest, her right leg seemed to be almost severed, her left severely crushed. "I felt the warm blood going out and I had the impression that my legs were completely cut off," she would remember months later. After a moment of shocked disbelief, she cried out: "Get this off me!" Her cries set in motion one of the most intensely observed rescue attempts New York had experienced in years.

The fate of Mrs. Gerney would depend in some significant degree on a doctor who as yet had no awareness of her predicament. Some forty blocks downtown, at Bellevue Hospital, on First Avenue and Twenty-sixth Street, a phone rang in the crowded doctors' station just behind the triage* desk, at the principal entry into the Emergency Department. It is here that emergency medical specialists make their initial judgments, separating all cases into one of three categories: "emergent" (gunshot wounds, heart attacks), "urgent" (broken bones, minor cuts) and "non-urgent" (sore throats, rashes).

The attending physician on duty immediately acknowledged the message and made his triage decision, "emergent," and set in motion the machinery to dispatch a physician in the mobile emergency rescue van (MERVan) to the scene. Whenever the Bellevue MERVan rolls, the first person to be apprised is Dr. Lewis Goldfrank, Director of the Emergency Department (ED) at Bellevue.

Goldfrank is a tall and vigorous man in his mid-forties who had been striding swiftly toward one of the trauma slots to look in on a seven-year-old girl who had been hit by a taxi and had just been brought in. As her stretcher was wheeled by, he could tell at a glance

---

*The word has been in the language since at least 1727 and has always meant "to sort things out according to quality." According to the *Oxford English Dictionary*, it was first used in reference to sorting wool (1727), then coffee beans (1825) and finally (1930) wounded people.

that the girl's color was good but her breathing was labored—sounding like the muted squealing associated with airway compromise. He swiftly assigned a team of physicians and nurses to the little girl's care, then paused momentarily to hear the specifics about the crane.

At least one person was trapped. Two others had just been freed, with superficial injuries. One of the city's Emergency Medical Services ambulances was on its way to the scene and police paramedics were even then burrowing their way toward Mrs. Gerney.

Goldfrank sought out the senior attending physician on duty. That was Kathleen Delaney, a Los Angeles native who, after progressing toward a Ph.D. in biochemistry at UCLA, had applied instead to medical school, and enrolled at Columbia University's College of Physicians and Surgeons. Now, nine years later, she was board certified in internal medicine and was just completing her second year of full-time service at Bellevue's Emergency Department.

"Kathy? Can you go with the MERVan to that crane accident?"

"Sure. Where?"

The attending physician at the doctors' station gave the particulars and sent word to the parking area, tucked into a courtyard behind a neighboring building, where the MERVan was parked, ready to go at a moment's notice. This kind of van, a small replica of an emergency room, extends the walls of Bellevue Hospital's Emergency Department into the streets. Two or three times a week, the MERVan is called on to augment the service that Emergency Medical Services (EMS) ambulances normally provide.

So successful has been the experience with such emergency medical services that the familiar term "DOA"—dead on arrival—has become something of a rarity on the Bellevue Emergency Department charts. Dr. Delaney ran out the sliding doors of the ambulance entrance and stepped into the MERVan, which then made a wide swing around the hospital drive and went howling off up First Avenue.

Goldfrank continued with his tasks—overseeing the work of the trauma team and pediatric staff working on the little girl. Pneumothorax. Collapsed lung. Possible injuries to the internal organs. But vital signs adequate and stable. Skilled hands were at work and there was no need for him to intervene. He looked in on the neighboring room. Here a passenger who had taken a tumble from a subway platform was being stitched up. Goldfrank looked at the chart: "AOB." Alcohol on breath. Fully three-quarters of the accidents that brought people to Bellevue's ED involved alcohol or drugs. Almost half involved both substances.

He went back past the doctors' station. It was 12:45 and Kathy Delaney was doubtless on the scene at Sixty-third Street by now. There were other reports that Goldfrank reviewed briefly at the triage desk—his preferred station for gauging the quality of care his service is delivering to the three hundred patients a day who present themselves to the Bellevue ED. Then he strode to his windowless office nearby. He took a minute to scan and approve proofs of a program for a conference on poisoning that he had organized with the toxicology team. As a ranking expert in the science of poisons, and as medical director of New York's Poison Control Center, he is considered one of the country's outstanding authorities on toxins. He has a redoubtable store of knowledge on poisons as varied as alcohol, aspirin, antidepressants, amphetamines, cocaine, nicotine, cyanide and myriad industrial and occupational agents, as well.

He also serves as the chief of both Bellevue's and New York University Hospital's emergency departments. As such, he sees more trauma and medical emergencies in a single week than many doctors see in a lifetime. From his special perspective as a toxicologist, he sometimes finds himself viewing the entire city as an organism vulnerable to systemic poisons. Any number of toxins contribute materially to the accident rate in the city and to premature deaths, as well. Such common poisons—especially alcohol, tobacco, cocaine and heroin—haunt Goldfrank and his forty attending physicians, the team that he has assembled during his seven years as director of Bellevue's ED.

As the city's leading organization in handling medical emergencies, Goldfrank's group of "critical care general practitioners" has become expert in confronting the unexpected. Or, as the doctor himself has written, quoting the White Queen in *Through the Looking Glass,* "we are able to believe six impossible things before breakfast." Now, the nearly unbelievable had happened yet again, and healing forces were mobilizing themselves to try to deal with it.

Even before the Bellevue MERVan arrived, New York Police Department Emergency Service units began to shore up the plywood under Mrs. Gerney. Paramedics, threading their way through the jumble of fractured plywood and twisted metal, did what they could to make her excruciating pain more bearable. Kathy Delaney would administer morphine when she got there. Ice was also passed down and was packed around the victim's wounds, banked up against the shattered wooden wall, behind which the crane's hard metal bore down with such relentless pressure on her thighs.

Meanwhile, a stoical Mrs. Gerney remained calm. Months later she told a jury, "I touched the crane. It was cold metal right next to me. I said, 'Can't you cut my legs off and take me out? I have two children. I have to live.'" Perhaps inspired by her own words, there was soon a widespread conviction on the scene that the best way to extricate the victim might well be by amputation. There was a general edginess among the onlookers, made worse by the obviously precarious angle at which the overturned crane sat.

Despite the growing tension, something of the relative calm of a Sunday spread through the streets as the sounds of traffic receded. Police closed off successively larger segments of the East Side to motor vehicles, to assure unobstructed routes for the rescue cranes that were on their way, and for a swift departure to Bellevue when and if Mrs. Gerney was freed. Everyone knew that Bellevue's microsurgery team had had amazing success in restoring severed hands and limbs, so long as the victim could get timely attention at the hospital. Awareness of this "miracle microsurgery team" on the one hand and impatience with the intractable nature of the overturned crane, on the other, began to unsettle the mood of the onlooking throng.

By now, some two thousand New Yorkers—doctors, reporters, paramedics, police, construction workers, public officials and others—were focusing all of their efforts on saving Brigitte Gerney's life.

There seemed to be a growing belief among police, fire and city Buildings Commission experts that the cranes that had been called to lift the upended machine would not be big enough. A search began for a still-larger crane.

Downtown, in the Bellevue Emergency Department, an acute myocardial infarction was striking a seventy-year-old merchant, whose wife had brought him in by taxi just before 1 P.M.

As Goldfrank studied the cardiogram with a medical resident, he could see evidence of a massive attack. As an educator as well as a supervisor of this sizable ED, Lewis Goldfrank advises frequently, observes constantly and occasionally acts as the primary physician himself. Today, he found himself in the role of educator-supervisor. The resident who had been assigned to the ailing merchant set up an IV and administered lidocaine in an attempt to prevent any ventricular irritability, irregularities of heart rhythm. As Goldfrank turned again to scan the tapes coming from the cardiac monitor, a nursing supervisor bent close to whisper into his ear. The telemetry center (which is a part of the doctors' station) had received an insistent message from the paramedic commander at the scene of the crane accident: surgeons were urgently wanted.

Goldfrank, satisfied that the old man was stabilized, and that his staff was doing essentially what he would have been doing had this been his own patient, went back to the doctors' station, where the telemetry unit and mobile phone base were located. Here he could study an electrocardiogram from a remote spot. And here, on more than one occasion, he had said a few words that had meant the difference between life and death to an unseen but gravely injured accident victim.

As he took another call from the EMS paramedics, it became clear that there was a crisis of communication. For years there had been a divergence of views between the paramedics and the doctors, made no less difficult by the fact that the paramedics had their own chain of command separate and sometimes independent of the emergency and in-hospital doctors who would have responsibility for the ultimate outcome of a case like Mrs. Gerney's. Given a thousand hours of health-care training, paramedics are able to do many things to save a victim of injury or sudden illness. They give artificial respiration, administer oxygen and/or adrenaline (epinephrine), apply splints or body boards and even intubate—insert an endotracheal tube into the throat of an unbreathing accident victim. But there are obviously areas into which they are not qualified to go, judgments that they are not equipped to render. These limitations define the critical role of the physician. Nevertheless, there is something of a swashbuckling esprit among the EMS personnel—something of a cocksureness that occasionally puts them directly at odds with more highly trained medical opinion.

Now, as the only communications came through the phone in a dress shop across the street from the accident, Goldfrank was soon aware that, without a direct link to Delaney, he had to rely on second- and third-hand reports, relayed via paramedics or emergency medical technicians, on information that did not come from Delaney at all. (The emergency medical technician [EMT] is one educational level lower than the paramedic—having had one hundred to two hundred hours of training.) In all cases, those interpreting the data and conveying the messages to Bellevue were paramedics or EMTs. Still, there seemed to be enough agreement among the various reports reaching

Goldfrank to indicate that Mrs. Gerney was in stable condition, with no indications that would force any precipitate action.

Among those on the scene there was a virtual consensus that amputation was the preferred course. As Goldfrank later recalled: "I told them on the telephone that we'd send a surgeon up there, but that field amputation had not been practiced in New York City since the Civil War era and we're not going to start now."

Sensing that his own presence was becoming virtually obligatory, Goldfrank turned over direction of the ED to his principal deputy, Neal Flomenbaum, and, with the acting director of the hospital and Bellevue's chief surgical resident, he went to the accident scene himself. A police escort whisked them north on an eerily silent East Side Drive. The authorities had ordered this major artery of Manhattan closed to through traffic from Thirtieth to Sixty-fifth Street to assure the speedy removal of Mrs. Gerney, once she was free.

# 2 "First, Do No Harm"

The Bellevue group was at the scene within ten minutes. "By that time," Goldfrank recalled later, "every television camera in the city was there. The afternoon was hot and sunny and there were thousands of people in the vicinity, including many political leaders." Among those leading citizens were the Mayor, Edward I. Koch, and Goldfrank's immediate boss, the president of the Health and Hospitals Corporation of the City of New York, lawyer John J. McLaughlin. They both had gone to college in Worcester, Massachusetts, in the sixties—McLaughlin at Holy Cross, Goldfrank at Clark University.

Goldfrank greeted McLaughlin, who accompanied him to the corner of Sixty-third and Third Avenue, where the Mayor had his command post in front of the same dress shop whose telephone was providing the link to Bellevue. Sensing Koch's wish for some clarification of the by-now frequently heard recommendation of amputation, Goldfrank told the Mayor that he was dubious about such a procedure, but he would come back with a report as soon as he, Dr. Delaney and their chief surgical resident had examined Mrs. Gerney. Then he was outfitted with a hard hat. Now he would have to slip down adjacent to Mrs. Gerney while Kathy Delaney went underneath. As he later recalled:

"I saw that the crane, as it tipped over, had smashed the plywood fence against her. The woman's legs were on one side, her body on the other and the crane was lying over this wall board on the woman.

" 'Hello, Mrs. Gerney, I'm Dr. Goldfrank from Bellevue,' I said. She was surprisingly calm, a lovely woman with an air of detachment that was tremendously impressive.

" 'Hello, doctor,' she responded, and it was clear that she was in complete command of her faculties and was evidently in a mood of calm resignation. She may even have been praying.

" 'We're just going to make a brief examination so we'll be better able to do all the right things for you just as soon as we can.'

" 'I understand,' she said.

"I couldn't determine how much direct pressure was on Mrs. Gerney. Her feet felt cold and clammy. There was certainly no motion there. And we couldn't be sure that the legs weren't partially or substantially severed.

"She was braced underneath and the paramedics had supported her with a backboard to prevent any injury to her neck and spinal cord. They also had established an intravenous line. That was all well done. Still, if the crane fell, she would certainly be killed. And I could see that there was just no suitable way even to consider amputation. Neither Delaney, David Livingston, the surgeon, nor I felt there was any justification at this point."

By the time Goldfrank got there, the two other cranes that had been called in had a line attached to the overturned machine. They had begun to exert slight upward pressure, to stabilize the crane. So, from the cramped confines of the small enclosure where he performed part of an examination, Goldfrank felt that the overturned machine was not threatening to fall any further. On that judgment—which he subsequently reviewed with various officials and experts on the scene—would depend most of what happened in the next four hours.

Two o'clock. " 'We're going back outside, Mrs. Gerney. Everybody's thoughts and hopes are with you. And it's all going to turn out for the best.'

" 'Thank you, doctor. I'm sure it will.'

"I wriggled my way out of there and began backing toward more solid earth. There was only that I-beam between us and the pit, I realized, and I hoped she wasn't thinking about that.

"I later learned that a police chaplain had administered last rites already, relaying the words through one of the police officers, who stayed near her throughout most of her ordeal."

Goldfrank would later meet that police officer. His name was Paul Ragonese, and, despite the severe cramping he suffered, he insisted on staying under the crane, to offer moral support to Mrs. Gerney. In fact, Ragonese had earned himself an enviable reputation as an officer who had repeatedly risked his own life to offer comfort and aid to others long before this latest confirmation of his unusual willingness to go to extraordinary lengths in the line of duty.

The Commissioner of Buildings for the City of New York, Charles Smith, was also on the scene. It was to him that Goldfrank found himself turning to establish what was the best chance of bringing this drama to a successful conclusion. For now, ultimate responsibility for the outcome had passed to Lewis Goldfrank.

Goldfrank and Smith discussed their options. "I knew from a medical standpoint that there should be no attempt at amputation; so our goal became to continue stabilizing Mrs. Gerney until the overturned crane could be hoisted and supported. The idea at that point was to tunnel from underneath, once the crane was secured against falling, then remove Mrs. Gerney from underneath. But they needed a better margin of safety than presently existed.

"We were joined by the heads of the police department and fire department emergency squads, who agreed with our general strategy—to play for time until a bigger crane got there.

"We got word that there was a 150-ton crane that would have to be brought into Manhattan from the South Bronx. It would require its own elaborate escort, its own sequence of blocked streets, and patience. Its cruising speed on city streets would be about four miles per hour."

They were told categorically by the Mayor, "Get the crane." To speed its entry into Manhattan the rest of the East Side Drive was blocked off southbound from the Bronx, snarling Manhattan traffic still further.

Meanwhile, forty traffic officers and five tow trucks, reinforced by scores of police detailed to the area, did their best to ameliorate the effects on traffic. Still, there was a massive traffic jam that created virtual gridlock from Thirty-fourth Street to Seventy-second Street between Fifth Avenue and the East River, and affected at least half of the densely populated island.

At Third Avenue and Sixty-third Street, a pair of motorcycle policemen—sleek in sunglasses, jodhpurs and shiny leather boots—stood by their idling machines, waiting for the moment when they would lead the way downtown to Bellevue. Reports kept circulating that as soon as the big crane arrived—a half hour or forty-five minutes longer—it would all be over. But the time kept stretching out.

Goldfrank and his team had sent one of these officers racing down to Bellevue a few minutes before with a sample of Mrs. Gerney's blood. At Bellevue's blood bank, technicians matched her blood with the donor blood of her general type available for transfusion to see if any of fifty possible antigen-antibody reactions might be provoked by giving Mrs. Gerney this new blood. They were satisfied that they had a good match, with no danger of a transfusion reaction. The motorcycle patrolman sped back uptown with a supply of this blood.

"We asked ourselves if she could stand another two hours of this," Goldfrank recalls. "Kathy Delaney got a second intravenous line established and more morphine was administered. We were also giving Mrs. Gerney oxygen. While we were waiting for the large crane I went back down and talked with her while one of the paramedics started giving her the typed and cross-matched blood that had just arrived. 'I know I'm doing okay,' she said. 'I'm more worried about you.'

" 'Thanks, Mrs. Gerney,' I said, somewhat at a loss for words. 'I think everything is going to be under control.' Her pulse and respiration looked surprisingly good. My gravest concern was now turning to the moment when we got her free of the weight. Lactic acid builds up—a natural breakdown product of the body's own glucose metabolism with limited oxygen—in the compromised part of the lower extremities. This toxin when liberated to the systemic blood supply could cause complications, such as severe acidosis and shock. We planned to enclose her extremities in medical anti-shock trousers (MAST) as soon as she was free, to support her mangled legs and to minimize both bleeding and the lactic-acid problem."

Meanwhile the media were everywhere, trying to get exclusive insight into the disaster, interviewing whoever seemed likely to provide a fresh angle. News had come that the operator of the overturned crane had been arrested for reckless endangerment. He had been operating the crane without a license. It was widely assumed that he had been doing so under instructions from the construction bosses, who wanted to get their project under way, even if all the required municipal licenses and approvals had not yet been granted. The young operator had complained of chest pains and dizziness and was taken to Lenox Hill Hospital's Emergency Service for treatment. His complaint proved not to be serious.

When Goldfrank returned from Mrs. Gerney, he joined the Mayor and other officials to meet yet again with reporters and cameramen. "I'm pretty used to it, but still the intensity of coverage that afternoon was something I hadn't experienced before," Goldfrank recalls. "I described the general situation and then came their specific questions: How's she doing? What are her chances? Are you going to amputate her extremities? I said, 'No, at this point she seems to be holding her own . . . she's doing remarkably well.'

"And the Mayor was then asked, 'How can someone operate a crane without a license? What kind of supervision are you giving on this?' And he said those responsible would be prosecuted.

"I then told the press that our entire microsurgery team, our orthopedic surgery and our trauma group were ready at Bellevue. I added that we had an experienced trauma surgeon, Dr. Livingston, our chief trauma surgical resident, at the scene; we were giving Mrs. Gerney blood; and we had all the medications we needed to care for her appropriately.

"Still, we were aware that, once the crane was lifted off, her clinical condition might dramatically deteriorate and her blood pressure might drop sharply. She could well bleed to death or suffer refractory shock.

"Each time after checking her out again we would talk with the Mayor, the buildings commissioner, the senior fire and police officials and the press when we came back to the command post. There was very little change during that long afternoon. She remained in amazingly stable condition."

At 3:55 the enormous crane arrived. It had to be stabilized and leveled, a task that dragged on for forty-five minutes. Then a hook slowly descended from the towering arm of the new machine and was gingerly secured to cables on the overturned crane. To make sure that there was no shifting as the larger crane started to lift, a second, smaller machine took up the tension from additional cables anchored on the overturned crane to keep it from slipping sideways as it was pulled upward.

"We were using the phone in the dress shop across the street to keep the hospital and the Health and Hospitals Corporation people downtown informed on our progress: making sure that the blood bank was ready with more of her typed and cross-matched blood; that the operating rooms and surgical teams were all ready; to let them know that we expected to be there in about forty-five minutes—as now, at last, the big crane was level. Various hospital cars and ambulances were waiting to rush downtown.

"The trick would be to get her gently into a scoop stretcher, which is like a wire netting, and move her out while maintaining the back support underneath her as the overturned crane was lifted off."

The engine of the giant crane strained and the cable eased upward, an inch. Another inch. A third inch. The paramedics were standing by with the scoop stretcher. Meanwhile, firemen from a rescue unit on the scene carefully removed fragments of concrete, sand and earth debris from beneath Mrs. Gerney.

At six minutes to six, she was free. During a somewhat chaotic final few moments, many hands lifted her into the special stretcher upon which lay the anti-shock trousers, which were then inflated. These MAST trousers proved their use in both the Korean and Vietnam wars in stabilizing any patient with massive blood loss in the abdomen, pelvis or lower extremities. Once inflated, they serve to maintain blood pressure by acting as a kind of coffer dam that keeps blood in the more vital central parts of the body and makes flow to the periphery more difficult.

She was pulled out, quickly placed in the back of a hospital ambulance. Kathy Delaney stayed on the scene to be available if anything had been overlooked or anyone injured in the final stages of the rescue. David Livingston plus two paramedics went downtown with Mrs. Gerney in the ambulance. "I went with one of her personal doctors, who had come to the scene, in a police car. It was only five minutes through those empty streets to Bellevue. The whole East Side had been held motionless, it seemed, for six hours so that we could make this trip at maximum speed. We took her directly to the Emergency Department trauma slot for quick x-rays and to reestablish her

IV lines, which had been pulled out when she was put into the scoop stretcher. Also, we got her cleared of all the remaining debris—small pieces of wood and masonry.

"It looked as though she did not have separation of the legs. There were pieces of free-hanging bone and comminuted fractures [bone that had been virtually pulverized], but no actual severance. And she had good pulses in her lower legs. She had what appeared to be reasonable tissue. Still, there were areas that looked as if they'd need grafting."

Just as suddenly as Brigitte Gerney had come into Lewis Goldfrank's life, she left it. Once he and his staff had stabilized her condition, had established that there was no abdominal bleeding and had prepared her for the next stage of her recovery, she was whisked away to the eleventh-floor operating room, where she was to spend four hours with a nine-person surgical team. The group included members of the microsurgery and reimplantation team of Bellevue-NYU as well as members of Bellevue's trauma unit, plus plastic surgeons from the NYU Institute of Reconstructive Plastic Surgery. Also among the specialists were vascular, trauma and orthopedic surgeons. The orthopedists won the particular admiration of Goldfrank and his Emergency Department colleagues for performing the structurally critical task—setting six-inch pins in the shattered major bones in her lower legs. That procedure restored these comminuted fibulas and tibias so that, in time, Brigitte Gerney could walk again.

Once she went upstairs, the first-floor Emergency Department shifted its attention to the next problems in critical care—which that Thursday evening included a patron of a midtown restaurant who had had a violent allergic reaction to a crabmeat salad. The young man arrived blue and apparently comatose, while a team of attending physicians, nurses and aides tried to insert a tube down his swollen throat and to get an intravenous epinephrine drip started.

"That's one of the peculiarities of this specialty," Goldfrank said later, after the high drama of the day had subsided. "You rarely hear the end of a story. You see someone in critical need of care—not breathing, cardiac arrest, traumatic injury—and you do everything you know how to do to meet the priorities of emergency medicine: ensure the airway; stop any bleeding; maintain circulation; keep the patient immobile; prevent shock and get them safely away from whatever is threatening their life. (For Mrs. Gerney, we were called on to do all six.) And it's only occasionally that you hear about the follow-up."

In the case of Brigitte Gerney, Goldfrank and others who played key roles in the rescue were recognized at special ceremonies at Gracie Mansion, the official mayoral residence, several weeks later.

And, a few weeks after that, a smiling Brigitte Gerney left Bellevue in a wheelchair, with every hope that she would ultimately resume a normal life.

It was a clear vindication of the medical principle first enunciated by Hippocrates, and which Lewis Goldfrank regards as the keystone of his profession: First of all, do no harm.

# 3

**Learning to Juggle**

When he comes through Grand Central shortly after 7 A.M. every morning, Lewis Goldfrank makes a conscious study of the numbers of pallets of corrugated cardboard he can spot—a rough index of the kind of day he'll be facing downtown at the Bellevue Emergency Department. When the weather turns bad, the homeless move into the ramps, hallways and waiting rooms of the terminal. They bring the cardboard with them.

There have been rumors that the pay luggage lockers here in the terminal are going to be removed. The official reason may well be to deny a hiding place to a terrorist's bomb. But the more urgent reason for removal will doubtless be to discourage the homeless, and all the unpleasant things their presence entails: filth, incessant coughing, offensive odors, sometimes abusive talk.

Goldfrank thinks about the homeless a lot, and tries to maintain in his own staff an attitude of warmth and compassion—a task sometimes difficult in an era when there are an estimated thirty-six thousand homeless in New York City, many of whom are alcoholic or quite mad. And, as Goldfrank remembers, even George Orwell, writing about French and British homeless of the thirties, said that the tramp was "abject, envious—with a jackal's character." The doctor well knows that Orwell's prejudice remains widespread. Yet almost every day Goldfrank and his staff find themselves in the role of the last friend on earth to scores of these American untouchables. His Bellevue team finds itself affirming worth where most of the rest of the world sees little if any merit whatsoever.

Goldfrank usually walks downtown, maintaining a steady hundred paces a minute to beat the traffic lights. At that rate it is hardly an eighteen-minute walk to Bellevue. The hospital is an imposing collection of buildings in various styles—all, in their way, memorializing the ideals of the age that gave birth to them. The most recent is a twenty-five-story concrete-and-glass tower, flanked by the red-brick structures of the early twentieth century. When those old buildings were built, a leading cause of hospitalization was abuse of alcohol. When the new building was completed in 1975, alcohol abuse continued near the top of the list of causes of illness and injury. Most of the other common diseases that filled the hospital in the old, red-brick era—rheumatic fever, tertiary syphilis, polio—have been conquered. Now new plagues have come to take their place: AIDS, hepatitis, intravenous drug abuse, lung cancer. Tuberculosis—then and now—remains a virulent presence.

He goes down an inclined walkway over which is suspended a bright banner announcing the 250th anniversary of Bellevue—the oldest municipal hospital in America. He enters what used to be a side door of the old Administration Building. A sign warns that no bicycles are allowed. This portal now serves as the hospital's main entrance. He is greeted by guards and staff members in the already busy main corridor that leads to the adjoining modern lobby of what is now the main building. He makes for a narrow doorway over which signs in English, Spanish and Chinese announce the Emergency Department.

At the end of this corridor there are a desk and a congestion of stretchers, filled with men and women of several races and ages. Just short of a reception area here, the doctor eases aside one of the stretchers that is blocking a door, with a quiet word to its occupant. Goldfrank pulls out a bunch of keys and unlocks the door. He walks past the small computer terminal that his secretary will use when she arrives and into a windowless inner chamber, lined with books and open file racks filled with clippings and papers. A desk against the far wall holds two multi-line telephone units, books and papers. Above it in rainbow hues ranges a bright array of children's art and photos of his wife, Susan, and of their children.

He slips his briefcase onto a small file cabinet, opens it and removes his lunch: a wedge of cheese, a few crackers and an orange. These go into a small refrigerator atop his files. He exchanges his tweed jacket for a white starched lab coat. He affixes his photo ID card to his tie. Without that, even he might well be challenged by some newcomer on the ubiquitous security staff that circulates in the rooms and corridors outside.

It is nearly 7:30. He checks his calendar, notes that a dermatologist is due in for the 8 A.M. staff meeting this morning, picks up his stethoscope and is ready to start his day.

There is motion everywhere—security guards moving down the corridors, the families of patients milling about near the waiting room. Police enter with a manacled prisoner, a striking young woman in tight-fitting jeans, shoeless. There is a sudden popping open of the ambulance entrance doors and the swift entry of a wheeled stretcher, on which is an unconscious man in a pin-striped suit. It disappears through the doors to the treatment rooms and trauma slots, and is followed briskly by a nurse and an attending physician.

Goldfrank has taken no more than three steps down the corridor toward the triage desk when he is accosted by a small woman of perhaps seventy with a bright blond wig and crimson lips. She is dressed in a number of layers of clothes and white running shoes.

"Hello, doctor. Do you recognize me, doctor?"

"Yes. Hello. How are you?"

She tells him. He has heard it before. Several times. She has high blood pressure, cellulitis, and other complaints, none of which seems new or unforeseen. And today she has a bad sore on her shin.

The doctor listens to her courteously, even though the triage nurse already has taken her medical history. The woman has been a daily visitor here for the past week and there's no denying that for someone in her triage category, "non-urgent," the wait can be a long one—sometimes several hours. Although she should be considered "urgent" with her several problems, she has never remained long enough to be examined. Because of the other seriously ill patients and her habit of impatient departure, she has put herself in a "non-urgent" category. She reminds the doctor that she lives in a nearby women's shelter and adds that she has only a subway token to her name. Now she is unable to climb back onto her stretcher. Someone has removed the footstool.

"We'll get to you as soon as possible," Goldfrank says. He offers to help her get back up on her stretcher. She declines. She just wants something for her sore leg, which is suffering from cellulitis, and something for her high blood pressure. "Try to be patient," he says.

Goldfrank checks the triage desk, which dominates this waiting-room area. Ranged about on either side of the adjacent doors to the ambulance port are a half-dozen stretchers on which are inert men.

A layman might assume they're drunk. That's only one possibility. There are others, which Goldfrank and his team consider also: hypo-glycemia, head trauma, central-nervous-system infection, a post-ictal (that is, post-seizure) state. Doctors here, almost as a matter of sur-vival, often develop into preternaturally swift diagnosticians—able sometimes to come to the right conclusion on the strength of evidence so subtle and fleeting that their junior colleagues are reduced to aston-ished silence.

Today the hospital is full. The wards upstairs to which "emergent" cases would be admitted have no beds empty. So all the patients that would have moved off the ED wards and corridors are still here, like planes stacked up awaiting take-off. Every new ED patient therefore finds the wait a bit longer than usual. One triage nurse goes over a stack of the charts from the night just ending. Four of those patients who came in have vanished, so she returns their preliminary charts to the bottom of the heap. It has been a chilly night. Perhaps they just came in to spend time in a warm room.

A dark-haired physician approaches Goldfrank. "Lewis, got a minute?" This is the senior attending physician, Dr. Robert Hessler, who has the pin-stripe-clad patient under examination. Hessler grew up in Montana, lives now with his wife and son in New Jersey, and finds the excitement of emergency medicine so intense that he cannot imagine any other pursuit presenting such intellectual challenges. He presents Goldfrank with such a challenge right now.

His patient is a fifty-four-year-old resident of Garden City who collapsed on the sidewalk on Park Avenue. The paramedics reported that he was twitching and foaming at the mouth and giving all the outer signs of a seizure. A helpful passerby administered first aid, and an EMS ambulance was fortunately on the very block where the incident took place.

The paramedics were in control even before the ictus, the dra-matic onset of the seizure, had fully run its course. The ambulance crew whisked the victim to the Bellevue ED within minutes, and shortly after wheeling him in, he began to regain normal conscious-ness.

The series of tests that Hessler administered established that it was probably not a stroke, as such, but a grand mal seizure.

As they speak, Goldfrank can see the patient in a room obliquely across the corridor from where the two doctors confer. He sits up and rubs the back of his neck, then describes a large circle with his head, as if to loosen constricted muscles.

"You're sure he had a seizure?" Goldfrank asks.

"Not much doubt." Hessler goes over the facts. A contusion on the back of the head. Abrasions on the arms, from the characteristic frantic elbowing gestures of epilepsy in mid-attack. A lacerated mouth, cut by the clenching teeth. In fact, he had nearly bitten through his own tongue. No. There was no doubt.

As they ponder the full meaning of their findings, a loud, high-pitched clangor rings through the corridor, the jangling steel of heavy-duty handcuffs and leg irons in motion.

"I'll sue! I'll sue every fuckin' one of ya!"

A suspect is shoved ahead of the arresting police officers. Blood seeps down from a saturated bandage on his cheek. The two doctors make way for the procession of police, paramedics and nurses toward the x-ray room. There is the cut cheek to be attended to, and another urgent matter.

The two doctors pick up the gist of the altercation. The suspect is accused of mugging a woman and taking her rings. He says he has done no such thing. He is being falsely accused. The victim says he forced her at gunpoint to give up four of her rings, and then, on the timely appearance of a patrol car, the suspect bolted. According to the victim, he probably swallowed her rings before the police caught, dropped and cuffed him. After he is stitched up, an x-ray will settle the question. "I'll sue for false arrest! I'll sue for medical malpractice!" The jingling recedes down the corridor and around the corner as Goldfrank and Hessler continue their brief conference.

At its conclusion they go into the examining room, where the patient now sits up on the examining table. "This is Dr. Goldfrank, Mr. Benton. How are you feeling now?" Hessler asks.

Benton and Goldfrank exchange nodded greetings.

"I'm still a bit woozy. Otherwise okay."

"You're clear about where you are?"

"Sure. Bellevue."

"Right. And why you're here?"

"Passed out, I guess."

"Right. You know what day this is?"

"Tuesday."

"And month?"

"March."

"What year?"

"Nineteen eighty-six."

"Okay. And your name?"

A faint smile plays across the patient's features. "Sure. Scott Benton."

Goldfrank studies the chart further as Hessler tells Benton what has been discovered. "It looks like a grand mal seizure."

"Grand mal?"

"Epilepsy."

"Epilepsy! I've never had anything like that in my life."

"Sometimes it hits just once in a lifetime. Other times, it's unpredictable. Could come three days in a row and then wait ten years."

"You're sure that's what it is—epilepsy?"

Hessler looks over to Goldfrank for reinforcement. "Yes, we are," Goldfrank says. "And we think you ought to be admitted for observation."

"Admitted? But I feel okay."

Hessler continues: "We'd feel better if we could be assured that you've suffered no brain damage. That it wasn't a stroke you had. That the medication we're going to suggest is the best for you."

"I don't want to be admitted. I'm supposed to be seeing someone on Forty-seventh Street right now."

"It's for your own good."

The patient remains silent as he turns over this remark in his head. Then his eyes meet those of the doctors. "Well, I don't want to be admitted and I don't want any report made to my employer, that's for damn sure."

Goldfrank later said that he could feel his pulse quicken at that unexpected remark. The doctors redoubled their efforts to persuade Benton to check in for further studies. Chances were, they said, that he might not suffer another such seizure in the immediate future. "On the other hand," Goldfrank told him, "you might have an attack right away. We can't predict."

Another silence, broken only by the sounds in the corridor of passing stretchers and the footfalls of patients making their way down the hall to the Walk-in Clinic, just opening for the day. The doctors realize that they have Benton's papers, in his wallet, which the paramedics have passed along.

"Who's your employer, Mr. Benton?" Hessler asks, glancing at the wallet.

"It's a transportation company."

"Transportation?"

"It's an airline, if you have to know."

"The paramedics picked up your wallet here. This is your ID, right?"

"Yeah," he says, reaching out for its return. His papers and wallet are given to him.

"Says you're a pilot."

"Right."

"Flying overseas?"

"Yeah, right. And I don't want to be admitted. I'm feeling fine. I'd like to go now."

There is another silence.

As Goldfrank said later, "It was a very bad moment for all of us. My concern was that this was one of those rare cases where the obligation of the physician to honor confidentiality in treating a patient runs right up against our responsibilities to society. His condition obviously might have catastrophic implications for hundreds of people."

Goldfrank, who has been leaning up against the cabinet in the small examination room, stands upright, takes a step forward and meets Benton's gaze. "Mr. Benton, with Dr. Hessler's concurrence, I'm going to have to say I don't think you should fly again until you have consulted with your airline's flight surgeon."

"I see."

"And I have to tell you that, quite aside from your own discussion with your flight surgeon, as director of this Emergency Department I feel obligated to make an independent report to your airline."

"Wait a minute. What about your obligation to a patient—confidentiality?"

"You make a valid point," Goldfrank responds. "And I truly wish I did not have to notify anyone. But there's no other option. You carry an enormous responsibility, and I'm sure you are more aware of the weight of that responsibility than anyone else. Still, I have to make a report."

"But how could it be epilepsy? I've never had epilepsy. Never. I'm sure it wasn't any such thing!"

"Wish we could agree with you. But we're sure. Of course your airline's medical people may disagree."

It is a taut exchange, one that leaves everyone feeling the worse for it. Finally, the pilot is given Dilantin and detailed instructions on how to treat himself.

"We know how you feel," Hessler offers.

"Clearly we don't want anything to jeopardize your well-being in any way—physical or financial," Goldfrank adds. "But we hope you can appreciate our dilemma."

Benton is grimacing and nodding. He lets out a slow sigh. There is another short silence. Finally he says, "Thank you, both. If it really was epilepsy, which I have to doubt, then I guess I see your point."

"You won't reconsider being admitted?" Hessler asks.

"No."

"You're going to sign yourself out, then?"

"Yes I am."

"It will be against our recommendation," Goldfrank says, "and in any case your flight surgeon will have to be notified, as we indicated."

"You really have to do that?"

"I'm afraid so."

"I've got a wife and four children—two still in college."

"I wish we had a choice."

So they shake hands and he signs his name at the bottom of the chart, on the line that says "against medical advice," and he is discharged.

When the doctors come back out into the corridor, an attending physician is clipping a fresh x-ray film to a glowing fluorescent viewing box. A small group gathers around to see it. The outraged prisoner of a few moments before, his cheek sutured up, is now silently studying the ceiling. There, clearly visible in the x-ray portrait of the rugae (muscle ridges) of the stomach, is a small cluster of circles—undeniable evidence that he has recently swallowed four objects that closely resemble finger rings.

A disturbance in one of the nine-by-nine-foot examination modules—the ID (Infectious Disease) and Rash Room. By some body language so swift it is hard for the visitor to detect it, a small group suddenly coalesces around a wheeled stretcher in the center of the room. The patient is the same woman Goldfrank encountered a half hour before: Mary Helen Neville, her chart says, lying on the table with a painful crater on her right leg. She is "undomiciled," and her phone number is listed, in deference to the needs of the computer, as (212) 000-0000.

She is suffering from cellulitis. The ulcer is about the size of the palm of your hand, with an angry-looking core. Miss Neville is whimpering as Dr. Goldfrank approaches. Her lament slowly dies away as he quietly inquires if she will object to serving briefly as center of attention during a teaching session that is about to commence. She has no objections.

"Good morning again, Miss Neville."

"Good morning, doctor."

"Looks like you have a real problem here. May I introduce my colleagues?" His objective is to instruct everyone in this small circle in the best way to handle such a case. Introductions of two medical students and a nurse practitioner are made as the doctor begins to clean the wound before him. Already a nurse's aide had removed the patient's running shoe and plum-colored sock on her right foot. Her past medical history and her vital signs are all listed on her chart. Her blood pressure is 160 over 110—high but not life-threatening—her pulse 88, her temperature 98 degrees. Nothing untoward there. The doctor says a few words, as notable for their soothing tone as for their content. She seems to become still calmer.

Her other foot is clad in an unmatching light gray sock. Goldfrank tells his small tutorial group, "The idea is to get her leg elevated, and drain the interstitial fluids [between the cells] to make it easier for her impaired veins to return blood to her heart. And then medication—adult diphtheria-tetanus toxoid and Hyper-Tet [human tetanus immune globulin]—provides an interim defense. Next, thiamine and antibiotics [dicloxacillin]. And then we'll want to make sure that the dressings are changed twice a day until the cellulitis clears up." As he later said, "It involved nothing fancy. It would have been simple for her to do for herself—if she had a home. Someplace where she could be free of the stresses she will feel on the street or even in a shelter. Somewhere where she could put up her feet and have good meals." But there is no such prospect for Miss Neville. So she is to be admitted despite the resistance of a house officer, who is informed of the impending admission.

Why admit someone if she won't take care of herself? But how can she take care of herself if she has no place to live? She'll do the same thing again anyway. So the discussion goes, briefly, until it is clear that the Emergency Department will admit her to the hospital. As Goldfrank later said, "The gravest error is to send someone home who should have been admitted. Still, the house officers call the Emergency Department 'the sieve.' We let anything through, they think."

Pleased to be the center of such intense regard, even if unaware of the hint of hostility toward her evidenced by the house officer, Miss Neville clearly is feeling better about herself.

As he finishes his explanations to the students, Goldfrank pats her gently on the shoulder. "This will be better in a few days, if you try to stay off your feet," he says. "So we're going to admit you to the hospital."

"But all I have is this subway token."

"You'll be billed based on your capacity to pay."

"Can't pay."

"Nobody's going to come chasing after you for what you don't have. Let's just get your leg cleared up."

"Thank you, doctor."

On Goldfrank's mind is the concern that if her condition is not treated at its present stage it could easily get much worse, could involve sepsis, osteomyelitis, gangrene, even the loss of her leg. He reflects that even from the viewpoint of the most hard-hearted budget watcher, treatment now is far preferable to what might ensue if Miss Neville's cellulitis and ulcer become so aggravated that massive doses of antibiotics, operations and extended hospitalization are required. Her lack of a home puts her at severe risk, and leaves society itself exposed to the possibilities of repeated costs of care.

"The problem with ignoring this kind of patient," Goldfrank has said, "is illustrated by what happens to the person wandering around with inadequate footwear. He gets a minor infection of the feet. If you don't care for that infection at that point—just a low-grade infection and cracks between the toes, for example—there's the beginning of a cellulitis or inflammation of the skin on the foot. If you try to treat the condition outside the hospital, what happens is that the person who has no home, or drinks too much, doesn't change his socks, if any, or shoes. The maceration of the feet gets worse, the infection becomes more consequential. Then the patient comes back two or three weeks later, when he wakes up from a stupor or is kicked out of his next hiding place. He comes into the Walk-in Clinic and you take off his shoe and you discover he doesn't have any muscle left. You're just down to the tendon and the bone. And he's got trenchfoot. So you will face the problems at one point or another.

"The problem we're seeing with the elderly homeless is just like Napoleon's surgeon, Dominique Jean Larrey, saw on the retreat from Moscow. The French infantry would battle the Russians, get half-frozen feet, then sit down at the campfire and get them half thawed. Before they could completely warm up, the attack would resume.

Larrey discovered that those who had partially thawed their feet
ended up with much more severe injury than those who walked on
frozen feet or totally thawed their feet. And what happens to the
derelicts today is that they have partially thawed feet and they're
suffering in the same way."

"I could understand the house officer's aversion," Goldfrank later
said. "We all have our likes and dislikes. But what I try to instill in our
staff is the belief that as care givers our responsibility is to see the
worth of each person as a human being and to give them all you've
got. And you can't just give them cold and detached attention—even
if it meets the technical standards of good care. There has to be some
feeling to it, some warmth. Even the down-and-out are very perceptive—of your beliefs, your philosophy, the way you treat them and the
way you look at them.

"Others call the Emergency Department 'the pit' because it's loud
and disorderly and oftentimes smelly and unpleasing to the eye. We
like to think it's more of a juggler's world. You have to be able to shift
from one thing to another, quickly adjust to the unexpected."

Even the Emergency Medicine Board examination acknowledges
the peculiar interrupted nature of the specialty. "The oral exams start
you out on one case, and then another is presented right on top of the
first and then a third one. You are asked to handle all three, but in
such a way as to sustain life even though you have a car-wreck victim,
a child dragged out of a swimming pool and a polydrug overdose all
coming at you simultaneously. Even on the exam you've got to work
with a team of people—nurses, radiologist, consultants from the various specialties . . ."

A young white male is being wheeled past as Goldfrank emerges
from the ID and Rash Room. He glances briefly at the man, and intuits
much of what will later become the medical history recorded on the
man's chart: "ETOH [alcohol] abuse. OD [Overdosing] Valiums and
PCP [angel dust]. Pt. found in street unresponsive. AOB [alcohol on
breath], beer can next to him. Pt. denies taking pills, suicidal idea-
tion . . ."

A much more voluble black male, thirty-one, is shouting now in
the corridor. He too has AOB, and is telling whoever will listen as he

hobbles down toward the Walk-in Clinic that a bus has run over his right foot. An x-ray fails to show anything broken.

Another patient has just come back from x-ray, accompanied by two police officers. He is being admitted to surgery. He's a twenty-nine-year-old black man named Wright Maccomb, who swallowed a key a few hours ago in the course of a police arrest. It is said to be a master key to the subway system and just now it is resting in his small intestine.

Candace Seixas, a thirty-two-year-old white female, passed out briefly at her work as a layout artist at the Acme Studios on Nineteenth Street. The triage nurse has noted "acute diffuse abdominal pain accompanied by nausea and dizziness" and judges her case emergent. Her boyfriend, who works with her, has brought her by taxi to the hospital. She is crying out in pain . . .

Hovering over the tumult of this Emergency Department is the spirit of Sir William Osler, the great clinician who served in the late nineteenth and early twentieth centuries successively at McGill University, the University of Pennsylvania, Johns Hopkins and Oxford. He is one of Goldfrank's heroes. It was Osler who urged his students, nearly a century ago, to "Learn to accept in silence the minor aggravations, cultivate the gift of taciturnity and consume your own smoke with an extra draught of hard work so that those about you may not be annoyed with the dust and soot of your complaints . . . we are here not to get all we can out of life for ourselves, but to try to make the lives of others happier." And the desirable frame of mind was "a calm equanimity. . . . How difficult to attain," he concluded, "yet how necessary, in success as in failure!"

# 4 A Morning's Rounds

Nearly thirty people are meeting in the narrow area that serves as the doctors' station, telemetry desk and paramedic communications area. Presiding are the two bearded elders of a conspicuously youthful group—Goldfrank, with a severe-looking graying beard, and Dr. Neal Lewin, Assistant Director of the Bellevue Emergency Department, with a neatly cropped brown beard. Lewin's bright gaze is sweeping the group, from behind his horn-rimmed glasses, conveying something of the restless energy that he brings to these gatherings.

An auburn-haired young woman takes a step forward to present the first case of the morning's rounds. Dr. Brenda Smart is English, about to be married and tired from working the 8 P.M. to 8 A.M. shift. She consults her notes and begins. She tells of a thirty-year-old female brought in by ambulance at 4 A.M. in a comatose state. The details come swiftly: The patient was a cab driver proceeding westbound on the Brooklyn Bridge toward Manhattan when she hit an ice patch and skidded into a truck. Her passenger, a thirty-four-year-old Hispanic male, was DOA at Bellevue. The driver gave all evidences of having ingested significant amounts of something—probably cocaine. There are obvious track—or needle—marks on both of her arms and even in her popliteal fossa [behind her knee]. A tattoo on one arm appears to have been placed there principally as a screen for needle marks. Her injuries included Le Fort Type 2 fractures of the face.

Several of the students look perplexed. Dr. Lewin, with a nod to Smart, offers a brief explanation: The French surgeon L. C. Le Fort dropped cadavers out of upper-story windows in the 1850s to study the consequences of falling from various heights. From these gruesome experiments emerged the nomenclature for three types of maxillary fractures. (The maxilla is the jawbone and the area that lies under the bottom half of the eyes and nose and the upper lip.)

"And you said this is a Le Fort Type 2, Brenda?"

"Yes. Involving the orbital rims and floors as well as the nose. And notable swelling of the entire midface area."

The case review proceeds. Meanwhile, there is a constant murmur of urgent-sounding activity around the perimeter of the meeting. And every time the adjoining door opens and a patient is wheeled down the corridor the air piston governing the door lets out a high-frequency whistle, which slowly dies away as the door closes. It's an audible reminder that neither sickness nor injury slows down for doctors' meetings. For a patient with a critical problem, a nurse records the vital signs, then passes the chart to a box from which one of the attending physicians or residents grabs it and immediately goes out to the patient. Thus there are frequent interruptions and inevitable informalities that tend to obscure rank and hierarchy during the meetings.

Just then there is an audible lamentation somewhere. Senior nurse Liz Reynolds leaves the group and moves swiftly down the corridor. She's attired in white pants and lab jacket and moves with a dancer's grace to a bay holding several wheeled stretchers. One is occupied by the moaner—an old man in a hospital gown and a beige snap-brim cap. She greets him with a smile and a calming word. His sheets are being changed by a nurse's aide who has been slow to note that the patient's legs, which are badly swollen, are tormenting him.

Reynolds gives the aide a soft reproof, then lends a hand in shifting the patient to one side while the old sheet is removed and the new one inserted. She says something else reassuring, and the moaning stops. The old man's worldly possessions are tucked away behind the elevated headboard of the stretcher. There is a clatter as two objects hit the concrete floor. As he turns to see what made the noise, his cap brushes off his head and also falls to the floor. Nurse and nurse's aide finish changing the sheet. Reynolds elevates his feet. She asks him at each turn if he's comfortable now. He says "yes," but keeps turning his head in an agitated way, trying to see the fate of his treasures—a pocket knife, a quarter and his beige cap.

He is settled now, and Reynolds drops to her knees to retrieve his possessions, which are quietly returned to the shelf under the stretcher with another encouraging word; then she goes back to the meeting.

There's a new face in the gathering this morning. It belongs to a well-built medical student who bears himself with a challenging jauntiness. "Did you have any interesting cases, Sidney?" Lewin asks him.

"No. All I had was uninteresting cases—garbage."

There is a momentary silence. Goldfrank's gaze fixes acutely on the new man. "Sidney," he says, putting an edge in his voice: "Osler said that there are no boring patients, only boring doctors."

The student drops his gaze to the floor. "I'm sure you weren't saying that a fellow human being was garbage, were you, Sidney?" Lewin asks.

"No, I guess I wasn't thinking."

"That's what we thought."

The reprimand is quickly past, but the student, on the day when he first gazes at the flyleaf on Goldfrank's text, *Toxicologic Emergencies*, of which Lewin is a co-author and co-editor, will doubtless have reason to remember this exchange. There is Osler's credo, adopted by Goldfrank and Lewin and by the specialty of emergency medicine:

In the hospital, we learn to scan gently our brother man, judging not, asking no questions, but meting out to all alike a hospitality worthy of the Hôtel Dieu and deeming ourselves honoured in being allowed to act as its dispensers.

Adherence to that venerable tradition sometimes puts the Emergency Department at odds with the rest of the hospital, as 75 percent of admissions come via the ED. It is a common perception upstairs that the Emergency Department overdoes its idealism, sometimes admits indiscriminately, heedless of the burdens such medical hospitality puts on house officers and other staff upstairs—and, of course, on the hospital budget.

The morning's rounds have entered their final phase: Brenda Smart is finished, now, with her description of the patient who had the critical accident on the bridge that morning. The cab driver remains in a coma and the outlook is guarded, although the normal computerized axial tomography (CAT) scan of her head is reassuring.

Now another physician, a dermatologist, is to present the case of a thirty-two-year-old obese male with a notable case of cellulitis—inflammation of the lower extremities so pronounced that to the untutored observer it might well appear to be elephantiasis. All thirty staff members follow the dermatologist down the hall to the treatment room.

A report from the triage desk draws Goldfrank from the room, so Lewin takes over the reassurance of the patient while the examination and discussion conclude. Then, as suddenly as the group materialized, they file out, returning to the meeting room, where the last few moments until nine o'clock will be devoted to an impromptu dermatological lecture drawing on the members of the group itself.

A nineteen-year-old college student has just been admitted and a medical history is being taken from her by Dr. Kathy Delaney, as Goldfrank looks on. The student is complaining of severe headaches and double vision. Those could be symptoms of a life-threatening disorder, such as a brain tumor.

Back in the meeting area, the discussion now centers on a blond young house officer (a resident) in an Icelandic sweater, who has drawn up his sleeve at the suggestion of the dermatologist. The visitor takes his ballpoint pen to point to a slight blemish on the house officer's arm. He describes how this slightly raised mole gives evidence of having penetrated below the epidermis into the dermis—which will make it a "junctional nevus." The Latin word *naevus*, the group is reminded, means "a mole on the body."

The visitor moves over to a young woman, a suntanned medical student, who says softly that she has been vacationing in Florida. He points out a pigmented striation on her forehead, then adds, somewhat to her chagrin, that it resembles strands of spaghetti. He goes on to a skin-blemished young man, who says with resignation, "I knew you'd get to me." The dermatologist enumerates the variety of blemishes readily visible—making a hurried listing, no doubt to minimize the young man's embarrassment.

As Goldfrank still has not returned, the dermatologist continues, demonstrating the depth of his training and illustrating the ease of finding teaching materials in this setting. No one is without interest, not even the healthy. For those who will but look, there is a wealth of instructive material at hand. The dermatology resident asks the young man in the Icelandic sweater to pull up his sleeve yet again. The resident notes a scattering of freckles on the young man's lower arm. "And what would you call these?" asks the visitor.

The young man says nothing for a moment. Then he brightens. "Hyperpigmented dots?" he asks, amid general laughter. That is all right, but not the wanted answer. He tries again: "Junctional nevi." A smile of pleasure on the resident's face. "Very good."

Goldfrank returns for just long enough to adjourn the meeting. As everyone scatters, he goes back down the corridor to where the young woman with double vision is now undergoing a full examination. Brenda Smart heads home, still shaken by the memories of the night.

$$\longleftarrow\!\!\!\bigwedge\!\!\!\longrightarrow$$

Kimberly Muti is nineteen and a sophomore at Pratt, planning to major in fine arts. As the charts would later recount, she is a healthy and attractive young woman, slender to the point of seeming to be anorexic. In response to Delaney's series of questions, she has negative answers to most—no excess use of drugs or alcohol, no blows to the head, no exposure to industrial toxins. She is a health-foods enthusiast and takes supplementary doses of vitamins. Delaney notes that her lips are scaled and cracked and there are small fissures at the corners of her mouth. Also she has had acne and it seems to be responding to vitamins and ointments that a local health-care clinic has prescribed.

Sensing a good teaching opportunity, Goldfrank involves other members of the department in the unraveling of the mystery that they confront. Among them, Kathy Delaney takes a particular interest. In fact, she has recently decided to develop her skills in toxicology, and there is a suspicion that this problem might involve a poison of some kind.

Delaney takes an ophthalmoscope from a shelf against the wall and approaches the young woman. "All right, Kim, just look this way, please." Peering through into the fundi, looking for the slight depression on the retina through which the optic-nerve fibers transport information from the retina's rods and cones to the brain, Delaney can see the actual brain tissue well enough to note papilledema—swelling of the optic nerve. This is a sign of generalized brain-tissue swelling, an ominous finding.

As Goldfrank later said, "We had a previously healthy young woman with several weeks of progressively worse complaints involving the musculo-skeletal system, integumentary system and central nervous system. She complained of pressure on the long bones, which made her moderately uncomfortable; her skin was dry and cracked, in places, and there were the headaches. A brain tumor could raise cerebrospinal fluid pressure. But the musculo-skeletal and cutaneous symptoms were not typical for a tumor. Still, we had not yet reached our unifying diagnosis, although a strong likelihood was emerging from the history."

"We think you ought to have some tests," Kathy Delaney says.

"It's that bad, is it?" Kim asks.

"No. Just that we need to make some tests, and the sooner the better."

She agrees. Now follows a succession of probes into the sensorium—the window of consciousness—of this young woman. Delaney moves her finger from one side of the student's field of vision to the other, looking carefully at the ocular movements.

Her reflexes are all normal and mental functions are, as the charts will note, "normal to routine testing." Then the young patient is taken upstairs to the seventh floor, where a CAT scan is done. That reveals no mass, no sign of a growth, but does confirm generalized cerebral swelling. So what they have is a pseudotumor, all the signs of such a growth without its actual presence. They had detected signs of swelling of the optic nerve by fundoscopic exam, and—based on the finger tests—"lazy sixth nerve" or sixth-nerve palsy, a condition that can account for the double vision. There is also every indication that the cerebrospinal fluid pressure is markedly elevated.

"Kim, we have to do a spinal tap and then we should have a clearer idea of what's causing your headaches and double vision."

The patient looks very small as she lies there and seems to be wishing that her mother—who is supposed to be coming in from Montclair—will get there soon. Dr. Delaney gives her hand an encouraging squeeze.

"We're very good at this sort of thing, so it won't be too uncomfortable—or take very long." So the young woman, following Delaney's directions, turns on her side and draws up her knees in a fetal position, while the doctor parts the hospital gown to reveal her back and swabs her lower spine with alcohol in the space between the third and fourth lumbar vertebrae. Kathy Delaney tells her it will all be over very quickly. From a lumbar-puncture tray nearby she takes a small needle with a bit of lidocaine—a local anesthetic—and raises a tiny mosquito-bite-size wheal with it. Then Delaney takes a long, hollow stainless-steel needle called a cannula, which encloses a pencil-lead-like inner stylette, and carefully strips off its outer sterile packaging.

The doctors have made sure of all the other indications. Everyone knows that under some circumstances a tap can result in a catastrophic drop in spinal-fluid pressure—in a worst case, it could draw the soft brain tissues forcefully against the firm, bony structure (the foramen magnum, Latin for "big hole") at the base of the skull and the top of the spinal cord. That would be somewhat like a mass of gelatin being sucked down a drain. But in this instance all are agreed that the procedure is the correct one.

Delaney deftly eases the long 22-gauge spinal needle into the space between the third and fourth vertebrae, aiming slightly toward Kim's head, taking care to maintain the needle horizontal until the spine is actually entered. There is a faintly audible "pop," and Delaney knows the needle has entered the subarachnoid space, as the terrain inside the middle membrane covering the spinal cord is called. (The term means, literally, the "under spiderweb" space. The membrane is said to be "spiderweb-like" because of its finely woven, fibrous texture.) She pulls out the inner cannula, and as she does so, a crystal clear fluid moves into the chamber of the needle, only to be glimpsed as a few drops come out while she attaches a three-way stopcock and a pressure gauge, called a manometer.

She collects less than an ounce of fluid in three different test tubes, for culture, chemical analysis and cell count. After a moment, she slowly removes the needle from Kim Muti's spinal column and puts a sterile dressing over the site of the puncture. The patient reports only a slight burning sensation.

There is an opening pressure of 320 millimeters of water, which means that the inner force of the spinal fluid pushing through the cannula has driven a column of water 12 inches up the gauge of the manometer, a height many times normal. The most common causes of such increased pressure are obstructions to the flow of cerebrospinal fluid, such as a growth of some sort. And, whether benign or malignant, such a tumor can be potentially lethal. But the CAT scan of the brain already indicated no mass lesions, no tumor. So the most likely diagnosis remains "pseudotumor."

Dolores Muti, Kim's mother, has arrived from New Jersey. She is a handsome woman in her late forties, dressed in a suède suit. Goldfrank talks with her a bit, eliciting further information on Kim's medical history.

"We knew that such pressure elevation leading to double vision could result from a variety of causes—obesity, for one. But Kim was very slender," Goldfrank later recalled, "and her mother confirmed that she had never had a weight problem. Diabetes, menstrual irregularity, oral contraceptives were other possibilities—but none of them applied here. Then the answer presented itself to me."

"Kim, how much vitamin A are you taking?"

The girl made a wry grimace. "Oh, pretty much."

"Like how much?"

"Maybe five or ten pills."

"A day?"

"Yes."

"Then we knew we had our answer," Goldfrank later said. "The pills she had been taking were 5000 IU (international units) each. She was taking up to 25,000 or 50,000 IU or more of vitamin A a day—probably in her own mind to keep her acne completely under control. But in fact that represents ten to twenty times a normal daily dosage. That was it. 'What you have, Kim, is acute vitamin A intoxication,' I told her."

She looked relieved and skeptical at the same time. "I thought it was something terrible—a brain tumor or something."

"For a moment there so did we."

"So what do I do now?"

"Stop the vitamin A and your headaches and double vision should clear up."

As Goldfrank later said, "We admitted her to the hospital to complete further studies of her visual deficits and observe for improvement of her headaches. And, sure enough, her main symptoms were cleared up in just a few days. In another couple of weeks her dryness of skin and skeletal aches and pains were gone, too."

In his standard work, *Toxicologic Emergencies*, Goldfrank has noted that megadoses of vitamin A and its close relatives pose a severe risk to young women, especially pregnant women, because of indications in animal research that central-nervous-system anomalies and birth defects are triggered by such overdosage. "Health food faddists," he has written, "continue to be exposed to the extensive advertisements of the vitamin industry. This continued interest must also lead physicians to think about vitamin A toxicity when faced with hepatic or cutaneous diseases [disorders of the liver or skin]."

Goldfrank's book notes that "many Americans are led to believe that vitamin A has an anticancer effect." Health enthusiasts suggest that the vitamin protects against such things as cigarette smoke and other "aromatic hydrocarbons" of the sort that cause cancer. But the side effects of such vitamin poisoning are numerous. "It has been suggested that only 5 to 10 percent of the 40 million Americans taking vitamins daily were prescribed those vitamins."

The tabular roster of vitamins and their associated disorders fills half a page in Goldfrank's book. Besides the vitamin A distress that Kim Muti suffered, these are among the other common manifestations of vitamin poisoning from overdosing:

- *Vitamin D*—headache, apathy, anorexia, nausea, vomiting, bone pain, high blood pressure, irregular heartbeat.
- *Vitamin E*—diarrhea, cramps, nausea, high blood pressure.
- *Vitamin B$_6$*—loss of coordination, numbness and tingling around the mouth, multiple sensory abnormalities of the peripheral nerves, lack of reflexes.
- *Vitamin C*—diarrhea, crystal formation in urine leading to kidney stones.

By the time Kim Muti had heard this catalog of unexpected possibilities, she gave every sign of being determined to change her dietary and supplement habits.

Mrs. Muti thanked the doctors effusively, and stayed to accompany her daughter upstairs.

Shortly before noon an ambulance brought in Pedro Monteverde, address unknown. His chart would provide a light moment later, when it came to Goldfrank's attention during a review of patient care.

The report said that Monteverde was a twenty-eight-year-old Hispanic ♂ who jumped into the Hudson River, apparently without self-destructive intent.

Treatment: T & R [treat and release].
Referred to: Medical clinic in 4 weeks.
Diagnosis: River exposure (Hudson).

# 5

# Doing a Decent Thing

In the main entry corridor at Bellevue there's an amputee in a wheel-chair checking the coin-return slots in a rank of vending machines for any odd change.

Lewis Goldfrank glimpses the wheelchair and its occupant as he moves quickly down the corridor. He runs a thumb over a cut in the palm of his right hand—the memorial of a screwdriver blade that gashed him as he assembled the bedframe his wife had built the day before, at his home in the country some thirty-five miles up the Hudson in Westchester.

His thoughts turn back to the slender black man, nodding and grimacing in the wheelchair. He can hardly suppress a slight feeling of apprehension. For the man in the wheelchair exemplifies a problem without a solution, and an ongoing reminder to the staff and the director of Bellevue's Emergency Department: here is where Manhattan's pariahs often come for respite.

Toby Wilts is a particularly vexing patient for the staff of the Emergency Department. He appears there as many as one hundred days a year, presents himself or is brought in by ambulance, typically in a state of acute intoxication, often enshrouded in such repellent filth that it is a small miracle that the staff, who by this time know him well, will respond in the unconditional way that reflects their professional commitment.

Goldfrank strides toward his own office, half a block ahead of the problem that seems likely to present itself once again at his threshold. A youth studies him as he passes. The boy is leaning up against the wall. "Excuse me, sir," the boy calls out, loud enough at first to sound challenging, but his words softened by a cagey grin. Goldfrank slows down slightly and makes contact with the boy's eyes. "Excuse me, sir, are you any relation to Abraham Lincoln?" Since he is about the same height, and has a full beard not dissimilar to that of the Rail Splitter, the question doubtless has been raised enough times to be familiar to Goldfrank. In any event, he answers with a broad smile, "No, but keep trying!" and continues on toward his office.

As Goldfrank reaches his office door, a wheeled stretcher partially blocks the way. On it lies a thirty-year-old black man in handcuffs and leg irons studying the passing foot traffic with a hooded intensity. Two nearby police officers—a man behind and a woman in front, leaning against a water cooler—keep an unblinking gaze on their prisoner, who shifts a leg, but not far. It is secured to the tubular steel members of the stretcher by one of the leg irons. Loss of a prisoner would mean severe disciplinary action, not just embarrassment, to the police watching him.

A glance beyond at the reception area tells Goldfrank it is a crowded morning. Soon he has a report from his second-in-command, Neal Lewin. One attending physician has overslept and another is home ill, leaving them short-handed. They agree to forgo the 8 A.M. staff meeting. Meanwhile, the ambulances have just brought in an AIDS victim. The man is in Trauma Slot No. 1, breathing again but barely revived after cardiac arrest in the ambulance on the way to the hospital. Goldfrank moves down toward the triage desk, and leans over to check the charts that assistant head nurse Sharda McGuire has stacked neatly on the edge of the counter there.

"Looks like Toby is going to be with us again," he tells her.

"Now?"

"Just saw him in the corridor coming this way."

She nods without saying anything further. He moves toward Trauma Slot 1.

Wilts wheels himself down the long hallway toward the ED, and it is noticeable that everyone moves away when he passes. He has adopted the tactics of the skunk, perhaps as a protection against violence. His life on the street, from all that Goldfrank and his team have pieced together from his many visits, is one of filth, noise, cold, hunger and self-induced delirium. And always the threat of violence. He is close to the portal, and a hospital policeman stands back a pace as he rolls toward the ED triage desk.

Dr. Hedva Shamir is the attending physician in Trauma Slot 1. She is a small woman with an air of cheery intensity. She smiles easily, with the sort of aplomb that the mother of three, who took off ten years to raise a family between finishing her internship and resuming her career, might be expected to display when confronted with three urgent but conflicting demands on her time and equanimity.

The patient from whose radial artery she has removed a sample for blood-gas determination is ashen. He is an extremely thin, balding man in his thirties, most of whose features are still obscured by the endotracheal tube and Ambu-bag over his nose and mouth. When the paramedics got to Jason Morelli in his apartment in the West Thirties, he was having a grand mal seizure, going alternately into spasms of contraction and of relaxation involving all of his major muscles. Such were the persistence and intensity of his seizures that a few minutes of that wild motion could be enough to do himself lasting injury.

Now Hedva Shamir already can see the confirmation of permanent damage from another source in the patient's body. There are cachexia (general emaciation), fever and a stiff neck, all of which can suggest infection (viral, fungal or bacterial), tuberculous meningitis or a brain tumor, among other afflictions. All are possibilities, in addition to the fact that the patient has AIDS. Quite aside from these clinical realities, there is the roommate's repeated assertion that Morelli is dying, that he wants to die and that he has doubtless overdosed himself on something to hasten his own death.

As Dr. Shamir worked to attach another tube to maintain the IV drip on the patient's left arm, the house officer representing the Medical Service joined the group working on Morelli. The house officer clearly did not relish the prospect of yet another AIDS victim admitted to the hospital. And there was a tacit question in the air: Why bother? If the patient has AIDS and indicates by his own actions that he does not want to live, why revive him?

Toby Wilts is in the reception area and he is highly audible. Without shrinking, Liz Reynolds makes a fast determination that he ought to be cleaned up and prepared for medical examination directly. With the help of nurse's aide Leon Fields, she wheels Wilts down to Room 2, where Fields moves with an athlete's smoothness to tidy Wilts up enough to make others willing to approach him. Fields is forty, a high school graduate who was trained as a cabinetmaker and later gravitated to hospital work. He has made a place for himself as a nurse's aide distinguished by unusual compassion. He says a word to the still-obstreperous Wilts, who grows less raucous. Fields and Reynolds shift Wilts to a stretcher and Fields starts to cut away the noisome trousers, which are so filled with excrement that they seem to be cemented to the man's body.

Back in Trauma Slot 1, the unspoken questions obtrude: Why work to revive someone on the brink of death anyway? In the absence of a DNR code, what is to be done in Morelli's case? DNR is medical shorthand for "Do Not Resuscitate," a code increasingly appended to the charts of terminally ill in-hospital patients by their private physicians or doctors who have had a long-standing relationship with them. What is to be done? For the director, the answer is unequivocal, and all of his attending physicians know it: the Emergency Department is in the business of saving lives. There is no other possibility without wreaking havoc on their procedures. Saving life is their absolute priority and they cannot qualify it without in some measure jeopardizing the standard of care they give to everyone who comes to them.

As Hedva Shamir moves the tubing to connect a new IV line, the plastic tube springs from her grasp, and, spurting with the crimson fluid, sends a small jet of Morelli's lifeblood arcing outward. Before striking the floor, however, it hits the trouser leg of the house officer, and then his shoes. His professional detachment disappears instantaneously. He grunts in dismay and dances backward away from the thin stream. Shamir grasps the tube, pinches it shut, says a brief word of apology and reattaches the tube to the IV line, containing the flow.

When the house officer comes back, after cleaning himself up, it is he who is ashen, but he who nevertheless has to take on the patient now, as the ED has admitted Morelli to the hospital. Morelli will pass into the care of this same house officer and his colleagues in the Department of Internal Medicine. A fortunate thing, an onlooker suggests, that Goldfrank himself was not in the line of fire. His still-healing wound on the palm of his hand could have given an avenue of entry to the stealthy AIDS virus, could it not? Goldfrank discounts the risk. "Possibly. But an exceedingly remote possibility."

He goes on to say, "All the earlier alarmist predictions on AIDS being casually transmitted turned out to be unwarranted panic. There is no evidence from any source that the virus can be passed along except by exchange of blood or other body fluids." Still, he agrees, open cuts of the sort he suffered from the screwdriver do argue for using surgical gloves when dealing with all patients and AIDS victims in particular—even though among the few hundred documented cases of health workers sticking themselves with needles or nicking themselves with scalpels or scissors in the presence of the AIDS virus there are no confirmed cases of AIDS.

Morelli is admitted, but does not survive the day. He had disseminated—or miliary—TB. "And he was probably already brain-dead by the time he got to us," Shamir later said. She shrugged sadly. "But we know our duty is to save life. And that is one of the good things about emergency medicine as a specialty. When we pass patients along—to Surgery, or Psychiatry, or Medicine or wherever—we're handing over living people. We've stabilized them, done the best we can. And it's a rarity to have someone die in our department. As a medical student at NYU, I thought I wanted to do pediatrics. But then I saw little children with cancer and it made me so sad to see their lives ending. That's when I switched to emergency medicine. Whatever else you can say, emergency medicine is about life—preserving it, defending it, restoring it. It's very satisfying."

Terry Haines, an attractive young bride, is a nurse practitioner who is able to relieve doctors of some of their responsibility, especially for the care of non-urgent cases. Now she is taking a history from a heavy-set Hispanic woman. A visitor, later scanning Haines's charts, marvels at the extreme economy of her notation. Among the abbreviations that appear: ♀ (female); ♂ (male); BRBPR (bright red blood per rectum). $G_7/P_5/SA_2$ (gravida [pregnant] 7 times; para [births] 5; spontaneous abortions 2; ↑ suggests alive; ↓ dead.

"Among the main tasks nurse practitioners have here," Haines explains later, "is stabilizing chronically ill patients. A lot of people with heart, lung and kidney complaints, for example, and asthma, arthritis and the kinds of things that go on and on."

Haines's new husband, James O'Connor, is also a nurse, who serves at Roosevelt Hospital and has an after-hours career that brings novelty and glamour into their lives. He has a degree in drama and has extensive acting experience. He's also a playwright who has had several of his productions produced in Los Angeles. "We live on Forty-third Street," Haines explains, "and have this apartment with a balcony, and sometimes when I get home I see Jim trying out a new script out there by speaking his parts to the passing winds."

Her work as a nurse practitioner is in a new area of medicine, for which she has trained by taking a master's degree at Pace University in a major especially geared to providing hospitals with graduates who can perform many tasks hitherto reserved to physicians—taking a comprehensive history, doing a complete physical exam, having laboratory tests worked up and preparing the patient for any additional medical attention that may be needed. Then she consults with an attending physician to work out their strategy for the patient.

Nurse practitioners have added essential staff strength to the health-care system and have been a part of Goldfrank's work since his earliest days running an emergency department at Morrisania Hospital in the Bronx.

Leon Fields finished his task, taking a carefully bundled plastic bag down the hall to the trash pickup station. "My job is to prep the patient for the doctors. I take the vital signs—pulse, temperature, respiration—and note it down. I get the patient comfortable for the nurses to take blood pressure and then for the doctors to get the examination going. Even if a person is disturbed or acting crazy, he's a real person and I try to see what the problem is and sort of act to calm him down if I can." In this case, Toby Wilts was nodding off, giving such a strong odor of alcohol on his breath that it was assumed that acute alcohol overdose—ETOH (EThyl alcOHol) abuse, as his chart would say—was involved once again.

At least, now that Fields had done his task, the little man was approachable. And Leon Fields could take pride in what he had done. "I really do like the work," he says quietly. "You're doing a decent thing in life and you're helping people."

In the adjoining Walk-in Clinic, head nurse Mary Dwyer, whose accent still lilts with the cadences of County Galway, tries to placate an outraged man who closely resembles those old portrait engravings of Franz Liszt.

"Where do you live, Mr. Jensen?"

"In the shelter . . . I'm a native of this city and I hope the city burns, including the Mayor."

"Why do you say that?"

"I've been sleeping on chairs for four nights, and I can't get any treatment and I'm itching like crazy."

"You can go up to Dermatology right now if you want. Let me give you a card." Mary Dwyer has already checked that Jensen is not louse-ridden. (If he had lice, it would be the job of the Emergency Department to clean him up.)

The patient mumbles his thanks and shuffles off.

Lying on a stretcher just down the hall from the Walk-in Clinic is a man with a plum-colored hand and two plum-colored feet, one of which resembles a club more than a foot, as it has no toes, and has formed itself into a tan and magenta cushion of scar tissue. A passerby on his way to the Walk-in leans over. "What happened to you, buddy?" he asks.

"It's from last winter."

"What is it?"

"Frostbite" is the laconic answer.

In another examination room a fragile-looking young blonde, who is a nurse, helps an old man up onto the table. His lower abdomen is grotesquely enlarged, and his shoulders and upper chest seem disproportionately delicate. He is suffering a grave liver disorder—cirrhosis so advanced that he resembles "a lemon on toothpicks," as the common phrase characterizes the appearance of those suffering his disorder.

In the midst of all the cross-talk and noise, Patrolman John Sisas, liaison with the 13th Precinct, makes his ongoing survey of the Emergency Department. His assignment is to see how many prisoners, how many cops and how many guards are in the department at a time as a precaution against too many weapons in the presence of possibly volatile patients or prisoners.

Now, however, except for the occasional shouted oath of Toby Wilts, there is no sense of instability or danger. In fact, office aide Ana Torres appears with a small dusky-looking sparrow that she has just discovered perched on the back of a chair in an examination room. She gingerly holds it cupped in her hands, which are clad in surgical gloves. She shows the small creature to Sisas and others, all of whom express amazement that the bird could get into a sealed building.

Leon Fields comes out, takes a look, and suggests how Torres can free the bird. He knows of a window that does open. Perhaps that is how the bird got in, in the first place. She has a sortie to make, first.

She goes down the hall, introducing the bird to all the staff members who are able to spare a moment to see it. Then, buoyed by the keen interest with which her discovery has been greeted, Ana Torres follows Fields to the window he knows about. It is half obscured by a cabinet. He reaches around, opens the window. She steps forward and releases the bird out into the parking area outside.

Toby Wilts surprises most of his old acquaintances in the Emergency Department because he is in his fifties and still alive. He suffers chronic and acute alcoholism and diabetes. Goldfrank has seen his decubitus ulcer on his backside grow until on occasions the bone is exposed. Wilts doesn't want surgery. He doesn't want psychiatric therapy. "Just care for me" seems to be his one message.

"After his stabilization in hospital, we've suggested he go to Goldwater Hospital on Roosevelt Island," Goldfrank says. "It's specifically for the disabled and chronically ill. But Toby doesn't want that. And he doesn't want to make use of shelters." So this exemplar of a manipulative patient, with no ability to care for himself, tests the Bellevue system as many as one hundred days a year to see if he can generate a concerned response from a dozen or so people, who hasten to help him.

Each day he spends on the wards at Bellevue costs about $600. Each visit to the ED, whether it leads to his admission or not, costs $125. So, adding up the twenty-five to thirty days when he is actually admitted plus the other days, in a year Toby Wilts will cost the City of New York some $30,000.

Goldfrank often reflects on the progression of events that has led to the present climate for the homeless. The hope in the 1960s was to get rid of the "snake pits," to use the new mood-altering drugs then being introduced to free mental patients from the bondage of forced incarceration. Back in their home environment, these patients would be supported and aided by families and by mental-health centers. For various reasons, however, these community resources have not been sufficient to provide the needed help—shelter, education, job development, continued social assistance.

So, many of these released mental patients have ended up instead in seedy single-room-occupancy hotels, which now are being taken over by real-estate operators, often for conversion to luxury housing. The result is that the high hopes of twenty-five years ago have been largely dashed by the stern realities that have supplanted them. And at Bellevue the tidal wave has been felt, as the number of released mental patients showing up in the Emergency Department has soared, year by year, almost, it seems, in proportion to the number of single-room-occupancy hotels that have been converted to more profitable housing uses.

"One of the worst days of the year for us is the day they run the New York City Marathon," Goldfrank says. "That is when 25,000 visiting runners come to town, looking for inexpensive housing. They take over all the rooms normally occupied by these down-and-out people. There's no place for them to stay, so, depending on the weather, they very often try the Emergency Departments of various city hospitals—Bellevue perhaps first of all."

In the Walk-in Clinic, Mary Dwyer is talking gently with a patient complaining of withdrawal symptoms, getting his medical history. "Are you a drug user?"

"Yes."

"Do you shoot up?"

"No more . . ."

Learning to interpret what truth lies behind those disclaiming words is one of the principal skills fostered in this kind of medicine. " 'No more' could mean 'not for the last six hours,' " Mary Dwyer says. "Or it could mean something else. You try to use past experience to arrive at a fair understanding. For example, heroin users overstate their dependency by three or four times." Heroin users don't want the doctors to give them less methadone than they will need, fearing less of a high or that they will suffer withdrawal pangs. So they ask for more than they need. Alcoholics, on the other hand, minimize their rate of use because those in respectable circumstances feel a stigma attaching to their drinking. Skid-row alcoholics, in the experience of the Bellevue Emergency Department, seem to be a little more frank, but still underestimate their intake. "So we multiply an alcoholic's estimate by two or three times," Dwyer says, "depending on what sort of life he's leading."

Down at the doctors' station, Hedva Shamir is writing up her charts, trying to bear down firmly enough so that she makes four legible copies. She sees many unclear copies of such records, as Goldfrank has asked her to head up a review of all the paperwork that attending physicians, residents and others create here. She hopes her own example will inspire her peers. She finishes up a report on Carlos Sanchez, a twenty-one-year-old messenger, who thought he had a blister on his foot but turned out to have stepped on a needle. An x-ray failed to disclose any FB—foreign body. He was given a diphtheria-tetanus shot, a Band-Aid, a kind word and a swift discharge.

One of the concerns of emergency medicine, as Goldfrank says, "is to maintain the herd immunity. Diphtheria has almost disappeared as a disease. But if we're not alert, especially in the derelict population, this old disease could come roaring back, and could sweep through a city like ours, as it is caused by a rapidly transmissible infectious bacterium that can inflame both the heart and the nervous system and often leads to death. So when we care for anyone, even though it doesn't register on society at large, we're taking advantage of the need for a tetanus booster by routinely giving a diphtheria booster, for which he may actually have a greater need. We're trying to defend all of us from the consequences of the lifestyle of some of us."

There's a ruddy-faced paramedic in the triage area, ready to go off duty, but looking as jubilant as an athlete who had just scored the winning point in a close-fought game. "That's two today, doctor!" he is saying to Dr. Shamir. "Four this week!"

Two saves for John Duval, a young paramedic who was trained here at Bellevue, and runs "bus" number 13-X, as he and his partner call their $80,000 Wheeled Coach ambulance. Their first save was the AIDS patient, who would live on for several hours. The second, just moments before, was the sort of save that sends a ripple of exhilaration throughout the whole ED.

"He fibbed on the sidewalk and now he's breathing on his own," Duval says. "It's really been our day!"

Duval has his own paperwork to attend to, so after a brief exchange, he goes off to complete his narrative of the sort of morning that will doubtless live in his memory for decades.

# 6

# A Shock and a Dive

It was 8:45 in the morning, fifteen minutes until the end of the shift for John Duval and his partner, Moe Padellan. Both were in their twenties, yet both were also experienced paramedics of the 1,500-person New York Emergency Medical Service (EMS). Their Ambulance 13-X was on the first shift: 1 A.M. to 9 A.M. They were moving slowly east on Twenty-sixth Street near Sixth Avenue—not far from the center of the territory for which they were responsible.

A call came over the police shortwave channel—cardiac arrest at Three, Three and Fifth. "We listen to the police channel," Duval later said, "because we like to 'buff' the calls—in other words, we don't always wait until Maspeth dispatches us if we can save a minute or two by moving right to the location."

In this case, there was a follow-up call from across the river in Maspeth, Queens, the Communications Control Center of EMS for the entire city. The center operates a highly sophisticated network that handles more than 700,000 calls a year. Now, 13-X heard that the same location had a "difficulty in breathing" report—presumably based on another passerby's 911 call. But, before this second report came in, Duval had radioed Maspeth that he and his partner were already on the way in response to the earlier call on the police channel.

Later, at Bellevue, the two paramedics told a number of staff members in the Emergency Department how they had made their second save of the day.

"It was the northwest corner of the intersection, and there was a crowd gathered around and two guys were already at work on the victim on the sidewalk."

The victim was well dressed and looked as if he had collapsed on his way to work. An attaché case lay close by, scuffed from its recent tumble. Duval and Padellan jumped out, Duval with the portable monitor and defibrillator (called a Lifepak) in hand, Padellan with an oxygen tank and endotracheal tube to insert in the victim's throat. The Lifepak is a briefcase-size monitoring and electro-stimulation device, with a small TV-like screen that displays an oscillographic picture of a heartbeat. After one thousand hours of training, paramedics like Duval and Padellan are adept at interpreting the various visual patterns. For example, a normal heartbeat produces regular spikes and troughs separated by short, flat recovery tracings. Asystole—no beat at all—generates an ominously flat line. And ventricular fibrillation—rhythmic anarchy—produces a chaotically writhing pattern.

"By the time we got there, these two guys, who later told us they had CPR (cardio-pulmonary resuscitation) training in Connecticut, had the old guy's shirt open and were administering CPR." One was giving mouth-to-mouth breathing, while the other was using the flat of his palm to push on the victim's sternum at regular intervals. The two had also made a hurried check of the victim's wallet. All they discovered was that his name was William Garrison, he was sixty-one years of age and lived about three blocks east of this very corner. They also gathered, from the scars on his chest, that he had undergone some kind of upper-thorax surgery in the fairly recent past—probably bypass surgery, they guessed. All of this information was conveyed in short, intense bursts as the Connecticut team kept up its rhythm until the new arrivals could take their places with hi-tech machinery to enhance the human means of life support—Duval with the Lifepak and Padellan with the oxygen and intubation gear.

The first thing Duval did was to set his equipment in its "quick-look" monitoring mode and then place the paddles on the victim's chest. As Duval and Padellan switched places with them, the original two reported that they had started immediately when the old guy had keeled over. They had been walking right behind him, in fact. He was still pink—a good sign when starting management of a cardiac arrest.

Padellan had first bagged the man—snapped on an oxygen mask—to increase oxygenation and now was trying to get the endotracheal tube established in the victim's throat, while Duval attempted to interpret what he was seeing on the small screen. He was seeing very little. The morning sun was streaming down directly on his equipment, overwhelming the screen with its radiance. He tried three hand positions to shield the screen so that he could distinguish the image.

Meanwhile Padellan was proceeding deliberately. In more than one case, a too-hasty insertion of a breathing tube has resulted in inadvertently pumping air down the esophagus and into the stomach, rather than into the lungs. In at least one instance, which all the paramedics had heard about in their training classes, the air pressure resulted in a burst stomach. So the team from 13-X was making sure of all its preparations before proceeding. On the other hand, any delay would lead to further risk of tissue and brain hypoxia—oxygen starvation.

Duval had the screen shaded now, and what he saw was alarming. "It was a pattern like worms—the line just wriggling all over the place." From his training and several years of experience he knew that he was witnessing ventricular fibrillation, the disorganized squirming of a heart that has lost its rhythmic memory.

The onlooking crowd, about fifty people, Duval later estimated, now was strangely hushed, even though the sounds of traffic and passersby continued to surround this small circle of hope.

"Moe, what's this guy weigh? Hundred eighty-five?"

"I'd say two hundred."

"So we give him three hundred watts?"

"Yeah." Padellan nodded as Duval reversed the paddles and applied an electrolyte gel to assure better contact and to protect the patient against an electrical burn. Duval placed one paddle under and to the left of the left nipple and the other just to the right of the sternum, on the upper center of William Garrison's bare chest. Duval had set his dials for a three-hundred-watt charge, and squeezed the trigger for an asynchronous shock.

There was an instant response. William Garrison opened his eyes, lifted his head and gave voice to acute annoyance. He brushed at the rubber tube that Padellan had installed but had not yet taped in place, pushing it entirely out of his throat. Garrison hitched himself up enough to eye both Padellan and then Duval. "What the hell are you guys trying to do to me?" he asked in a strong baritone voice.

"Behind me I could hear the crowd gasp," Duval later remembered. "I probably gasped myself. I never had a save like this one. It was amazing."

One of the first things Garrison saw, in his revived state, was a boldly lettered pin that Duval wears on the front of his shirt. "Don't Panic," it says.

"We told him to relax. We wouldn't tube him. We'd give him one hundred percent oxygen by face mask. We told him he had had a problem—we never want to alarm someone by giving too much detail—and we had been trained to deal with it. And just relax and we would take care of everything. So he looked at us a moment, shrugged, and relaxed, just as we hoped he would."

Governing everything that was done that morning was a sixty-page pocket-size booklet written by the Medical Advisory Committee of New York City, which was headed by Goldfrank's principal deputy, Neal Flomenbaum. This is the document that assures standard quality of care in such extreme cases as Garrison's. Following the next step in this *Paramedic Treatment Protocols* for ventricular fibrillation, Duval and Padellan established an IV infusion with an 18-gauge intravenous catheter, drew some blood and fed in 5 percent sugar water at the rate of 50 to 100 cubic centimeters an hour, to keep a vein open. This way, there is a ready access to the venous system if you need it for quick medication, should conditions suddenly change, as they sometimes do in such cases.

To get this IV drip started, Duval had to take a pair of extremely sharp shears. "They're only about two-inch blades, but they can cut through a penny. I used them to cut his coat sleeve, to get to a vein where we could establish the IV. It was too bad to cut up that nice suit jacket, but that's all we could do. Couldn't make him sit up and struggle out of his coat."

Finally, as they lifted Garrison into the rear of their ambulance, the two paramedics gave him a bolus of one milligram per kilogram of body weight of lidocaine, which worked out to about 90 milligrams for someone of his two-hundred-pound weight. This bolus and the IV drip of lidocaine that followed would defend against the heart's tendency to go back into ventricular fibrillation.

"We were ready to roll," Duval later recalled, "when this guy comes running up to us telling us to wait a minute, where did we think we were taking his brother? We were glad to see him because we could get some more medical history. He told us his brother had gone to his periodontist that morning and, some months before, had had double bypass surgery.

"He wanted to know what had happened, and we were sort of

vague. Didn't want to get him all upset. So we put him up front, where he wouldn't have to see if his brother went into fibrillation again on the way to the hospital. Which he didn't."

Ambulance 13-X headed toward Bellevue. En route, Duval transmitted via a special electronic device a rhythm strip showing the pattern of heartbeat that Garrison was now sustaining. By the time they arrived at Bellevue, Dr. Robert Hessler had already had a chance to study the ECG at the telemetry desk, where it was received and printed out. It was quick work to oversee Garrison's admission to the Emergency Ward.

It was to Dr. Dana Gage that the paperwork on 13-X's dramatic resuscitation would ultimately be routed. She would see the Garrison chart, among many others, and from them she could derive a sense of how the medical needs of those who fall critically ill in New York are being met.

She came to Bellevue, after finishing her residency in internal medicine at Montefiore Hospital in the North Bronx, to spend a year's time. She ultimately became involved in developing the paramedic training program at Bellevue and decided to follow through, as the opportunity arose, organizing additional training programs in years that followed.

You hear, from Dr. Gage and from others, that there has been a genuine revolution in health care in the last thirty years or so. "The experience of the Korean War," she says, "showed that if you can get people quickly moved to an emergency doctor's care, you dramatically improve a badly injured patient's chances.

"What began with traumatic injury over the past ten years has been applied to other areas of medicine: heart attack, acute allergic reaction, stroke, drug overdose. And we've taken what we've learned and extended it far beyond the walls of the hospital." Powers that used to be reserved to the doctor, with years of training and experience, are now passing into the hands of young people who, as Dana Gage says, "are extremely dedicated to the work that they do and are constantly having to operate outside a stable environment in poor conditions, with little support—and yet somehow manage to use their training, instincts and desire to serve with a remarkable degree of accuracy. I know the hardships they deal with constantly. And I'm always amazed that they don't burn out more frequently than they do."

Down at the Emergency Department telemetry desk, John Duval hears that his patient is now judged to be in fair condition in the hospital's cardiac-care unit. He tells his story several more times, and receives the congratulations of many others. He describes his feelings about the EMS service:

"We make fifteen hundred calls a year and maybe 10 percent of them demand our utmost skills—I mean really call on us to do ACLS or even BCLS." Those are acronyms for "advanced cardiac life support" and "basic cardiac life support." Only a few years ago, these techniques were known only to a limited number of highly qualified doctors. So it seems understandable that today's paramedics take on something of the cockiness of jet pilots.

"But sometimes you don't really get much cooperation. I mean we're a hot team, Moe and me, more highly motivated than most. We pride ourselves on being on the scene of an incident in the fastest time—under five minutes in our area—a lot of times under three minutes. We got to this guy this morning in about two minutes. Lucky there wasn't much traffic in our way. Sometimes they won't make way for you. I had to tap a cab on the bumper one time. Guy had me blocked, wouldn't go through a red light and there I was right on his tail. Howling on the siren. Still didn't move. So I rapped him on the bumper. Didn't move. I got so desperate I got out and ran up to him and shouted at him to get moving. He didn't speak English! But at least he finally moved."

Dr. Goldfrank cannot help observing the tendency of ambulances throughout Manhattan to bring two kinds of patients, especially, to Bellevue—those at death's door and those without any financial resources. There are some institutions in the city that will not accept a patient without a credit card or signs of medical insurance sufficient to assure payment for whatever hospital costs are incurred.

Bellevue was the first hospital in the city to put ambulances on the street—five horse-drawn vehicles in 1869. Now it is one of some twenty hospitals in New York City that have ambulances based there. Yet it is the hospital of choice for poisonings—including drug overdoses—both because of Goldfrank's eminence as a toxicologist and the hospital's close links to the New York City Poison Control Center. It is also the preferred destination in cases of traumatic injury, particularly involving gunshot or mutilation, as its trauma teams are so adept at reversing the results of such violent events. And it is the place preferred by the police or fire department when their fellows are struck down in the line of duty—occasionally they bring in gravely wounded police or firemen by helicopter to the Thirty-fourth Street heliport for quick transit by ambulance a thousand feet to the Twenty-seventh Street Emergency Department ambulance portal.

On the third floor of Bellevue's Administration Building, down the hall from Dana Gage's office, a CPR course is beginning, and a lecturer is telling a new class of Emergency Department nurses about arterial blood gases, acid-base disturbances and the means by which resuscitation can be improved. They are told the signs that will indicate whether a victim is suffering heart- or lung-related difficulty and just how carbon dioxide in the blood relates to relative acidity or alkalinity in the body.

"You're not dead unless you're warm and dead," the instructor is saying, and a score of pencils busily make notes. Meanwhile, the characteristic wailing sound of a siren marks another urgent problem at the Emergency Department entrance below.

A Hispanic grandmother and grandson are wheeled in together. Neither is conscious. Both have been in cardiac arrest. The boy has burns on his fingers and lower arms—signs, the paramedics think, that he had been playing with matches in his grandmother's crowded apartment on the Lower East Side. The place was ablaze when firemen extricated the pair and loaded them into an EMS ambulance at the scene.

The boy is about six. He is diaphoretic—sweating profusely—and coughs fitfully under the oxygen bag clamped to his face. His grandmother, perhaps seventy, has similar symptoms, but fewer signs of life. She is not moving at all. Both are covered with soot and are rushed along the entry corridor past the ranks of wheeled stretchers awaiting other incoming patients, through the double doors and into the trauma section of the Emergency Department. "Doctor in the slot," the loudspeakers intone, and several white-coated forms rapidly converge on the pair.

Soon there are still other victims coming in—a pair of firemen from the same blaze brought in a Fire Department car, both suffering from smoke inhalation. They too come into the trauma slot. Within minutes, arterial-blood-gas and carbon-monoxide analyses are being run on all the victims to assess the impact of the fire and the victim's body condition. Before arrival in the Emergency Department the paramedics, following the protocol drawn up by Bellevue doctors and other emergency medicine specialists, have administered 100 percent oxygen and 50 cc. of 50 percent dextrose in water and have drawn blood samples to further define the patients' problems.

The firemen are successfully treated on the spot with 100 percent oxygen via a non-rebreathing mask, which quickly removes their small quantities of carbon monoxide. But the soot in their upper and lower airways is substantial.

The boy and the old woman are not looking good, however. Few have survived carbon-monoxide-induced cardiac arrest. The boy has 35 percent of his hemoglobin bound to carbon monoxide, the grandmother 60 percent. Normal is 2 percent.

Now, the battle for their lives is focused on the basic unit of their well-being, the cell. As Goldfrank later explained it to medical students and residents, carbon monoxide has no smell and no taste and is nonirritating, and it has a voracious affinity for hemoglobin. It forms a stable compound with the blood that, paradoxically, makes it vividly red at the same time that it denies the hemoglobin its capacity to carry life-sustaining oxygen to the cells. As he has written in his text, *Toxicologic Emergencies,* "Inhalation of smoke from one cigaret—which includes substantial amounts of carbon monoxide—reduces the amount of oxygen available to the tissues by approximately 8 percent. This is the equivalent of going from sea level to 4000 feet altitude." In the intense environment of a smoky fire, the victims may have suffered the equivalent of a journey to perhaps 30,000 feet altitude—or losing half their red blood cells.

The doctor goes on to explain that carbon monoxide binds to myoglobin, hemoglobin and all the iron-carrying molecules of the mitochondria—the factory and powerhouse elements of the cells. So the elemental transactions, without which life cannot proceed, are interrupted and finally halted by the furtive invasion of carbon monoxide. "Since the brain and the heart are the organs with the highest metabolic rates, they are most sensitive to hypoxia (oxygen shortage) and account for the major toxic manifestations." Insult to the heart can result in any number of irregularities in the heartbeat, owing potentially to both muscular and electrical failure. Depending on the past history of the patient, very low levels of carbon monoxide can produce severe poisoning of the heart.

Swelling of the brain (cerebral edema) is another result of such poisoning, and can ultimately show up as loss of vision, inability to adjust either to brightness or darkness and hemorrhaging of the retina.

"The half-life of carboxyhemoglobin," Goldfrank has written, "is as much as five hours when the patient is exposed to air (21% oxygen) at normal barometric pressure—one atmosphere. This means that it takes five hours for the level to fall by one half. At three atmospheres, if you give 100% oxygen, the half-life is just 23 minutes—and there have been many instances of dramatic reversal of symptoms like coma, neurological deficits of various kinds and things more difficult to characterize like memory impairment and personality disorders." But there is no hyperbaric chamber, where such therapy can be given, at Bellevue. The facility that Bellevue—and actually the poor of the entire city—depend on is on City Island, a fourteen-mile trip from the Emergency Department. The hyperbaric oxygen committee of the EMS system of New York—of which Goldfrank has been a member since its inception—drew up a protocol for transferring patients to City Island as there are no multiple-place chambers in New York hospitals willing to care for indigent victims. The main chamber is on City Island because that unit was developed to treat sea divers with the bends—nitrogen bubbles in the blood caused by too-rapid resurfacing. Such divers can be easily rushed to City Island from anywhere in the region.

Shortly after the arrival of the now-comatose pair, arrangements had been made to whisk them to City Island, in hopes of reversing what seemed the unmistakable signs of impending death.

The FDR Drive was again blocked off and a police escort sped the pair, along with two Bellevue teams of doctors and nurses, in a swift convoy of EMS ambulances to City Island. All of this happened within an hour of the time the boy and his grandmother were rescued from their burning apartment.

While awaiting the results of the "dive," Goldfrank reminded his staff—in another of the informal teaching sessions that constantly take place in the rooms and corridors of the Emergency Department—that hyperbaric oxygen can reverse the poisonous effects of carbon monoxide, but it also can cause problems. It can complicate upper-respiratory-tract infections such as sinusitis. Those with asthma or chronic obstructive lung disease may suffer collapsed lungs. Other complications include ear discomfort, vision deficits and even oxygen toxicity of the central nervous system and lungs.

When the little boy and his grandmother came out of the tank, they were admitted to Bronx Municipal Hospital, the back-up institution for hyperbaric therapy, because of its proximity to City Island. For both boy and grandmother, carbon-monoxide levels were sharply down, but neither responded to any stimulus. Both appeared to be brain dead.

Nevertheless, Goldfrank and his team were unwilling to give up. There had been reports of dramatic improvements in other, similar cases, although probably in victims less ill than this pair. As Goldfrank says, "There's no way to practice emergency medicine without being an optimist. And there's no way to survive without making decisions. If you're not decisive, people worsen or die. Hamlet would have quit emergency medicine after the first few weeks."

The boy and his grandmother were taken back to the hyperbaric chamber the next day for another dive. But the additional treatment resulted in no improvement in their condition. Shortly thereafter, both died.

When Goldfrank gets back to his office after attending a meeting upstairs, there are several messages. One asks him to call Dr. Richard Weisman, the clinical pharmacist and computer expert who directs the staff at the New York City Poison Control Center, which is located across First Avenue from Bellevue, in the Department of Health building. Three other messages are from radio stations, and one from Channel 4 news. Goldfrank returns Weisman's call first.

"Hello, Richard."

"Yes, Lewis, we just heard there's been another cyanide incident. . . . It's Tylenol again. Girl up in Yonkers. Radio station just called."

"Looking for a reaction?"

"Yes. I referred them to you," Weisman says, in deference to Goldfrank as medical director of the Poison Control Center.

The last time there was an over-the-counter-medication scare the calls came in at the rate of 2,500 a day, nearly overwhelming the staff of the Poison Control Center. It was the director's hunch that this one wouldn't precipitate quite such a reaction. But they would know in a few hours. Meanwhile, the media were calling before he could return their calls. In twenty minutes he had given two on-the-air radio and TV interviews.

# 7 ⌁

# A Question of Poison

A three-year-old boy has drunk about two ounces of vanilla extract. The voice on the phone is not calm. It's one of the attending physicians at Lincoln Hospital's Emergency Department calling to the Poison Control Center. Dr. Weisman takes the call himself.

The center is a small suite of offices on the ground floor of the New York City Health Department building: one narrow room with four desks, computers and microfiche files for the pharmacists on duty to consult, plus four other adjoining rooms—Weisman's office, his deputy's office, the toxicology fellows' room and file and supply room.

As Dr. Weisman listens, he types in a few phrases on an IBM PC before him. He designed the system he is now using. Through a laser-disk mass storage unit attached to his computer, he has within his reach the entire contents of a thousand books and periodicals on all the varieties of toxins that are used as ingredients in various drugs, foods and household products, and the most recent research findings on these substances. The great source of current information comes from Denver, from the Poisindex data base at the Rocky Mountain Poison Center. Weisman's advanced interface now makes it possible to gather essential patient data while scanning this trove of information on various toxins at regular intervals to update the patient's record.

Weisman dials a number that alerts Dr. Roberto Bellini, a toxicology fellow in training at the New York City Poison Control Center/ Bellevue Hospital. Bellini occupies an adjoining office when he is not over in the Emergency Department or making toxicology rounds. He is a debonair young doctor whose exotic background equips him to handle a variety of unexpected crises. Born in Kuala Lumpur, fluent in several languages, he chose Bellevue after finishing medical school in Cincinnati and a residency in emergency medicine "because it is unique in all the world," he says, "for its caliber of care and variety of learning opportunities." After a moment, Weisman has linked Bellini and the Lincoln Emergency Department caller.

"What are the child's vital signs?" Bellini asks.

The attending physician responds: "Blood pressure 70 over 40; pulse 88 beats per minute; respiration 10 breaths per minute. Temperature normal." The child is drowsy.

Up on the screen of the small computer flashes a whole protocol for treating this misadventure. Vanilla extract, the computer reveals, is mostly ethyl alcohol. Because of the child's small size, the effects of even a few ounces of alcohol can be devastating.

"You should start an intravenous," Bellini says, "and give both dextrose and naloxone, just in case the drowsiness is due to an opioid. And it would be a good idea to put the child on a cardiac monitor and watch his vital signs carefully. And give him activated charcoal and a cathartic." The dextrose will treat potential hypoglycemia caused by alcohol or any other substance that might be involved—a "coingestant."

The activated charcoal, mixed into a cup of water, will make a slurry that can adsorb almost any toxin, once it is swallowed. Each particle in suspension can grasp to itself five or six times its own weight of most poisons it might encounter in the boy's stomach. (Alcohol, however, is poorly adsorbed.)

The attending physician acknowledges all of these suggestions as Bellini goes on to urge that all routine blood studies be drawn and that an immediate blood-alcohol-level reading be determined.

It becomes clear that the child will have to be admitted to the pediatric intensive care unit, because there is already evidence of central-nervous-system depression, in the drowsiness and low rate of breathing. As Bellini continues, "With that amount of alcohol he may very well need respiratory support. You'll also have to monitor the child's blood-glucose levels closely, watching out for hypoglycemia, until the alcohol is completely metabolized."

"How long will the child be intoxicated?" the attending physician asks.

"How much does he weigh?"

"About fourteen kilos . . . say thirty pounds."

Bellini asks Weisman's best estimate. The pharmacist does a few calculations and shows the physician his results. "At least ten hours," Bellini says, if the child really drank two ounces of vanilla extract. It would be 35 percent alcohol—that's 70 proof. "We'll know better when you get the first blood-alcohol figures."

A colleague, Dr. Mary Ann Howland, also a clinical pharmacist, and a professor of clinical pharmacy at St. John's University—in addition to being a consultant to the Poison Control Center—is handling another call.

A doctor out in New Jersey has a thirty-one-year-old woman with worsening numbness and prickling sensation in both her hands and feet. She reports that her gynecologist had suggested that she take large doses of pyridoxine—vitamin $B_6$—to treat fluid retention. The patient has been taking two and a half grams a day.

Dr. Howland says she will get some information if the doctor will hold on for a moment. A few key strokes on her PC and she has a voluminous listing before her. She reads through several pages of data. "Before 1983," she tells the doctor, "the literature said that only the fat-soluble vitamins could produce toxic effects in humans." That is, vitamins A, D and E. It used to be thought, she continues, that even large doses of water-soluble vitamins (such as $B_6$ or C) would be rapidly cleared from the body. "It is now recognized," she goes on, "that megadoses of pyridoxine are responsible for a newly recognized sensory neuropathy."

Today there is widespread "orthomolecular" use of massive doses of the vitamin as a remedy for premenstrual tension and water retention, Mary Ann Howland tells the caller. Often people take twenty to a thousand times the recommended daily allowance.

"How much has to be ingested before you see toxic effects?" the physician asks.

"In all of the cases of this kind that we have seen or have been reported, the daily dose has exceeded two grams per day. Patients sometimes are unable to walk or they develop a staggering gait. Victims have lost the sense of touch, don't react to high temperature, pinprick or vibration in arms or legs. These symptoms can go on for several months after you stop the megadose intake. And it can take many more months for complete recovery."

As these calls are concluded, others start coming in at the four stations here, where pharmacists receive queries from throughout the New York metropolitan area and are backed up by physicians on the scene, like Bellini, or others throughout the city and surrounding suburbs who voluntarily make themselves available for telephone consultation around the clock—an obligation that Goldfrank and his chief aides, Flomenbaum and Lewin, assume every night. So each is on call one night in three.

For Goldfrank, the Poison Control Center represents a triumph of interagency cooperation. When he came to Bellevue, in 1979, the Poison Control Center had been operating for some twenty-five years without physicians having been involved in its day-to-day operations for a number of years. It had grown in that time to become the largest such center in the country, receiving sixty thousand calls a year. Shortly after his arrival, Goldfrank took a leading role in reorganizing the center, for which he continues to serve as chief medical consultant.

Owing to Goldfrank's skills as a grantsman and a maker of consensus, the center received substantial seed funds under a national grant to upgrade poison control centers. As a result, the center now is able to give twenty-four-hour poison-control information by means of a staff of specialists in pharmacy, pharmacology and in medicine. In addition, the center disseminates a great deal of poison-prevention information and is an active site for training and research, involving both pharmacists and physicians. And in one small particular, Goldfrank's tenacity has an unusual monument—a memorable telephone number.

Hoping that the number 212-POISONS could be assigned to the center, Goldfrank asked the phone company if that might be possible. The answer was no. The number already belonged to a client and could not be reassigned without his consent. Who was the client? The phone company could not divulge that information. So Goldfrank began calling the number, hoping to persuade its owner to relinquish it for the public good. "I never got an answer. I would let it ring ten or twelve times, and no one ever picked it up. I tried at all hours, with the same result. So then I asked my secretary if she would try, whenever she had a moment to spare. For more than a year we kept trying. For more than a year, the same result."

Then one day, after they had almost given up, there was an answer, and the discovery was made that the number was in an unused office at Celanese Corporation. "That was a rare stroke of luck," Goldfrank recalls, "because for many years my father had been a vice president for Celanese. Through him, an approach was made to top management, and that's how we finally got the number." Today, a handsome lapel pin that has been widely distributed, and which Goldfrank himself wears, shows a glowing red apple on which is imposed a skull and a phone. Below, on a white, heraldic banner, the legend "212 P-O-I-S-O-N-S."

A call comes in from CBS. They've received a news bulletin: A young woman has just died from cyanide introduced into an Excedrin capsule. What should New Yorkers know to protect themselves? Weisman refers the caller to Goldfrank, then quickly calls over to the Emergency Department to alert the director to the news and the expected deluge of inquiries.

There's another call that Dr. Howland handles. She hears the symptoms recited, then asks, "Urine the color of sea blue, would you say?" To an affirmative response she replies, "Then the child has ingested a product containing methylene blue . . ."

"Poison Control Center?"

"Yes."

"I just took an Excedrin and I've got this funny feeling. Could it be cyanide?"

"May we have your name, telephone number and address, please?"

The caller complies.

"Is there any sign of tampering with your capsules?"

"There might have been."

"Please hold on for Dr. Bellini."

Bellini picks up his extension and introduces himself.

"What are your symptoms?"

"I feel sort of dizzy."

"Anything else? Any sweating or faintness?"

"Not exactly. I'm sort of damp and feel a little woozy."

"Then I'd suggest you get to an emergency room. Want us to call an ambulance?"

"No. I'm close enough to walk. My friends will bring me in. Is it cyanide?"

"Probably not. But it won't hurt to check it out. Those symptoms are significant enough so you should come to the hospital. Then we'll decide."

"It could be cyanide, then?"

"It's possible."

"Oh God."

"But stay calm. Please bring your capsules with you, just in case. There's good treatment available. You'll be in plenty of time."

The calls about the new cyanide scare start coming in in ones and twos. Most can be handled with a reassuring word or two, based on the symptoms. There's a progression of symptoms that Goldfrank has set forth that acts as a screen to determine if there really might be cyanide involved. First comes headache, then faintness, then diaphoresis—heavy sweating. Following those might come vertigo, excitement or apprehension, drowsiness, arching of the back and lockjaw, elevated temperature. And finally, the signs of imminent collapse: convulsions, unconsciousness, paralysis, coma and then death.

In helping to develop clear and helpful advice for such callers, Goldfrank, Weisman, Howland and the other staff members are constantly refining their methods of presentation so that highly complex material can be rapidly assimilated by those who do not have much scientific background.

Nevertheless, there is awareness among these professionals that the populace is awash in toxins—as many as 500,000 identifiable poisons, Goldfrank has said, are a part of modern life. Each has its own method of working, each a preferred line of defense.

He has listed an imposing summary of drug and poison exposure in his book, *Toxicologic Emergencies:*

- 20 million people in the United States are alcoholics.
- 10 million are cocaine dependent.
- 8 million persons are poisoned a year.
- 7.5 million use sedative-hypnotics (such as barbiturates or Valium).
- 500,000 are heroin or other opioid addicts.
- More than 2.8 billion drug orders and prescriptions are written a year.

Some sense of the complexity of handling individual cases may be gained by considering, as Goldfrank and his colleagues do every day, the "half-life" phenomenon. As the doctor uses the term, "half-life" is the time it takes your body to eliminate half the dose of a drug or poison you ingest. As Goldfrank has written, "If the half-life of a drug could easily be predicted, one might be able to determine when the drug would no longer be toxic to the patient. When a drug is removed at a constant rate, it is said to obey 'first-order kinetics.' " In other words, the kidneys will get rid of the poison in such a way that on successive cycles, or half-lives, you get a decrease from 100 percent to 50 percent, 25 percent, 12.5 percent, 6.25 percent, 3.12 percent and 1.56 percent.

"Unfortunately," Goldfrank says, "many drugs that behave that way in normal concentrations have a different pattern in toxic doses. When a person is overdosed—his enzymes saturated or his transfer mechanism overloaded—the elimination rate becomes a problem in 'zero-order kinetics.' " That is to say, the rate of elimination may well become independent of how much you have ingested.

Some common drugs like aspirin or alcohol switch from a first-order to a zero-order pattern as soon as the metabolizing enzyme sites become saturated. Then there may be new toxins created by your body as it tries to get rid of the old toxins, so that normal metabolism is blocked or slowed, sometimes dangerously.

"Thus," Goldfrank concludes, "overdose kinetics are complex and patients with abnormally high amounts of any drug have to be watched carefully."

Cyanide calls are coming in at a regular rate, and Bellini has been asked by Goldfrank if he will meet a camera crew from Channel 11, which is already on its way.

As Bellini answers other calls, his thoughts turn back to another cyanide scare, of a different sort entirely. A year before at a small jewelry plant in midtown, a pair of young men got into an odd sort of macho duel. One, a heavily muscled Armenian, and the other, a mustached Greek, were on their coffee break when one apparently dared the other to drop a twenty-five-gram pellet of crystalline sodium cyanide compound into his coffee. The chemical is used in precious-metals fabrication as a cleansing agent. The Armenian took the dare, and within minutes after he had sipped just a few swallows of the toxic brew, he was already experiencing sharp headaches. By the time the ambulance arrived, he was sweating profusely.

"When he got to us," Bellini recalled, "we knew we had to act quickly." Already the paramedics, responding to Goldfrank's advice from the Bellevue telemetry desk, and suggestions from the Poison Center (all three linked up by radiophone) had started him on a breathing mask, so when he arrived at the Emergency Department he was getting 100 percent oxygen. As Goldfrank has noted, mouth-to-mouth breathing in such poisonings can be dangerous. There is at least one case in the literature that suggests that secondary cyanide poisoning can be induced by such resuscitation attempts. In one instance, a man in Wyoming who tried to revive his dog, which had accidentally eaten some cyanide poison pellets, had himself developed a numbing of the lips and shortness of breath. CPR in such circumstances is now considered a very high-risk procedure.

The Armenian victim already had apparently ingested a lethal dose of the poison. He had alarmingly irregular heartbeats—ventricular arrhythmias and multiple premature ventricular contractions—when he arrived. The doctors worked side by side with an antidote kit to try to reverse the effects of the poison. The victim looked extremely ruddy and was short of breath, despite the enriched oxygen that he was breathing. What happens, Goldfrank has explained, is that the body suffocates because cyanide interferes with the ability of the individual cell to receive the oxygen that each pulse of fresh blood brings. Normally, hemoglobin picks up fresh oxygen in the lungs and each heartbeat sends it on to cells throughout the body. The exchange to the cells depends on off-loading from the hemoglobin to the part of each cell that acts as the loading dock—the mitochondria. But the cyanide blocks this transfer point, so even though you are breathing, and your blood is picking up the oxygen from the lungs, giving you that ruddy glow, your cells are still starving for oxygen, because of this blocked reception point. So you pant breathlessly, sweat profusely, have seizures, ventricular arrhythmias—erratic contractions of the thick muscular wall of the heart—and then you die.

The antidote kit removes the cyanide from this key obstruction point. To accomplish this life-saving reversal of an otherwise lethal progression of events, the doctors next broke an ampule of amyl nitrite—the same stuff that homosexuals, especially, prize as "poppers," which supposedly intensifies their sexual pleasure. "Pearls" of this substance were broken onto a gauze sponge, and held under one of the ports of the oxygen mask. Next swiftly followed a series of IV injections that had the effect of plucking the cyanide out of its obstructing position and then binding it to a transformed hemoglobin molecule. But there was a possible hazard here. The hemoglobin is converted to methemoglobin, which reduces the oxygen-carrying capacity of the blood. Should the hemoglobin, as reflected in the red-blood-cell count, be below a favorable level, death can occur as surely as if caused by the cyanide itself. So fire victims, who may have breathed considerable carbon monoxide and cyanide simultaneously, are at special risk because the cyanide antidote can aggravate the effects of carbon-monoxide poisoning. In such cases, the hyperbaric oxygen chamber is one possibility. So at best amyl nitrite, sodium nitrite and then sodium thiosulfate represent a difficult therapy, in which the patient is poisoned one way in order to get rid of a still more dangerous poison. Once the cyanide removal is under way, the doctors administer another part of the antidote kit—sodium thiosulfate—to render the extracted cyanide inactive and to promote its excretion into the urine. If they get successfully to this point, they have won the battle. But the war for unobstructed oxygen transport may continue to rage on.

For the Armenian jeweler, the doctors had to go back and repeat the whole procedure, which meant keeping the patient under observation for twenty-four hours. "He was saved all right," Bellini recalls, "but at first it looked as though he wouldn't be."

The prospect of treating numerous cyanide victims simultaneously remains one of the least-welcome scenarios, yet after the Bhopal disaster in India, Goldfrank called his Poison Control Center colleagues together to make contingency plans for any similar toxic-gas release in the New York area. Within months of that planning session, they had a near-miss that still arouses apprehensions years later:

Near Kennedy Airport in Queens there is a metallurgy plant that uses quantities of cyanide products in its processing activities, and keeps a sizable inventory of sodium and potassium cyanide stored in granular form in fifty-five-gallon drums. A fire broke out in the plant and raged briefly out of control, raising apprehensions that, when water was used to douse the flames, it might well react with the crystalline cyanide compounds and form hydrocyanic acid, putting the lives of thousands of surrounding residents at risk.

"We at the Poison Control Center and the city Environmental Protection Agency coordinated closely with the Department of Environmental Health, the police and the fire department in making a survey of the number of antidote kits available for treatment," Goldfrank said later. "A worst-case projection suggested that there would be as many as fifteen hundred victims, but we could only verify a few hundred kits in New York—typically there are a half-dozen at each of the major hospitals. So Richard Weisman got on the phone to Eli Lilly in Indianapolis, which makes the kits, and they reacted admirably— they sent sixty kits in their own private Lear jet, just so that we would be covered if the fire spread. We had the extra kits within three and a half hours of our call."

Meanwhile, Dr. Weisman and his staff were alerting every hospital in the area of the fire to their patients' possible exposure to catastrophic cyanide poisoning. Weisman had already managed to get all available extra kits shifted into the half-dozen closest hospitals—Elmhurst, Queens General and other hospitals within a ten-mile radius. Then Goldfrank and his team had made the difficult assumptions on the probable course of events if the fire took its most unfavorable course and there were as many as fifteen hundred victims of highly concentrated fumes. Exposure to one hundred milligrams of cyanide gas is probably irreversibly lethal. That's equivalent to about one-fiftieth of a teaspoon. "But those with lesser doses were likely to be major beneficiaries of the antidote, if we could get kits in sufficient numbers," Goldfrank recalls.

Evacuation plans were being advanced by city authorities as the fire fighters succeeded in dousing the blaze before the flames got to the cyanide bins, so the crisis passed.

Still, in the minds of the Poison Control Center, that was a warning, one that made urgent a close study of the city's toxicologic defenses against a real catastrophe that might arise from any of a number of cyanide-related toxins (primary poisons or metabolites) that have become a part of daily life.

The range of such poisons that have become essential in the United States is disquieting. All the plastics in most packaging and lightweight gears and mechanical assemblies, for two examples, depend directly on cyanides of various compounds for their manufacture. So does the photo industry. There are at least sixteen different kinds of such cyanides, each deadly to the human body, each essential for the comforts and ease that we enjoy today.

In addition to cyanide, there are many other common poisons at large. The miracle drugs, which in small doses enhance health or suppress microorganisms, in larger doses can kill—and there are at least twenty-five such substances. Then there are nearly a hundred chemicals that are used as building blocks for still more complex compounds; nearly 150 pesticides, any number of plasticizing compounds, solvents, heavy metals, and catalysts, each of which poses its own threat to human health. Some kill enzymes in the body. Others suppress the immune system. Still others act as the trigger that sets off cancer, or act to erode the lining of the nerves, or damage the liver or destroy the lungs' ability to function.

Against this ever-increasing tide of substances the toxicologists oppose all their knowledge and skill, and now, as in Richard Weisman's laser-disk device, they increasingly call on the computer to help marshal the defensive facts quickly enough to save lives.

"What we try to keep in mind," says Roberto Bellini, "is that kids younger than five years old are responsible for half the calls we get. And those typically involve ingesting just one substance. It's relatively easy to field that kind of question. Then, of the half that involve adults, perhaps half are deliberate poisonings, suicide attempts. And the other half are so-called recreational. Someone makes a bad trip on some kind of drug. It's a battle. And it's exciting. And I love it."

The phone is ringing at his elbow again. "Dr. Bellini . . . And you're using capsules? . . . But no sign of tampering? . . . Headache? Any sweating? . . . Well, then, I'd say you're not at risk. If you start sweating, or have shortness of breath, give us a call back. Otherwise, I'd say relax." His phone is ringing again.

As he takes another call the Channel 11 team appears at his door and, as he nods his consent, the camera crew begins immediately filming the young doctor as he reviews the symptoms of cyanide poisoning with yet another caller.

√\~

Across the street, Lewis Goldfrank has just received a call from an old college chum, Lloyd Schnell, now a busy corporate attorney who remains a devoted friend and admirer. Schnell wonders if the Poison Control Center is being inundated by all the cyanide excitement. No more excitement, Goldfrank responds, than when he and Schnell were engaged together in raising funds for civil rights organizations during the era of James Meredith's integration of the University of Mississippi Law School in the sixties. They reminisce about that for a while, and as Schnell rings off Goldfrank gathers up his papers and makes ready to leave. As he does so, he sees a reprint from which the words "Worcester Foundation" leap to his eye—enough to put him into a retrospective mood as he sets out for Grand Central. For the Worcester Foundation was one institution that played a key role in helping him define his own purposes in life, a major influence in forming his aspirations to become a medical student.

# 8

# The Making of a Medical Student

Lewis Goldfrank was born just a few months before America's entry into the Second World War. He was a large and precocious child, who gave delight to his parents from his earliest years.

His mother, Helen Colodny Goldfrank, cherished the ideal of "a golden, rural childhood" for Lewis and his two younger sisters. Helen and her husband, Herbert, came very close to matching the ideal, given the fact that they found a house surrounded by woods and farmlands within thirty-five miles of Manhattan in the town of Thornwood. It was a minor miracle that they discovered a place in a rural setting, inexpensive yet close enough to the city to permit them to meet mortgage and commuting costs for $60 a month. Here they raised their family.

Herbert Goldfrank was a department manager with the international chemical concern Stein Hall and Company. Drawing on his own interests in chemistry and botany, he was soon to develop into an innovative marketing engineer—able to find new uses for old botanical and chemical products and finding natural resins such as locust bean, guar and gum arabic suitable for unexpected tasks—like ice-crystal preventives for ice cream, strengtheners for paper pulp and additives for assisting oil recovery from deep-earth wells. When Lewis was two, Herbert went off to war with the U.S. Army, serving in the European theater.

Helen Goldfrank spent her youth learning her trade as a writer on various small left-wing newspapers and magazines. When she wrote her first story, the printers could not spell Colodny, but liked her story. So they gave her a new name, Helen Kay, which she kept as her pseudonym for all her writing and professional life.

She spent two years at Time Inc., first at *Time* magazine and then at *Fortune*. She left there to go to Cleveland, to act as assistant editor of the newspaper of the Central Trades and Labor Council of the CIO.

After Lewis was born in 1941, she helped found and edited a monthly journal for the State, County and Municipal Workers Union (CIO) and continued to work there while Herbert was overseas. During the last year of Herbert's army service she was on the staff of the CIO *News* in Washington, D.C.

Lewis was born in September, 1941, when his mother, Helen, was twenty-eight. "It seemed like an advanced age, then, to have your firstborn," she recalls. He was able to speak at fourteen months and to read by the time he started school. His mother went back to work when he was still quite small and managed a successful career and motherhood at the same time—a pattern at least thirty years ahead of its time.

Both the Goldfranks had strong family traditions of social consciousness and concern for those less fortunate than they. Lewis's maternal grandfather was caught at age thirteen by the secret police in Russia with duplicating chemicals on his hands—considered to be absolute proof of his involvement in producing anti-czarist publications. He was sentenced to exile in Siberia, when his father—an affluent cheese broker in Pinsk—managed to secure his release on condition that he emigrate. He did so, leaving his family behind.

He prospered in the United States and continued his keen interest in left-wing politics, devoting a good portion of his time to labor-union organizing. He earned a good living as a pharmacist, but found himself drawn to radical causes. His membership in the Communist Party brought him to the attention of government authorities, notably the FBI. In view of what happened later, it seems likely that a dossier on the family was begun in some government bureau as early as the 1920s.

On his paternal side, Lewis's forebears were among the ten founding families of the Society for Ethical Culture, which was founded in reaction to overly formal Judaism and Christianity, sparked by Felix Adler, the heir-presumptive to his father's post as rabbi of Temple Emanu-El on Fifth Avenue in New York. Young Adler rebelled against "priestly" religion and fostered a movement that, beginning in 1876, drew on the prophetic tradition to emphasize a faith founded on "deeds, not creeds."

The values of this movement pervaded the Goldfrank household from Lewis's earliest years. Among its key concepts is the ascription of worth to every human being. As one of the leaders of the society has written, "Of course, we do not mean that every man and woman is valuable to society. There are drunkards, thieves and murderers. What we do mean is that even in the most depraved of men there is latent worth that may be brought out."*

When young Lewis was of Sunday-school age, the Goldfranks and a half-dozen other Westchester families founded the Ethical Culture Society of Northern Westchester and it was there as well as at home that the boy heard the major goals of Ethical Culture set forth in various ways: First, no creeds, no priests, no pomp or elaborate ceremony. Second, belief in the infinite and unique worth of every human being. Third, the determination to conduct your own life so as to foster the maximum potential in others, hence in yourself. Fourth, an insistence on personal integrity without any pledge to specific beliefs—as in the existence of God, a life hereafter or posthumous punishment or reward. And finally a strong faith in the power of man to shape his own destiny.

After Herbert's return from the war in 1945, Lewis came to know, and later to emulate, his father's talent for drawing out latent skills in those he knew and worked with. It was during the immediate postwar period, when paper mills were unable to satisfy the clamor for their product, that Herbert Goldfrank helped discover that locust-bean gum chemically helped the slurry of wood fibers (the "liquor") bind together with more tenacity so that paper-making machines could run 15 percent faster, thus increasing output at modest additional cost. This was the era, too, when the family expanded, with the birth of two younger children—daughters Debbie and Joan.

Under the pen name Helen Kay, Lewis's mother had turned to writing children's books as her son grew up, and he himself was the subject of several of them. Among the titles were *Apple Pie for Lewis,* *One Mitten Lewis,* and *The Magic Mitt.* The first was to be the cause of an unexpected and painful turn of events for the family. It was a successful book, and was selected by the State Department as one of

*Ethics as a Religion,* by David Saville Muzzey, Frederick Ungar Co., New York, 1951, p. 265.

the exemplar works that would demonstrate to others how Americans live. So it was added to the official list of especially recommended titles and was placed in U.S. embassy libraries around the world.

But, in their indefatigable investigation of possible sources of subversion and covert Communist influence, investigators for the subcommittee of the junior senator from Wisconsin, Joseph R. McCarthy, came across author Helen Kay's name. She had served as a courier for funds that were taken from various sources, some apparently linked with Communist elements, and given clandestinely to labor groups in Germany in 1934, shortly after Hitler rose to the position of Chancellor. These investigators included Roy Cohn and G. David Schine, two young men whose world-wide junkets in behalf of the subcommittee drew considerable publicity in the tumult surrounding the senator's flamboyant disclosures of subversive activities in all walks of American life. Somewhere, in some government file—perhaps FBI, perhaps elsewhere—there was a dossier on the travels in behalf of anti-Nazi groups by the young Mrs. Goldfrank, whose great aplomb and extremely youthful appearance made it possible for her to pass as a sixteen-year-old even though at the time she was twenty-one.

As she explains today, she undertook those trips with the understanding and support of her family, and with full sympathy for the goal of helping in the fight against Nazism. "We felt very much that we were striking a blow against Hitler by conveying those funds," she says.

Twenty years passed. One day there was a stranger at the door of their house in Thornwood.

"Mrs. Goldfrank?"

"Yes."

"This is for you," the stranger said, handing her an envelope.

As she reminisced, "Of course it was all a bolt from the blue. I looked at the papers and it was a subpoena from the U.S. Senate. The marshal, or whatever he was, was already on his way back down the walk when I called out after him. 'But who's going to look after the children? I have a baby, you know.' Our third child—Joan—was just fourteen months old at the time. 'Lady,' he said, 'that's your problem.'"

Three days later she had to appear in Washington. The local press prominently featured the wire reports on the McCarthy subcommittee's subpoena. And there was a sharp change in the local environment: some of the Goldfranks' neighbors turned immediately against the family. A few expressed sympathy and solidarity. But, for the family, the essential mood was one of increasing foreboding.

Helen had considerable trouble getting a lawyer. One well-known attorney in Washington suggested that she go ask Senator McCarthy what he wanted. Another, a leading law professor and a friend of her father's, explained the importance of the Fifth Amendment in times of hysteria. But the first didn't want to go with her to face the McCarthy subcommittee.

She was accompanied by the second lawyer, who told her that if she tried to explain, if she went beyond the familiar words of the Fifth Amendment's provisions against self-incrimination, she would no longer have constitutional protection and could be cited for contempt of Congress forthwith.

So, at the hearings themselves, Helen knew that there was essentially no defense against the demands that she name names of those with whom she had been associated in years long past. As with many others summoned before the committee, she came to know that the accusation was the verdict. As a result of her brief appearance, and the categorization of her as "another Fifth-Amendment Communist," she passed into a kind of limbo.

It was an especially trying time for Lewis, as his classmates were not above taunting him and trying to provoke him to fight. He withdrew into himself, which was not difficult in view of the isolation in which the family house sat. There were open fields around it, and one neighbor, in fact, had a herd of sixty cattle. During this time Lewis and his father worked intensively in the fields and gardens—bringing wild shrubs and trees out of the woods and transplanting them. He learned about root pruning, grafting and fertilizing and the ways of many different plants—among them mountain laurel, pine, hemlock, rhododendron—as well as the many flowers, fruit trees and vegetables in the family gardens, which he also helped to tend. His interest in growing things, kindled in those difficult years, remained an active enthusiasm that would one day come to fruition in his expertise as a toxicologist.

Under the stress of the times, the Goldfranks steeled themselves as a number of anonymous threats came by phone and mail. Herbert Goldfrank hired an off-duty New York City policeman to guard their house against arsonists' threats. For a period one friendly neighbor appointed himself their guardian; he pitched a tent on their front lawn and occupied this outpost during the night, as defense against clandestine marauders.

"One bright September day ponies began to appear at the foot of Nannyhagen Road. . . ." So began Helen Kay's *A Pony for Winter*, a children's book she wrote about the arrival at the Goldfrank place of a little black pony named Mollie.

Mollie was one of the small steeds that provided pony rides during the summer at the Playland amusement park in Rye. Now, as fall approaches, their owner planned to board them at various places in northern Westchester. Both the Goldfrank parents agreed to Debbie's pleas that they take in a pony named Mollie until the next spring.

While the girls had Mollie to take their minds off the tensions of the time, Lewis turned increasingly to his stamp collection and to his favorite sport, baseball, where his size and quickness helped establish him as a fine batsman in the Thornwood Little League. His father recalls that his average was a lofty .576 in the summer of 1954, when his team, the Giants, advanced to the final game of the season—for the championship. Lewis awoke the morning of the big game with a sharp pain in his stomach. He skipped all his meals, and sat with a basketball thrust into his stomach, hoping to stifle the pain enough to play. His basketball treatment gave only slight relief. Still, he suited up for the game and arrived at the field looking wan and telling the coach how poorly he was feeling.

The coach left him on the bench until the sixth inning. The Giants were locked in a scoreless tie with the Indians and it looked as if there never would be a score.

"I could hardly stand, the pain was so intense," Lewis recalls, "but the coach told me to come in as a pinch hitter. So I did. I don't know how, but I got hold of a pitch and drove it over their center fielder's head. It should have been a home run but I could not run, so I only got to second base by the time the ball was retrieved and thrown back into the infield. The next batter hit another long ball. Again, it should have been a home run. But I was walking around third base and just barely moving by the time I got home. I scored and we won, one to nothing."

Lewis collapsed on home plate. He was rushed to Grasslands Hospital in the neighboring hamlet of Eastview and there was treated by a surgeon who was called from a country club dinner-dance, arriving in the examination room in black tie. He successfully removed Lewis's inflamed appendix. As there was no children's ward in those days at the hospital, Lewis was put into a ward alongside prisoners, where he became the virtual mascot of the ward. He learned to play cards from his fellow patients, who were manacled to their beds. "It was my first experience of prisoners," he recalls.

Meanwhile, the headlines in the local paper, the *Patent Trader*, said of twelve-year-old Lewis, "Wins Game, Loses Appendix," suggesting that already the climate of opinion among their neighbors was swinging in their favor. His mother's troubles continued to haunt him and the family, but gradually the passage of time helped soften the impact of the publicity and the notoriety.

While caring for her growing family, Helen went on writing children's books, and Herbert continued traveling in behalf of his company. But, because of the McCarthy hearings, neither of them was able to leave the country, as they were refused renewal passports. They didn't get their passports back until 1960.

There was some irony in Lewis's graduation from junior high school, when he received an award for academic excellence in history from the Daughters of the American Revolution, an award his son Andrew would also win twenty-two years later at *his* graduation.

Besides exhibiting prowess on the baseball diamond, Lewis was becoming an able scholar, and, when he was given a high school reading list in ninth grade at Pleasantville High School, he was delighted to discover that virtually every one of the required works—from Plato's *Republic* to Mark Twain's *Huckleberry Finn*—could be found in his parents' library. Among his intellectually inclined classmates, the attitude toward the Goldfrank family increasingly became one of admiration for their cultural attainments and indifference toward their controversial political views. In fact, at this time, Helen had begun a correspondence with Pablo Picasso, with a view to writing a book about all of Picasso's pictures of children. When the Goldfranks again were free to travel abroad, one of their first trips was to meet with Picasso to plan the art for Helen's book, *Picasso's World of Children*, which Doubleday brought out in time for Christmas, 1965, in a large, sumptuous format commanding the then-considerable price of $25.

As Lewis's academic success continued, a family friend, Albert Q. Maisel, a science writer for *Reader's Digest*, suggested that Lewis apply for one of the much-sought-after study grants for an intensive research science course established by the Worcester Foundation for Experimental Biology in association with St. Mark's School in Massachusetts. The foundation was then in its second decade, having been set up by two Clark University scientists, Hudson Hoagland and Gregory Pincus, with a particular emphasis on the fields of biochemistry and physiology. It was through the work of the Worcester Foundation that oral contraception was developed.

Lewis won a spot in the two-month program in the summer of 1958. "It was the first time I had been away from home," he recalls, "my first experience in a dormitory setting. And my first exposure to the strict atmosphere of a boarding school." Even though the regimen chafed somewhat, the intellectual excitement of being among world-class scientists and the stimulus of learning advanced laboratory techniques and the habit of scientific meticulousness awakened in him a burning ambition to become a scientist himself. It also acquainted him with the scientific strength of Clark University and inspired him to apply for admission there.

In these difficult years, one of the most admiring of Lewis's schoolmates was a petite blonde, Susan Harrington, who was two years behind him in school. She was as able a scholar as he, and both earned exemption—as honor students—from attendance at study hall. As they strolled the grounds of Pleasantville High School together, they began a romance that would culminate in their marriage six years later.

He graduated near the top of his high school class in 1959 and was accepted at Clark, where he enrolled the following fall. Clark had been founded, like Johns Hopkins, as a graduate school, by Jonas Gilman Clark, who made a fortune supplying forty-niners during the great California Gold Rush. Its first president was drawn from Johns Hopkins, and, like Johns Hopkins, Clark later added undergraduate instruction to combine elements of both college and university under its roofs.

Lewis was a serious student and an active member of the student council. During his junior and senior years he was the winner of a National Science Foundation grant to serve as a research assistant at Clark. Again his grades put him near the top of his class. He was also something of a firebrand, speaking out against fraternities and in favor of civil rights—using his persuasive powers to rally support for the Freedom Rides and sit-ins then sweeping the South. His activism sat well with most of his classmates, and he was elected president of the student body in his junior year. A student editor summed up

Goldfrank's performance at the end of his senior year by noting that, while he was not the most popular president, he nevertheless would "own up to his failures," and had earned general respect for his "decisive and energetic leadership." He was a Phi Beta Kappa graduate with honors in chemistry.

During his undergraduate years he also won a grant from N. W. Pirie, a celebrated British biochemist, who had backing from Rockefeller Foundation funds, to serve as a research assistant. Lewis had first learned of Pirie's work at the Worcester Foundation. Goldfrank spent the summer of 1962 in England, at Harpenden, Hertfordshire, the oldest continuously cultivated agricultural research station in the world. He worked on leaf-protein studies and absorbed the philosophy of Pirie, an unusually quotable scientific gadfly and a refreshingly down-to-earth and inventive researcher.

Pirie's combination of scientific and social goals had a ring of solid practicality, of a sort that his young visitor found highly congenial. Of the many provocative plans that Pirie developed, one that particularly caught Lewis's imagination was the extraction of useful protein from such rampant weed growth as water hyacinth, which is one of the most prolifically efficient converters of sun power into potential food power on earth. It is also the bane of many tropical and subtropical societies, as it clogs waterways. Pirie's plan includes the use of a low-cost, gigantic food blender that can pulp and then pelletize such foodstuff and render it as palatable as an alfalfa cube or tablet.

As he worked alongside his lively mentor, Lewis came to appreciate the costs of originality, as set forth by Pirie in a *Scientific American* article, "Orthodox and Unorthodox Methods of Meeting World Food Needs."* Wrote Pirie: "The innovator must expect to run into trouble. When someone made the old comment that genius was an infinite capacity for taking pains, Samuel Butler replied: 'It isn't. It is an infinite capacity for getting into trouble and for staying in trouble for as long as the genius lasts.' "

It was almost as if the great satirist had been foreseeing the future for Lewis Goldfrank.

*February, 1967.

# 9 $\wedge\!\!\!\backslash\!\!\backslash\!\!\backslash\!\!\backslash\!\!\sqrt{\ }$

# Becoming an Emergency Doctor

"Our Admissions Committee has given me the pleasant task of inform-
ing you of your acceptance as a student at the Johns Hopkins School
of Medicine." So read the letter Lewis received in November, 1962, in
response to his application for early admission.

Shortly thereafter he also got good news from a half-dozen other
leading medical schools. But Lewis already had his heart set on Johns
Hopkins. This was the institution that bore the stamp of Sir William
Osler. As one of its founding faculty members, Osler helped lead a
revolution in medical education in the 1890s. Hopkins became the
cradle of the modern, clinical approach to the instruction of medical
students.

As Osler had said: "The natural method of teaching the student
begins with the patient, continues with the patient, and ends with the
patient, using books and lectures as tools, as means to an end.... The
whole art of medicine is in observation ... but to educate the eye to
see, the ear to hear and the finger to feel takes time, and to make a
beginning, to start a man on the right path, is all that we can do."

In the minds of many, the distinction Osler lent to Johns Hopkins
made it the most prestigious medical institution in the U.S.—some
said in the world.

Goldfrank departed for Baltimore, eager to become part of such
a tradition. The honeymoon was not long and not happy.

"From Day One I began to think it had been a major mistake," he
recalls. "As part of my orientation I was given a tour of the hospital
by a third-year student. When we got into the Emergency Room, my

mentor drew my attention to an old black man lying on a stretcher, his head badly lacerated, waiting to be sewn up. My guide turned to me. 'See that old guy?' he said. 'How'd you like to sew him up?'

"I was shocked by his question. I had studied inorganic chemistry in college. I knew molecules, not anatomy. I couldn't tell a vein from an artery and he must have known it. I could have done the old man great damage. But that didn't seem to count. It was very disillusioning."

At Clark, Goldfrank had worked against fraternities on the grounds that they were exclusionary and had no place in an institution devoted to the pursuit of knowledge. He discovered that at Hopkins there was a strong fraternity–social club influence—indistinguishable in his mind from undergraduate fraternities.

He was strongly affected by the obvious poverty he saw in the sections of Baltimore surrounding the medical school and the seeming indifference of his fellow students and faculty members. He'd raise the question in discussions, but felt himself being tuned out. He got repeated warnings against becoming "too political." Still, he persisted in drawing others into conversations about how students might work toward social or political solutions to the problems that seemed to come right up to their doorsteps—poor housing, bad nutrition, alcoholism and skimpy health care. But there was a sense that he was breaching some unspoken tradition. He could not understand how such apparent refusal to observe what was going on right in their own immediate environment could square with Osler's remark that "the whole art of medicine is in observation."

Instruction was organized on the quarter system. Ten weeks of concentrated effort were spent on each subject, which meant five days a week from 9 to 5 and Saturdays until 1 P.M. They started with biochemistry. Goldfrank and a bio-medical engineering student, Bob Scobey, teamed up on a project that took them most of that quarter—analyzing the metabolism of frogs. As their description in their term paper said, to prepare the necessary enzymes from spleen, liver, kidney, pancreas, and brain, the frogs, species *Rana temporaria*, "were sacrificed," and their mitochondria—the "power plant" within the cells—extracted by a series of mincing, centrifuging, spinning and filtering actions.

"We wound up with lots of frogs' legs, and after we were through in the lab, we would take the frogs' legs over to the Scobey apartment and have a feast. I developed quite a taste for frogs' legs," Goldfrank recalls.

He and his partner turned in their paper with the imposing title, "The Hydrolysis of Adenosine Triphosphate by Tissue Homogenates of *Rana temporaria.*"

Just a week later came President Kennedy's assassination. "There were cheers and parties in the dormitories. I felt very much estranged and I went to Washington during the funeral just as the previous August—right before enrolling—I had gone to the Lincoln Memorial to hear the Martin Luther King 'I have a dream' speech. Both were great experiences, but neither was something I could share with my classmates."

The next two quarters were devoted to anatomy, and, as Goldfrank recalls, "there was a tremendous sense of intellectual excitement in the course material itself, but there was little rapport with students or faculty. They just had no response to the things that seemed to me most urgent and most worthy of discussion."

There were warnings. "They told me I had to get better grades. That I better halt my 'outside activities' and stop wearing turtlenecks. It became clear that I wasn't fitting in. They knew it and so did I."

By now it was early spring, and the quarter was devoted mostly to physiology, with some genetics and—on Saturdays—the history and philosophy of medicine. Goldfrank felt particular discomfort in several of his classes, as the teachers ridiculed the answers of a black classmate, the first black student admitted to the school. "When he stood to recite," Goldfrank remembers, "the instructors often just ripped him apart. It was obvious to me that they weren't giving the student a fair chance. And I told the instructors and my classmates that I thought so—more than once.

"What evolved was an argumentative, provocative relationship with many of the people at Johns Hopkins. And it was an era of that kind of confrontation—so much ferment and protest on campuses all over the nation, really. But not there. Not at that institution."

It has been said that it's harder to flunk out of than to get into medical school. Nevertheless, shortly after his return home for summer vacation that June, Lewis received a letter from the dean. Based on his record—five passes, two fails, one of which was physiology—the dean wrote that "the committee has voted that you be required to withdraw from school. We all wish you success in your future endeavors," the letter concluded.

It was too late to make a transition to another institution for the coming school year. In any case, Goldfrank went ahead with a summer research job he had already secured at the Geigy Chemical Company in Ardsley, near his home. He went to talk with admissions people at several medical schools, including NYU, that summer. At those institutions he was told that, as he had had difficulty adjusting to life at Johns Hopkins, it was very likely that he would find it equally difficult to adjust to life at any other medical school. Therefore, none of those institutions felt it could offer him a place.

"One evening that summer, I was discussing my setback with a young couple that my parents had befriended. He was an IBM physicist and she a lawyer, and they seemed very interested and sympathetic. Her father came to visit them later that summer at their place in Chappaqua. They told him all about my predicament and then invited me to meet him. He was Dr. Chaim Perelman, the chairman of the department of philosophy at the University of Brussels."

Dr. Perelman had a seat on the Medical School board, and encouraged Goldfrank to apply. "It was the Free University, which meant that anyone could enroll, but not everyone could survive." Furthermore, Brussels had a reputation for feistiness of its own, being one of the few universities that chose to close rather than submit to the Nazis during the Occupation. Dr. Perelman felt sure that Lewis would find the academic climate compatible.

"My French was pretty good—reading and writing knowledge, anyway. My French teacher at Clark had been a Quaker, and I was much impressed with both subject and teacher." Admission to the Medical School was to a seven-year course, and included additional courses not on the American syllabus—among them philosophy.

It was to be an eventful summer, as Lewis married his high-school sweetheart, Susan Harrington, and they went off to begin a new life together in a new country.

Both worked hard at their French, Susan learning hers on the job. She was to be the support of them both for that first year. "But every job I applied for said they needed bilingual skills, French and Flemish, which I didn't have. Finally I was beginning to despair," she recalls. That was the day she called at the offices of the First National City Bank of New York. "They asked me how I could be bilingual and I said English and German. I had known German as a child, when my grandparents, who fled Vienna as Hitler came in, lived with us. And I became proficient in German during my formal studies in high school. Then they asked how much pay I would need. I told them I needed to take home $200 a month."

She was hired on the spot and within a year she could take dictation in French or English. Lewis, meanwhile, plunged into his studies with sustained intensity. He was granted second-year standing (because of credits from his year at Johns Hopkins) in most areas, but not in anatomy. The curriculum in Brussels was much more detailed in this subject, and devoted two years to it. In any event, he rose to and stayed near the top of his classes throughout the six years of study in Brussels, which made him eligible for scholarship aid from his second year on.

The young couple supported themselves on a budget of $3,600 a year: Susan's earnings plus some help from home and the scholarship aid. They did other jobs, as well—Susan as a movie director's assistant in dubbing in English dialogue in Belgian films, and Lewis as an unusually tall Santa Claus in Brussels department stores at Christmastime. They also translated French scientific papers into English for their professors at the university, and the papers went on to various journals both in America and Britain.

To get to the various jobs and to school, they had an old black 1951 Volkswagen bug, prone to frequent collapse, with which they braved Brussels traffic and its sometimes perplexing combination of traffic circles and trolley cars.

As the years passed, the Goldfrank family developed as they discovered that, because of their great success at their studies, both Lewis and Susan could attend school on a full-time basis. She had enrolled in psychology and ultimately won a master's degree. It meant getting by on four hours of sleep a night, while the demands of daughter Michelle, son Andrew and her own studies all competed for her time. The presence of a crèche attached to the university—a day-care center with a professional staff of *puericultrices* (child-care specialists) as well as a pediatrician in frequent attendance—made possible a schedule that would have otherwise daunted even the most energetic young parents.

They lived in a three-story brownstone in an unfashionable part of town, populated with Moroccan, Portuguese, Spanish and Belgian neighbors. They rented the top floor of their place to a fellow student, thus reducing their monthly rental to just $40. Lewis and Susan rose at six, got the children dressed and fed and off to crèche and elementary school in the little black VW. "Sometimes, if we were blocked in at curbside, we had to lift the car out to get on our way," Lewis recalls.

Susan's classes started at seven, Lewis's at seven-thirty. There was a two-hour midday break. Lewis would walk home past the great central market, where a janitor in the pathology lab at the Medical School also was a greengrocer. He always saved choice vegetables—especially eggplant—for the Goldfranks, and he, like almost all of their other townsmen, went out of his way to be helpful to the young American couple. A loaf of bread, some slices of cheese, and a vegetable or two, and Lewis and Susan had lunch, then a quick nap, and back to classes at two. They picked up the children after the last classes, at six, spent the evening preparing their supper, getting the children bathed, then reading to them and playing with them until bedtime. The children unwittingly assimilated their elders' scientific studies, drawing and redrawing bones, muscles, cells, molecules on their parents' laps as they studied. After the children were put to bed, Lewis and Susan typically studied on and on—until well past midnight and often until 2 A.M.

Despite their heavy family responsibilities, they developed a close circle of friends, among whom were two of the outstanding Belgian medical students. As they became senior students, Goldfrank formed a student committee with them that petitioned the faculty leadership for reforms. The trio framed a strongly worded protest about the curriculum—the burden of which was that students were required to do the work of technicians, to take blood samples and do routine tests. The three expressed concern about the intermittent instruction by house officers and attending physicians at the hospital affiliated with the Medical School (St. Gilles) as well as at the University Hospital.

The group led a successful campaign that quickly won concessions from the faculty and improved conditions so that medical students were able to use their hours in hospital service more as students and less as technicians. "The whole thing was a battle against the ex-cathedra style of teaching, the rote-recitation approach to knowledge."

Goldfrank won his M.D. *magna cum laude*, in 1970. He looks back on those days fondly, and stays in touch with his Belgian friends, Drs. Françoise Janssen and Jean-Marie Bouton, both of whom are in pediatric critical-care medicine in Brussels.

Lewis, Susan and family returned to the United States, where Goldfrank accepted a post at Mt. Sinai Hospital in Hartford. He had been recruited by the medical director, Stanley Bernstein, who had toured European medical schools looking for especially well-qualified students.

Goldfrank worked in pediatrics at the University of Connecticut Health Center in Hartford for three months and then in internal medicine at Mt. Sinai Hospital. "At the time I thought I wanted to specialize in pediatrics. But one case in particular made me change my mind. A little Hispanic boy with leukemia. Lots of spinal taps. Lots of hopeful moments. But he kept slipping away. I had a hard time getting to sleep at night. If you like to work in a hospital, as I did, but couldn't forget your pediatric cases—which I couldn't—you begin to wonder. I was better at coming to terms with adult deaths, but I couldn't accept it in kids. I wasn't ready for that. Never would be. Instead, I decided to continue in adult internal medicine.

"But, before the year was out, I knew that Mt. Sinai wasn't really the best situation for me, and Dr. Bernstein agreed to help me find a more stimulating spot. He very kindly offered to recommend me to Montefiore Hospital in the Bronx, where his old friend and my future mentor, Dr. David Hamerman, was the chairman of the Department of Medicine."

Montefiore was named after the best-known Jewish philanthropist of the nineteenth century, who was also the lord high sheriff of London, Sir Moses Montefiore. Montefiore Hospital in the Bronx was founded at just the time that the Statue of Liberty was being erected in New York Harbor.

Its rich tradition and its strong teaching absorbed Goldfrank's energies for two years, at the end of which he sat for the arduous exam for board certification in internal medicine, which he passed in September, 1973. A few months earlier he had accepted a position at neighboring Morrisania Hospital, named after the part of the Bronx that used to be the leafy preserve of the family of Gouverneur Morris, one of the architects of the U.S. Constitution.

Morrisania and Montefiore were linked by an affiliation arrangement set in motion by the Health and Hospitals Corporation of New York, the $2 billion agency that runs the sixteen municipal hospitals and—except for the federal government—is the largest health-care organization in the land. To attract quality staff to its various hospitals, HHC, as it is called, went to prominent private and voluntary hospitals and medical schools and worked out arrangements to link their academic strengths with the teaching opportunities presented by the municipal hospitals.

Thus in the late sixties Morrisania became affiliated with Montefiore, Bronx Municipal with Albert Einstein College of Medicine, and Bellevue with New York University Hospital.

Because of their affiliation, Montefiore sent Goldfrank to Morrisania as a resident in internal medicine for a tour of duty in 1971. In that role, he served in the Emergency Room on a regular rotation. He was tapped to become chief resident at Montefiore for 1972–73. During that year, he proposed to the Montefiore Hospital administration and to Dr. Hamerman that he take over the Morrisania Emergency Room and try to create an environment that would assure better care for the patients and better teaching opportunities for the young interns and residents. Both Montefiore and the Morrisania affiliation administration gave him the go-ahead. And so, at age thirty-one, just three years out of medical school, he became head of the department. At the time, in all of New York City, there were only two or three other doctors who were acting as directors of emergency services, a position without particular cachet in the eyes of most of the medical world.

As the hospital served an acutely impoverished area with no other health services available, the emergency service became a focal point for a great deal of public attention. So, for Goldfrank, there was quick recognition.

"Morrisania Hospital is located in the center of what has been termed a necropolis or city of death," he wrote while acting as the director of that Emergency Department in 1976. "Between 30 and 40 percent of the 400,000 people who live in this part of the South Bronx are on welfare.

"Violence and poverty are everywhere. Every day five adults are bitten by wild dogs that roam the streets in packs. Sexual assault is a common occurrence; daily one adult or child is brought to the Emergency Room for the assessment of rape. Merchants working behind bullet-proof glass serve as quiet reminders that the gun and the gunshot wound are also a part of the daily life of the South Bronx. Thousands of people here get their sole running water from open fire hydrants. During the winter there are continuous fires, and it is no wonder: 50 percent of the homes have inadequate heating. In our catchment area 30 out of every thousand newborn babies die (the city average is less than 20 deaths per thousand). Tuberculosis is not a disease of the past, not when 37.8 cases per hundred thousand are found (the national average is 14.3 cases per hundred thousand). Malnutrition and kwashiorkor—chronic protein deficiency—are diseases of Central Africa and Southeast Asia. But they are also the causes of pediatric admissions in the South Bronx. Septic abortion is common. Women use potassium permanganate, quinine, herbal concoctions, willow sticks and needles—doing themselves permanent damage, even killing themselves . . ."

The raging plague of heroin addiction in the South Bronx first awakened Goldfrank's interest in toxicology, and shortly after making the acquaintance of a former *New York Times* writer, Peter Frishauf, who had revived *Hospital Physician* magazine, Goldfrank began contributing case histories drawn from his Morrisania experiences. Those early articles bore titles such as "Stop the Noises!," "The Homewrecker" and "The Aromatic Vegetarian." These would form the nucleus of Goldfrank's toxicology text, which has, after three editions, grown from 180 to 929 pages.

During this time, the Emergency Room at Morrisania was kept from destructive chaos by the intimidating presence of a six-foot-eight-inch head nurse, Carol Temple, whose commanding manner was usually enough to cow the drunks, the addicts and the disorderly, who were always in residence there. But even her dominating spirit did not avail against the harsh realities of the street or the arrival of sudden death.

One relatively quiet afternoon, a police car roared up to the ambulance entrance and two policemen dislodged from its rear seat the limp form of a cabby who had been shot during a holdup. Goldfrank and his fellows worked intensively on the man, but there was no pulse, no blood pressure, no breath coming from his lungs. When the suction catheter began pulling cerebral cortex from his mouth, they gave up.

Among the strong impressions that remained on Goldfrank was the matter-of-fact way in which two young surgeons examined the body, trying calmly to determine which were the entrance and which the exit wounds. At the other end of the stretcher a young student volunteer, only a few weeks from his first year in medical school, watched as the dead man's head bobbed around, his eyes partially open. There was the sense that just a moment ago there had been breath, and quick perception and perhaps laughter. Now there was the inert form, an empty vessel, the ruin of hope and aspiration.

Goldfrank could offer the student a word of comfort, of the sort he would have been all too glad to have heard during his own stint at the dissecting tables in medical school.

Among his other memories was of an evening when five people came in in rapid succession, comatose from heroin overdoses. "The ability to handle all these people simultaneously was quite beyond me. But I was taught by a group of Caribbean nurses—True Samms, Margaret Chase, and Norma Weir were three I particularly remember. They reminded me that this was only the beginning and that, if I could handle these problems, I'd be able to do the tasks of even greater difficulty, such as caring simultaneously for ten children with neo-natal tetanus—as they had done in their home countries of Jamaica and Trinidad.

"These were my first tangible lessons in forming a team on which a nurse plays a key role in the provision of quality care. These nurses were revered by all of us doctors on the house staff—saving us from mistakes that would have resulted in fatalities. What they invariably did, when one of us made an error, was to gently move the hand, or call for a senior attending physician to come in, saving us from catastrophic error. They were the ones who first taught me that if the patient looks sick, the patient *is* sick, no matter what the laboratory data say.

"For example, there was a forty-five-year-old woman who came to the Morrisania Emergency Services almost daily, with severe chest pain. We thought it was Munchausen's syndrome—imaginary illness. She had been admitted to the hospital many times, with no findings. And she had been often discharged from the Emergency Service, with no findings. On this particular day, the house officer was about to discharge her once again when Norma Weir heard the patient say that she was dying and Weir thought she *looked* as if she were dying. Her electrocardiogram was taken again, and this time showed the lethal arrhythmia of ventricular fibrillation. Because of Weir's medical skills, the patient didn't go out the door to die. But she did die in the hospital a few hours later. That impressed all of us deeply."

Meanwhile, the new field of emergency medicine had already grown substantially. As Goldfrank settled in at Morrisania, he was aware that the Emergency Room had become the most frequently used part of the hospital. There was a nationwide trend clearly evident: In 1960, 42 million patients visited hospital emergency services. By 1977, the figure had grown to 76 million. By 1990, it is expected to reach 100 million.

But he also became aware that there was a tendency to resort to cold, detached science. "We were told, for example, by forensic pathologists, that carbon monoxide poisoning is manifest in the pink nail beds and ruddy mucosa of victims. But when we had victims of fire, with carbon monoxide inhalation, we couldn't distinguish who had pink nail beds because anyone alive has pink there. And anyone alive also has red mucosa—as on the outside of your lips. And then it struck us that the pathologists only dealt with corpses, and sure, if you look at a corpse with those signs, carbon monoxide has caused them. But that's no use in diagnosis on a living patient."

Much the same sort of remoteness from real-life conditions caused the proliferation of "subspecialty" clinics. So, if you had a rash and an asthmatic condition, you would be shunted to two separate clinics, neither of which knew what the other was doing. In fact, when Goldfrank got to Morrisania, there were more than fifty such subspecialty clinics.

When he was appointed associate director of ambulatory services, Goldfrank teamed up with the director, Dr. Mutya San Agustin, and together they managed to get the total number of clinics reduced to just twenty-three, with the result that there was more continuity of treatment, more of a personal bond between doctor and patient—in a word, more warmth and responsiveness to the needs of the poor. There was more awareness of such problems as losing a day's pay if they missed work while seeking health care.

The Health and Hospitals Corporation closed Morrisania in 1976 because of the completion of the North Central Bronx Hospital, which was built adjacent to Montefiore and linked to it not merely by affiliation as Morrisania had been, but also physically. Its Emergency Department was planned by Goldfrank and nursing supervisor True Samms—systems, traffic, triage arrangements, staffing and the like. At the time, there was still limited prestige accorded such work. The general principle animating most such services was "when in doubt, discharge the patient." Goldfrank proposed to supplant that approach with another spirit: "If in doubt, admit and observe." Within a few months Montefiore asked him if he would care to take on the additional responsibility for their ED as well.

Simultaneously, he was advancing through the academic hierarchy of Albert Einstein College of Medicine—first as assistant professor of internal medicine and then associate professor.

As the young specialty of emergency medicine continued to burgeon—reflecting the virtual disappearance of the general practitioner and the increasing mobility of American life—there were inevitable struggles over turf. Departments of surgery, internal medicine, pediatrics and psychiatry, especially, often resisted the increased self-assurance of the young doctors in emergency medicine who followed a "horizontal" kind of medicine as opposed to the "vertical" specializations.

So, as emergency medicine came "up from the pit," Goldfrank came up with it—successively as director of emergency services at North Central Bronx Hospital and then at Montefiore. Meanwhile, at Bellevue, Neal Lewin was serving as acting director of the Emergency Department.

Today, Neal Lewin calls himself "the Ed McMahon of the Emergency Department," characteristically making light of himself. Somewhat mercurial in manner, Lewin did not always have the close rapport with Goldfrank that now distinguishes their association. Lewin took his undergraduate degree at NYU–University Heights and then went to SUNY–Downstate Medical School. Thereafter he did his internship and residency in internal medicine at NYU–Bellevue. He later became board certified in emergency medicine after five years of experience.

"I was acting director here at Bellevue back in the late seventies, while the search was going on for a permanent director." Lewin had his private practice, also, which precluded putting himself forward as a candidate for the permanent post, which was soon to be offered to Goldfrank.

"I knew Lewis, or knew of him, back then. And the way we actually met was sort of unusual. My grandmother had been in a car accident. She was in her eighties and was admitted to North Central Bronx Hospital, where Goldfrank was directing the Emergency Room. She had a hip fracture and multiple injuries."

Lewin and his brother-in-law, Dr. Melvyn Damast, both looked in on the patient, and did not feel entirely comfortable with the attention she was getting. "So my first encounter with Goldfrank was not the most amicable. It wasn't hostile, but it wasn't warm, either. Here I was trying to get my grandmother to another hospital, while she was in the Emergency Department, not fully stabilized, so it's not hard to understand why we weren't in perfect sympathy."

As Lewin tells it, he and his brother-in-law, thinking that their grandmother really ought to be across the street at Montefiore, where she had been an active volunteer for years, got approval for her to be transferred there. Then they waited impatiently for the transfer to be effected. They got tired of waiting.

"There's a tunnel connecting the hospitals. In bad weather it's used as a passageway between the adjoining institutions. Well, we wheeled her over to the intensive care unit at Montefiore ourselves."

It doubtless speaks well for the temperament of both men that they nevertheless became close friends and mutually admiring colleagues.

"Time went on," Lewin recalls, "and I heard a rumor that it was going to be Goldfrank at Bellevue, but no one had told me. And one day I saw him in the corridor at Bellevue and I went up to him and re-introduced myself and asked him frontally, 'Are you coming here? And what's your plan?'"

There was no firm agreement between Goldfrank and Bellevue. As Lewin was later to learn, Goldfrank was still negotiating the degree of autonomy that the Emergency Department would have in admitting patients. He was seeking assurances that a major commitment would be made to upgrading the entire approach to emergency planning at Bellevue, including specialized training. Goldfrank's plans for training developed into the entity that became the Emergency Care Institute.

"So, as it became clear that he would be taking over and bringing in his own team, I went on with what I was doing until the transition. I was in charge of scheduling—never a popular job—and people would say, 'Oh, no, I can't work weekends,' or 'I can't work holidays,' and I would say, 'All right, you're better off working someplace else where your needs can be met.' People were slipping notes under my door, 'I quit!' So we were depending on sessions doctors [physicians hired on an hourly basis], and I was really ready for Goldfrank to take over, as I had had enough of the attitude of people saying, 'Who's Neal Lewin? Oh, he's some guy at Bellevue who keeps firing people.'"

As the new regime came in, it was clear that there was a strong similarity between the two doctors. Lewin's strong family bonds echo Goldfrank's own feelings. Lewin spends as much time as possible with his two small sons, who, with his wife, Gail, a graphics designer, live in a comfortable old house in the Edgemont section of Westchester. "But I didn't want mindless affluence, didn't want fancy cars or an extravagant lifestyle that doctors can so easily attain."

Earlier, Lewin had worked at St. Mark's free clinic, which helped clarify his attitude toward service. "If I had wanted to go into business . . . group practice or some other sort of medical enterprise, I could make two, three times as much. But there wouldn't be the same sense of service, of making a difference in people's lives."

He has always looked to Bellevue "as a kind of model of public service." As Goldfrank came in and set up his own style of operation, opportunities for intellectual and professional growth opened up, and Lewin has responded to these opportunities fully. He has become an increasingly greater factor in the evolution of *Toxicologic Emergencies*. In fact, Lewin himself has contributed to no fewer than fifteen chapters, and his wife, Gail, designed the cover for the second edition. He is principal author of the chapters on cocaine, caffeine, herbal preparations and arthropods. (The last involves creatures that bite, ranging from those that crawl, bite and settle down [body lice] to those that crawl, bite and leave swiftly [wood ticks, bedbugs, assassin bugs].)

Both strong family men, both nurtured in the tradition of the Ethical Culture Society, both active patient advocates, Lewin and Goldfrank have formed a highly effective working team. Each is free to follow his own bent but both converge on the absolute commitment to render quality care in a compassionate setting.

As for Lewin's grandmother, "We recently celebrated her ninetieth birthday at our home. It was a tremendous party. There were poems and all the generations gathered, and it was really a splendid occasion."

Catherine O'Boyle played a key role in Goldfrank's decision to come to Bellevue in 1979. She agreed to accept the job of associate director of the department, drawing on her extensive experience in emergency care.

She had started at Bellevue in 1967 when the Emergency Room was at the back door of the hospital and few of the modern standards for treating traumatic injury and cardiac arrest had yet emerged. She rose to become nursing director and administrator of the Emergency Room. In this role she met on a regular basis with the Hospital Administration Committee as de facto head of the department while it was still under the purview of the committee. Then a part-time medical director was named. And finally, in 1979, a full-time director was appointed. Until then, the medical staff was predominantly made up of those who could be summoned as emergency cases came in, but who did not maintain an ongoing presence in the department.

By 1970, when a part-time medical director had been appointed at Bellevue, and during the time when the new building was being built, there was considerable development in the field of emergency medicine itself, paralleling developments that were taking place at Bellevue. As Kay O'Boyle recalls, "Until the late sixties, most ambulances were hearses, really. And nationwide half the ambulance services were run by morticians because they owned the vehicles that could transport people. In those days, training of ambulance operators was neither intensive nor broad. I've seen figures that only about 8 percent had advanced first-aid competence. So nationwide many of the heart and trauma cases were lost that today are routinely saved."

Then the Highway Safety Act of 1966 began the process that has resulted in the upgrading of ambulance service, the emergence of trained paramedics and emergency medical technicians. Meanwhile, in 1970, O'Boyle had overseen a newly established Emergency Department building, which had a sixteen-bed intensive care unit, two trauma slots and sixteen examination rooms in the adult emergency service, in addition to walk-in examination facilities and a twelve-bed observation unit. But there was still no ongoing medical direction of a consistent sort.

In the seventies, further legislation provided for better ambulance and 911-style rapid-response organizations like the Emergency Medical Services. Such ambulance corps, staffed with paramedics and emergency medical technicians with advanced training, have done much to improve the survival chances of badly injured or acutely ill patients.

"And while that was going on nationwide," Kay O'Boyle says, "here at Bellevue, by the mid seventies, Dr. Lewin and I had the responsibility of running a sizable emergency operation, but with none of the authority to maintain standards that would have been equivalent to what we see today. There wasn't a problem with major injuries or acute illness like heart attacks. But our category two of today—urgent patients—were not taken care of nearly as well as now.

"When Lewis Goldfrank came to Bellevue, he was like the missing part of a puzzle, who made it all work, and what we have today has evolved from the cooperation of the people who have dedicated a large part of their lives to the Emergency Department. Each of them has left a part of himself here over the years, and the memory of each of our predecessors has done much to give us our sense of tradition and commitment."

Shortly after Goldfrank's arrival, Appleton-Century-Crofts published Kay O'Boyle's book, written in collaboration with several nursing co-authors from Bellevue and Hunter College–Bellevue School of Nursing. The book is titled *Emergency Care—The First 24 Hours,* and in it all of her experience came to flower. It's a six-hundred-page work that sets forth for nurses and doctors everywhere the "Ten Commandments of Emergency Health Care," which were framed in Goldfrank's earlier days and then refined by various staff additions over the years:

- Do no harm.
- Do first things first. Is the patient breathing? Bleeding? Thinking?
- Make sure each diagnosis is correct, setting an appropriate therapeutic endpoint.
- Remember that patients with head injuries require expert care.
- Bear in mind that alcoholics present atypical manifestations of typical diseases. They should be examined thoroughly **in the nude**.
- Before instituting therapy, make sure that blood gases and electrolytes are normal.
- The borderline patient needs reevaluation by someone else, even if the data are normal. "If the patient looks sick, the patient is sick."
- Anticipate drug interactions and reactions.
- Be humble and conservative.
- Be a patient advocate.

Kay O'Boyle's reputation over the years has been such that a producer for one of the television networks spent several months cultivating her acquaintance, telling her he planned to develop a television drama. When he showed her a script for the forthcoming *Kay O'Brien,* she discovered that the main character, based loosely on her own life and work, had become a young female surgeon. And Bellevue had become the butt of an endless series of complaints from the heroine and others as they went about their work. O'Boyle protested loudly, and the producers responded by changing the name of the institution from Bellevue to "Manhattan General." The series opened on CBS in the fall of 1986. What did O'Boyle think of it? "I haven't had a chance to look at it yet," she said some months after its premiere.

Today, every time Goldfrank is described as director of Bellevue's Emergency Room he quietly corrects the phrase. It's "Emergency Department," not "Emergency Room." The physical realities at Bellevue strongly support his point. Far from being a single room, the department has in fact some sixty rooms, occupying slightly more than a half-acre (22,000 square feet). Today, there are forty attending physicians and scores of nurse practitioners, nurses, nurse's aides, and other staff members who populate the many chambers—trauma rooms, observation rooms, examination rooms, waiting rooms, laboratories. There's a rash room, a cardiac resuscitation room, a large room only for the most urgent surgery, a room for radiographic procedures with a variety of devices. And, when the art of the men and women and the science of the rooms fail, there is a room for keeping those who have not survived and who are destined for the City Morgue.

But this is the department that has to do with sudden battles to save life, with especially rapid treatment to preserve and restore the breath of life in every patient who passes through its doors. It is that absolute dedication that seems to charge the very air with excitement.

# 10 The Carriage Trade

A stretch limousine comes swiftly down the FDR Drive, speeds off onto the exit at Thirty-fourth Street, and darts into the roadway leading to the Bellevue Emergency Department entrance. It is nearly 2 A.M. The big car brakes sharply and comes to an abrupt stop. Out of the limo step a couple in their fifties, in evening dress, supporting between them a young woman in her twenties. The young woman is whimpering softly as they guide her along the entryway leading into the Emergency Department. They lift her onto one of the wheeled stretchers queued there at the entrance. Their chauffeur, after starting the trio on their way, thinks better of leaving the large car idling unattended and quickly nips back to turn off the motor. A hospital policeman tells him he'll have to move the car. He gets back in and heads around toward First Avenue, to await his clients there.

The trio in the hospital seem a bit uncomfortable. As assistant head nurse Elisabeth Weber later recalled, "You could see a look of apprehension on their faces as they saw the people in our waiting area that night. There was one old regular who calls himself Eighth Street Eddie and a weeping Haitian couple and a Chinese man with abdominal pain and a prisoner in handcuffs—looking really furious—being guarded by two policemen. And other regulars—Berenice, who has sickle-cell anemia, and several homeless . . . It wasn't an unusual night at all."

Elisabeth Weber is a fresh-faced young woman who came to New York from Cleveland just three years ago. At the time she had already had four years' experience in pediatrics, medical intensive care and emergency medicine. "I always wanted to live in New York, and had wanted to work at Bellevue for as long as I can remember." She enrolled in graduate school at NYU and started at Bellevue at the same time. She has had two promotions since joining the Emergency Department staff. "I came here because it is probably the most famous hospital anywhere and because of its reputation as a place that doesn't reject any patient," she says.

Elisabeth Weber quickly notes the facts: Mr. and Mrs. Newell Carberry of Atlanta have brought their daughter Stacey in with what looks like a compound fracture of the forearm. As she was leaving a wedding reception at a midtown hotel with her parents, the girl tripped over the hem of her gown on the sidewalk outside in such a way that her arm came whipping down onto a fireplug. Stacey Carberry is not in the clearest frame of mind and appears to have had a great deal to drink. Her injury has left her right forearm looking misshapen.

Weber makes an immediate judgment. The break, severe as it is, is clearly not life-threatening. But it does merit immediate splinting and evaluation. As she feels for a pulse she picks up a microphone at her elbow. "Doctor to the triage desk," she says. She suggests that the parents take a seat in the waiting area as she and an orderly move the stretcher to the trauma section. Dr. Robert Hessler appears, greets the patient and begins to feel for the girl's radial pulse as he examines her arm. He and Weber get a splint in place while she gives him the girl's vital signs and as much of her history as she is able to convey, which isn't much. The patient is crying softly but not responding in a comprehensible way to Hessler's questions. So Weber returns to the parents to get the rest of their daughter's medical history. Allergic to penicillin. No loss of consciousness. No excessive drug or alcohol use. As best she can recall, Mrs. Carberry estimates the date of her daughter's last tetanus shot and the last time she ate.

Elisabeth Weber disappears with this supplementary information. Then they wait. As they do so, the Carberrys look uneasy. The surrounding populace includes an old man clad in discolored Glen-plaid slacks and disintegrating black basketball shoes, who snores loudly on his stretcher. A few seats down from the couple, two balding young men with diamond earrings seem to be comforting each other. Weber has already seen the smaller of the two; he has slight shortness of breath and blueberry-colored splotches on his arm—suggesting *Pneumocystis carinii* pneumonia as well as Kaposi's sarcoma. Almost certainly AIDS. A chic Hispanic woman with green Louis-heel shoes and intense brown eyes draws languidly on a cigarette while a young black next to her sits with his head swathed in a giant turbanlike bandage, around which Walkman earphones are clamped. On the walls are various announcements, including a "Bill of Patients' Rights" and a poster announcing special prenatal nutrition classes. There is also a warning about leaving your personal property unattended.

Weber senses the discomfort of the elegant couple as she goes about her other tasks. She thinks anew of the tradition that she serves—of the unconditional kindly acceptance of all who seek care and healing. And she hopes somehow the Carberrys will sense the spirit of the place that to the undiscerning eye might well be obscured by surface appearances.

When he sees x-rays of the young woman's fractured arm, Hessler calls for an orthopedic resident. He has assured himself that there are no other injuries. As the orthopedic resident busies himself with Stacey Carberry's arm, yet another luxury auto draws up to the Emergency Department entrance, this one bearing diplomatic plates. The patient who appears in the entry hallway is dressed in a splendidly tailored double-breasted navy blue suit. He cannot speak, so thickened is his tongue. From time to time he laughs maniacally. His pockets yield no information whatsoever about his identity. The driver who brought him in is from one of the UN legations, but now he's done his job and wishes to get away. He knows only that the patient has had too much saki to drink, and was discovered, long after the rest of the company had departed, passed out behind a couch. They loaded him into the legation's auto for a swift trip downtown. Weber thinks this is an emergent case, and moves the stretcher right into Room 6, where nurse Jeanne Delaney begins taking the vital signs, and Dr. Toni Field quickly comes in to begin ministering to the anonymous diplomat—readying a twelve-lead electrocardiogram for use once Jeanne Delaney has monitored his respiratory rate and pulse.

Elsewhere, Elisabeth Weber sees that Eighth Street Eddie is now asleep on his stretcher, as he has been on many previous occasions in this Emergency Department. His usual haunt is about a mile down-town, at Broadway and Eighth Street, hence his self-styled sobriquet. He had been asleep right on Broadway itself, not on the sidewalk, at 1:20 A.M. "He's in his late fifties, and he's undomiciled, as we say. And he would never hurt anybody," Weber says, and yet he is one of those repeat patients who tax the equanimity of Bellevue every day. Weber shudders when the police refer to him as a "dirt ball," as they did when they roughly brought him in earlier tonight.

Weber has tender feelings that she has learned to cover up. "I remember one night—it was last February—they brought in an old derelict. His pants were entirely soaked in urine and I remember we were getting his vital signs and trying to clean him up when I saw that he had four or five Valentine chocolates in his pockets. And I could see, behind all the stubble of beard and filth, that he once had been a fine-looking man. I started to get sort of teary, looking at those soaked chocolates and thinking of the life he was making for himself. Then one of the others—I forget who it was—said, 'Come on, Elisabeth, for goodness' sake, nobody sent him those candies. He stole 'em!' It was probably the truth, and it made me feel less teary."

The Carberrys are pacing the hallway outside the trauma section. Elisabeth Weber has a report for them. Both bones of the forearm—radius and ulna—are broken, but the young woman is going to be fine. "The orthopedic resident will be out in a few minutes." Meanwhile, Weber offers a copy of a magazine-size booklet, which she suggests that the waiting parents might like to browse. "Thank you," says Newell Carberry as he sits down to read a copy of *The NYU Physician*, devoted to the subject of Bellevue as it marks its 250th anniversary.

The booklet describes one of the most renowned medical teaching institutions in the country. Bellevue was founded in 1736 as an almshouse in lower Manhattan. It formed an alliance with New York University in 1847, a bond that has been intact ever since.

By the turn of the century, Cornell and Columbia universities had also entered into relationships with Bellevue, until there were three "divisions," one linked with each of the universities. There was an ambiance of intense clinical teaching almost as soon as that style of medical instruction arose in this country—following the lead of Dr. William Osler and his colleagues at the newly founded Johns Hopkins Medical School in Baltimore in the 1890s.

In the days when the three universities staffed the three divisions, there was keen competition among them. "Once, three men came into Bellevue in a coma and vomiting, having drunk an unknown solution," the magazine reports. "Each division took responsibility for one of these victims." There was special delight in Division Three—NYU's division—when one of their attending physicians was the first to figure out that what the trio had ingested was an aniline contained in shoe polish.

As the hospital changed and evolved, it grew dramatically. It had as many as three thousand beds by the late thirties and became even more firmly linked in the minds of New Yorkers with its municipal hospital system, of which it remained the flagship. The intensity of the academic atmosphere clearly imprinted itself on its students, as there are more than a thousand NYU medical alumni today currently holding faculty positions or deanships at medical schools across the country. Only Harvard has more.

By the fifties, Bellevue had declined considerably in size and somewhat in quality, as its physical plant deteriorated and other, more modern hospitals were built in the city. By the late sixties, both Columbia and Cornell had withdrawn from their affiliation with Bellevue, leaving NYU to meet on its own the needs for staffing the hospital. Meanwhile, plans for a modern replacement for the old red-brick complex of structures advanced—but slowly. And it wasn't until the mid-seventies that the new twenty-one-story, twelve-hundred-bed structure that is today the main building was completed and occupied. The success of the newly built University Hospital, which adjoins Bellevue, enabled NYU to establish a teaching faculty at Bellevue, as well as giving it the allure to attract distinguished new faculty members. All of this made positions on Bellevue's house staff eagerly sought-after by the best-qualified new medical-school graduates.

Former NYU Medical School dean Lewis Thomas has written, "I regard it still, as when I first walked through the unhinged doors of the old building, as the most distinguished hospital in the country, with the most devoted professional staff. If I were to be taken sick in a taxicab with something serious, or struck down on a New York street, I would want to be taken there."

The orthopedist has finished. Stacey Carberry has had her arm set, x-rayed again to confirm correct alignment of her bones, and enclosed in a lightweight cast. The resident comes out to tell the Carberrys that there are minimally displaced fractures of both arm-bones but that they will easily be reduced with the long-arm cast that he has put on Stacey. They'll need to fill two prescriptions for painkillers for their daughter. They have a noticeably altered bearing. Perhaps the booklet has persuaded them that they have chanced upon one of the best places anywhere for the urgent treatment that their daughter needed. Elisabeth Weber, from her station across the reception area, where she is taking a medical history from a distraught woman who has been raped, can see the obvious signs of reassurance on the Atlanta couple's faces. She can sense the doctors are telling them that they can see their daughter now in the cast room.

Later, after the rape victim has been taken to a treatment room for evaluation, the Carberrys see Elisabeth Weber in the corridor and come out to offer their thanks. For Weber, there is the unusual pleasure of knowing not only that the patient has been well attended but that the hospital will receive full recompense for its efforts.

"We just want to thank you for everything," Newell Carberry says. "It's quite a place, this hospital." Elisabeth Weber is very pleased to have made a convert. She thanks the Carberrys, wishes Stacey well, and goes on about her business. Within a few minutes their chauffeur appears, having been located by Carberry and escorted in from First Avenue. The small party, pushing the young woman in a wheelchair, make their way out the front corridor and head back uptown.

An Emergency Medical Services medical technician, twenty-eight-year-old Sean Kroptkin, is waiting, shirtless, in one of the examination rooms as the attending physician jots a phrase on the chart lying on the small desk there: Tenderness over the trapezius and latissimus muscles of the shoulder and upper back. The attending physician is Dr. Toni Field. Dr. Field, who is board certified in internal medicine as well as emergency medicine, and who has been at Bellevue for six years, has wanted to be a doctor ever since she was two, when she used up all her mother's Band-Aids to cover a playmate's chicken pox scars.

"So you have discomfort here?" She touches the left side of his back.

"Yes."

"How'd it happen?"

"We had this real heavy MI (myocardial infarction) in the East Thirties. Fifth-floor walk-up. Guy must have weighed nearly three hundred. Had to lift up the stretcher to get around the stairway post . . ."

"That's when it happened?"

"Yeah. It felt like it just sort of popped out on me."

Field studies the back for a moment. "All right. I'm going to suggest an ice pack, which we'll get you. I'd keep it iced for twenty-four hours, then warm soaks."

"Okay. Then I got my own doctor out in Rego Park."

"All right, Sean," Field says. "How did your MI case work out?"

"Great!" He slips his shirt and jacket back on and leaves. Field finishes up his chart.

"You never know what surprise is going to come next," she says pensively to a visitor. One of her favorite memories involves the subway. As she was leaving work one day, she saw an EMS ambulance stop abruptly at Lexington and Thirty-third and its crew dart down the stairs to the subway station below. "I heard someone say 'cardiac,' so I followed, thinking I might be able to help." When she got to the scene, a man in his forties was seated on the floor, his back up against the wall, conscious but obviously frightened. The paramedics dropped to their knees. Field went up and showed a pair of transit police at the scene her ID and then moved forward to identify herself to the paramedics at work on the man.

" 'I'm Dr. Field. Perhaps I can help,' I told them. And at that moment the patient's face lit up and he said, 'Dr. *Toni* Field?' and I said, 'Yes,' somewhat startled. 'You saw me at Bellevue last year. Said I had something like mitral valve prolapse.'

"He was right. I had forgotten, but I looked it up later. His chest pain resolved. His condition was not serious. But it gave me a peculiar feeling—a good feeling—to discover that this cold city still has a lot of human warmth in it and a lot of people who care for each other, despite the stereotype of the heartless metropolis."

Dr. Roberto Bellini is checking his charts at the physicians' desk near the telemetry area at 4 A.M. when they bring in a woman who has been robbed and beaten. Amanda Mathis is about forty-five and beautiful. She has pearly white teeth and luxuriant red hair. But now her fair features are badly misshapen. She has deep, irregular gouges in the soft tissues of her face.

Bellini sees them wheeling her in and catches Weber's imploring glance. He moves forward to begin his examination. The paramedics have the story, as best they can piece it together. The woman was found in the East Eighties, where she had apparently been dumped out of a car or a taxi. The indications are such that Bellini assumes that she has received a dangerous overdose of at least one substance. So he chooses to give naloxone (a narcotic blocker or antagonist that reverses the severely depressive effects of such substances as heroin). He also gives oxygen, and then dextrose (in a 50 percent solution in water, administered intravenously) and thiamine. The dextrose will act against hypoglycemia (low blood sugar), while the thiamine (vitamin $B_1$) will counter the possible nutritional deficit an alcoholic is likely to suffer.

Bellini learns from her halting answers to his questions, as she regains consciousness, that she has been visiting New York City as sales representative for a group of regional magazines. Earlier that evening she had stopped in the lounge of her sumptuous Manhattan hotel, where she struck up an acquaintance with several well-spoken strangers at the bar.

That's all she can remember when she awakens under the bright glare of the lights in Room 7. Even though the neurological exam shows that her thinking is unimpaired, her pupils are of unequal size. Then, as she shifts on the examination table, her left side moves with reluctance. She has lost motor strength on that side of her body, a sign of a probable brain injury.

Missing are her purse with all her credit cards, her watch, her rings. Besides her many gruesome-looking facial cuts she has a broken arm. Once she begins to come around, Bellini gives Goldfrank a telephone call at home, despite the hour—a practice the director encourages in difficult cases. They agree that her awakening is not the result of the antidotes. More likely it is the lucid interval after a major injury to the brain. She may well have suffered a contusion, a subdural hematoma or a more serious epidural hematoma—often missed because of alcoholic stupor. But, in any event, some such cause must be at work here, because of her left-sided hemiparesis (partial paralysis on one side).

After their consultation, Bellini orders a CAT scan, so the patient is whisked to the seventh floor. A large subdural hematoma—an abnormal pool of blood between her brain and the thick dura mater lining the inner surface of her skull—shows up. Given the urgent need for immediate surgery, she is admitted to the neurosurgical service, and moves out of Bellini's ken.

"Later we learned that the neurosurgeons had to operate right away to relieve the life-threatening pressure," Bellini reported. Whether she was a daily imbiber or a periodic binge drinker, there was also evidence that the patient had done herself substantial injury over the years.

Goldfrank later studied over the charts and came to his own understanding of Amanda Mathis's plight: "In our work-up of her blood sample we could see the abnormalities in the atypically large and multi-segmented white blood cells as well as their diminished numbers." Normally, he explains, white blood cells have two to five tiny sausage-shaped segments linked by little filaments. Folic acid—literally meaning "leaf" acid, one of the B vitamins that was first found in spinach—is metabolized and excreted by alcoholics, and not replaced, because of their negligent eating habits. So their white blood cells, starved for folic acid, develop five to eight of these small segments, signs of atypical cells. Goldfrank continues:

"Red blood cells are also reduced in number as a direct result of alcohol's toxic effects on bone marrow and blood loss from acute inflammation of the stomach (gastritis). Then her platelet count—reflecting her blood's power to clot—was also very low. In addition to these abnormalities, her white blood cells doubtless will function poorly in the fight against infection. And, beyond that, many of her liver cells will have been destroyed and those that remain will have limited stores of glycogen—the precursor of glucose, which supplies energy. So that's why we give glucose. Those hepatic cells that survive perform poorly, often resulting in inefficient metabolism, which can lead to yellowing of the skin (jaundice), an indication of liver damage or failure.

"As a result of the damage she had suffered, Miss Mathis will probably be prone to seizures in the future. Alcoholics often suffer vitamin depletion, particularly vitamin $B_1$ (thiamine). Thus their nerves no longer receive and transmit information such as pain, touch or temperature very well. In addition, critical cells are destroyed in the brain by the absence of thiamine. Therefore, the patient has trouble interpreting sensation in her environment. In short, she has trouble feeling and thinking. Memory, particularly of recent events, is often dramatically impaired. The classic Wernicke-Korsakoff's syndrome is often encountered in such patients. They can recall remote things, like who pitched the gopher ball to Bobby Thompson in the 1951 Dodgers-Giants playoff, but don't recall meeting you on admission ten minutes before. Other vitamins, such as vitamin C, are also depleted. In fact, the only people I've ever seen with scurvy were alcoholics.

"Then the chronic use of alcohol alters the functioning of the gastrointestinal tract. Bleeding is common. Heartburn and hemorrhage due to irritation of the stomach—gastritis—also occur. And there are the nutritional consequences of using drink as 'food'—lack of vital proteins, vitamins and minerals. Poor nerve and muscle function follows and this affects even the heart muscle."

At any rate, after ten days and an outlay of $7,000, which her medical insurance largely covered, Amanda Mathis was released and went on back home. As well as her story could be pieced together, she had gladly set out with her new friends to go uptown to see a late-night show. On the way there, she lost consciousness. Unknown persons apparently robbed and beat her and left her unconscious for some 911 Samaritan to find.

Seven-thirty A.M.—Elisabeth Weber's tour of duty is over and she collects her handbag from the wheeled bins that fit under the counters in the doctors' station, a secure place for the valuables of the staff during their shift. She bids Roberto Bellini and Toni Field, both bent over their paperwork, goodnight and heads down the corridor, where Eighth Street Eddie still slumbers. She walks across town to the Lexington Avenue subway. As she later recalled, going uptown that morning was more disquieting than anything that had happened at work. An irrational rider, a man in his forties, suddenly began shouting obscenities at his fellow passengers, and no one was bold enough to try to shut him up. "I see these people on the streets, and I'm as afraid as anyone else. It's very upsetting. But then the same sort of person shows up at Bellevue for help, and that never gives me a qualm. I'm not sure I understand why."

There's a call to the doctors' station from an embassy uptown. Bellini takes it. "Do we have an unidentified black male admitted about 2:30 A.M. in a car with diplomatic plates?" Toni Field looks up from her charts. "Yes. He's sleeping it off there in the recovery area."

Bellini turns back to his caller. "All right," he says and rings off. "His embassy is coming to get him," he reports.

As Lewis Goldfrank comes through Grand Central at about the same time, he notices that the expected removal of pay lockers has been completed. Another element in the web of life for the homeless of Manhattan has ceased to exist.

# 11 A Midday Arrival

The boy was a seventeen-year-old black youth who stood six feet three. He appeared alone, with a bad laceration on his right arm. That had been five days ago. Now it is Tuesday morning and Dr. Hedva Shamir is inconsolable.

"Hedva, would you stop crying?"

"I can't." But she does stop, and then she can tell what has so crushed her spirits. It's the boy. His right forearm had been badly ripped up. He had accidentally put his hand through a window, he told Shamir as she applied pressure with multiple compresses that previous Thursday morning.

"At the time I wondered if it was really a suicide attempt. 'Are you a righty or a lefty?' I asked him. 'Lefty,' he said, and that should have been my first real clue. The second was that he was seventeen and homeless. Lived on the street. I called the plastic-surgery resident and he got the boy sutured up and then we had him admitted to Psychiatry, even though he denied any suicidal intent. But I still felt strongly that it had been a suicide attempt and that he ought to be watched until he was no longer so depressed. And then there were a thousand other things on my mind, and I forgot about him.

"They kept him on suicide watch from Thursday through Sunday, watching him continuously. Then he was taken off and he hanged himself Monday." There is a long silence as Shamir tries to regain her composure, but it is one of those cases that is especially difficult. "And that wasn't even the worst I've seen lately, even though it was bad.

"There was this other case, a twenty-year-old Hispanic brought in by the police with a methadone overdose. He had swallowed some other tablets besides, so we tubed and lavaged him: an endotracheal tube so he could breathe; an orogastric tube to remove the stomach contents. It was almost ten liters of water before we thought we recovered all the tablets. Then we gave him activated charcoal to soak up—adsorb or bind—whatever had already gone into solution. And of course we gave IV naloxone and dextrose and thiamine, all the things you usually do to reverse or treat the overdose. As he came around he wanted to shake hands, to thank me for what I had done. But how could I shake his hand? I heard, as we were working on him, from the police that they had taken his girlfriend to the medical examiner's office. The morgue. She was dead. He knifed her to death. A sixteen-year-old girl. And he wanted to shake hands. I knew I couldn't shake his hand. How could you shake a hand that's murdered a sixteen-year-old girl? I've got girls at home. Three daughters of my own." There is another silence, and the sounds of the gathering of the morning staff meeting come down the hall.

"I never have problems, even with AIDS patients. It's a wonderful human gesture, shaking hands. But I wasn't going to shake his!" She falls silent for a moment. "It was a very difficult thing. And now this news. It's so sad."

Lewis Goldfrank appears and with a subtle nod of the head lets Shamir know that their morning rounds are about to begin. She gathers up her papers and moves toward their conference area.

On everybody's mind this morning is the case of the Italian businessman who arrived the day before at Kennedy International Airport. On his way into town, coming through the Queens Midtown Tunnel, he started experiencing acute searing chest pain radiating to his back. He was brought right to Bellevue by the cabby. The diagnosis was a pulmonary embolism, a wedgelike obstruction of a major blood vessel supplying the lungs. After admission to the Emergency Ward he was treated accordingly by the Internal Medicine house staff. He was put on an intravenous heparin drip, to thin his blood and dissolve the clot. But then, quite suddenly, he died. Instead of an embolism, it was discovered that he had a dissected aorta* of such magnitude

*A condition in which the wall of the aorta is dangerously weakened, often by high blood pressure.

that the displacement of this principal artery of the heart obscured the actual condition and gave the semblance of a pulmonary embolism.

As they reconstructed what had happened, Goldfrank and Lewin led the discussion, in that self-critical manner that they hoped would arm everyone for any similar case in the future. What was known? An embolus (blood clot) had presumably formed in the veins of the patient's legs or pelvis during the long transatlantic flight. According to the diagnosis of yesterday, such a clot blocked part of the pulmonary vascular bed in his lung. It should have led to shortness of breath, an increased heart rate and abnormalities that an electrocardiogram would reveal.

But now, as they looked at the documentation, they could see that signs of a blood clot in the lung had been missing. There had been no shortness of breath, no rapid heartbeat—or tachycardia—a normal ECG and no deficit of oxygen in the arterial blood gases.

Everyone is studying the floor as Goldfrank concludes the discussion: "Remember, everyone, that vital signs are the hallmark of emergency medicine. They rarely mislead. But if we get harried, or overwhelmed with other cases—and I know that this happened at a particularly busy time—still, we have to be alert to think actively. To make no assumption. To think and rethink so that we make sure we don't neglect any clues." There is a moment of silence as complete as can be amid the clatter of constant arrivals and departures.

After a review of other current cases, there is a brief discussion of crack, the newly prevalent form of cocaine that seems to be entrapping the very young, especially. An eleven-year-old boy has been admitted suffering severe pulmonary distress from his first try at crack. Neal Lewin reminds the group that, smoked in this form, the drug is rapidly absorbed into the lungs after being inhaled, "like a bolus IV," a sharp jolt of a drug dumped into the system in a large batch—or "bolus"—rather than gradually introduced in the manner of snorting by applying the cocaine powder to the nostrils. Gradually, the staff is hearing more about the manifestations of the drug, and as their knowledge accumulates, a new awareness will emerge of the lethal effects on the heart, brain and other organs that are showing up in the wake of the epidemic-like spread of this new form of powerful and easily obtainable cocaine.

A medical student is asked if she has had any interesting case experiences she would like to discuss. She says, "No, I've been on vacation."

"Where?" Lewin asks.

"Cape Cod."

A lively discussion of the medical consequences of a vacation on Cape Cod immediately gets under way. "What if you had symptoms of gastroenteritis?" Goldfrank asks. "What could we infer as to etiology?" Several voices volunteer that clams or oysters could be responsible. There is also the possibility of hepatitis or similar viral diseases because of the affinity these shellfish have for sewage.

Goldfrank continues to ask about other possible sources of diseases, thus extending the horizons of his staff and students and awakening them to the realization that there is virtually no subject that does not have medical implications, and very likely emergency medical implications.

The exchange grows more lively as other possibilities are raised. Goldfrank, Lewin and Richard Weisman all have lectured and written extensively on fish toxins, so there is a quick recapitulation of the bivalves (oysters and clams) and the dinoflagellates (single-cell organisms) for the benefit of the students here this morning:

Healthy clams and oysters live on these tiny plankton-like organisms. But these same single-cell organisms can go off on a population explosion—blooming and becoming so profuse that they form the "red tide," and then can even affect people walking by on the beach, through the aerosol-like effects of wave action, which lofts these tiny toxin-laden creatures into the air. It can have an allergic effect on the unwary—causing a cough, runny nose and itchy eyes.

"What could be the consequences of walking your dog in the woods?" Goldfrank asks. "If you were out near Truro, your dog might pick up a tick—they've been isolated at the national park there—harboring one of the *Rickettsia* organisms that set off high fever and severe headaches, which are indicative of what?" Several voices chime in: "Rocky Mountain spotted fever."

"Right," says Goldfrank, "which is not endemic to the Rocky Mountains, by the way, but is most commonly an East Coast disease." He looks around the gathering briefly. "What about a bite from a tick followed by the onset of flu-like symptoms, with arthritis and lassitude?" He gets several identical responses to this query, as well—Lyme disease. The discussion ranges over other disorders that one could suffer from sojourning on Cape Cod—babesiosis, a feverish anemia, from a teardrop-shaped organism that infests red blood cells; tularemia, a glandular fever; and even, given the heavily homosexual populace of some areas on Cape Cod, AIDS.

Neal Lewin offers a word of encouragement to the medical student: "Lucky thing you got back healthy." Everyone laughs.

The last segment of the meeting is devoted to a slide presentation from one of the NYU-Bellevue dermatology residents, which goes on for about fifteen minutes, describing various skin cancers and disorders caused by both sun and foliage, such as poison ivy. Before he can finish, an urgent call summons the lecturer away. Lewin has the last word. "And you had hardly scratched the surface." There is a muted round of laughter and the group adjourns, swiftly scattering to various parts of the building.

A question has come up for Goldfrank's consideration: Herbal tea sales in the Bellevue newsstands have been stopped, as such mixtures are potentially dangerous and notorious for various contaminants that neither administration nor staff thinks appropriate for a health-care center to purvey. In addition, does herbal therapy have a place in the midst of sophisticated modern medicine? Now there is a related issue: why shouldn't the sale of cigarettes also be banned?

No one questions that the sale should be banned, but there are practical considerations that finally argue against a ban. As the discussion developed, it was argued that if the Medical Board forced the newsstands out of the cigarette business, that would not stop smoking among patients in the hospital. Instead it was thought to be likely that such an act would merely create a black market, with cigarettes selling for as much as $1 each. So the more pragmatic members on the Medical Board prevailed and decided not to prohibit their sale. Goldfrank understood but still couldn't agree with the decision. "I thought it was worth taking a stand and trying to make a change in people's habits."

Dr. Stephen Waxman serves as medical director of the Walk-in Clinic. He studied at the University of Pennsylvania, went to medical school at SUNY–Downstate Medical Center and completed his residency at North Central Bronx Hospital, where he met and worked with Goldfrank.

"I went on a trip after finishing my residency and then came back to New York and applied for a position as an attending physician in the Emergency Room at Lenox Hill Hospital. They asked for references, and among them I listed Goldfrank.

"When they telephoned him to check on me, he learned I was back and looking for work, so he called me and urged me to consider Bellevue before accepting another position."

Waxman wound up taking a post at Bellevue instead of Lenox Hill. That was in 1979. His arrival at Bellevue lives vividly in his memory.

"On my first day, I was coming downtown on the subway—the Lexington Avenue line. At Forty-second Street, the train stopped and the doors opened and remained open. There on the platform I could see a group of people gathering around an old man who was sitting propped up against a column. I went forward, identified myself as a doctor and asked if I could help. The man was about seventy, pale and short of breath. As I was examining him, the ambulance crew arrived with oxygen and a stretcher. I asked them where they would be taking the patient. 'Bellevue,' they said. So I said, 'Great—let me ride along. This is my first day on the staff there and I guess this will be my first patient.'"

From that first day, Waxman was impressed by the conspicuous sense of dedication he saw. "Doctors come and go, but the nurses endure." Their style combines a practical mixture of strong idealism and tough-mindedness.

There's some sort of disturbance in the Walk-in Clinic. The clear voice of the head nurse—Mary Dwyer—can be heard. As she later explained, a woman came in wanting a painkiller for her backache. "Not before you see a doctor," Dwyer told her, and the woman sat down, sputtering angrily. A few minutes later she came back to Dwyer, virtually begging for a pill. Based on her vital signs and a history of long-standing pain, Dwyer did not think her condition urgent. "If her temperature had been high, or if her heart rate rapid, or if she had been giving signs of withdrawal symptoms, she would of course have had priority. But you have to be alert. Sometimes people try to prowl around to get needles or syringes from the cabinets, so we know to keep on our toes. We see people who fake pain," she says.

Says Dr. Waxman: "Here there are people suffering almost everywhere you look. And a patient in real pain gets our immediate attention. At the same time, we get very good at spotting the fakes, the ones who are playing the angles to get some painkillers. One person told me he had acute kidney-stone pain. So we looked for blood in the urine. There's a microscope right there," he says, indicating a small lab across the hall from his office, "so I can readily check the blood in urine. Some people bite their lips, then spit into their sample. You see epithelial cells in microscopic examination." (Such cells are typical of the lip and mouth, not internal organs such as kidney or urinary bladder.)

Waxman has also detected different-shaped red blood cells in such examinations, leading him to conclude that animal blood has been put in some of the samples he's been presented with. And he has even seen the blush of pink that came, as was discovered after a bit of detective work, from red crêpe paper which someone had dipped into a specimen jar. So there is a degree of skepticism if a walk-in patient claims an urgent need for painkilling medication but doesn't present vital signs that support the claim. Mary Dwyer has also felt just this skepticism.

"I told one woman to wait her turn the other day. Someone would be with her in a few minutes. Then she started shouting. Called me a white mothereffer, and made such an outburst it disturbed the other patients in the waiting room. So I turned back to my work, and I could hear something breaking—like a chair being torn off its mounting. And that's exactly what it was. I heard someone call out, 'Mary! Watch out!' and here comes a part of the chair flying toward me! That's when I shouted 'HP!' and the hospital police came running." The irate patient was quickly subdued. The chair missed.

Dwyer worries about the waiting time for patients. But the Walk-in is closed from midnight to 8 A.M., a reflection of budgetary restrictions imposed by a cost-conscious Health and Hospitals Corporation. And, as services are restricted, waiting times tend to stretch longer, and angry patients become more frequent. It is a growing problem.

While Goldfrank lunches at his desk, he sees the memo about Dr. Bellini. Bellini has tendered his resignation to accept a new position in Tampa, to be an attending physician in an emergency department there, with a major role in their Poison Center.

Goldfrank has a stoic response. Congratulations to the departing one and confidence that outstanding new candidates will present themselves for consideration, choosing Bellevue for its enormous variety even if it does not as yet have a residency program equivalent to those at some forty other leading teaching hospitals around the nation.

Besides that academic problem there is a financial one. Goldfrank is aware that his attending physicians are paid perhaps half or a third of what they might make elsewhere, and that the rigors of life here are daunting to all but the most energetic applicants. There's also the unresolved problem of residencies. Each attending physician might legitimately expect to act as teacher to a few highly trained residents, themselves on the brink of becoming attending physicians. But, as presently constituted, no residency in emergency medicine exists at NYU-Bellevue, because of resistance on the part of the establishment to this new specialty. (Emergency medicine has been recognized as a distinct specialty, the twenty-third so recognized, only since 1979.)

A summer program that Goldfrank has developed along with the head of volunteers at Bellevue, Project Health Care, has much to do with developing ultimate applicants for service. College students come in for the June-to-August period as volunteers—working in the various emergency areas. He checks over a stack of evaluations written by the students about their experience. One, submitted by a young Dartmouth student, Frances Gmur, catches the flavor of the department:

My first rotation was in the Emergency Ward, where all of the severe trauma and life-or-death cases are treated. Feeling a bit nervous, I walked in and introduced myself to the head nurse and others. Tom, the head nurse, was very nice, explained a few things to me, and then quite calmly said, "Well, you've picked a fine day to start: both trauma slots are full." Thrilled to be in on this, but not quite sure what to expect, I poked my head in the door he indicated. Just then they were pronouncing a thirty-year-old black woman dead after trying to resuscitate her for an hour. Right beside her, in the other slot, was a longshoreman who had been crushed by falling cargo, and both his arm and leg on one side were severed. They were rushing him up to the operating room to try to reattach the limbs. Later I heard that they could not save either limb.

She added:

The saddest event of my summer was the little girl, nine years old, who had been hit by a car and had gone into complete respiratory and cardiac arrest. They tried for over an hour everything that they could to bring her back. A team of about fifteen doctors was working on her. They opened up her chest and massaged the heart. They shocked her. They put in a breathing tube, drilled a hole in her skull to drain the blood. But nothing worked, despite slight encouraging signs. The worst part was that I went in later when everyone was gone, because I wanted to see what they did with her. She was just lying there, not covered or anything. They had sewed her back up, including sewing her mouth shut. Her eyes were open though one was very swollen. It was a sight I will never forget.

Goldfrank turns to page proof of the third edition of his book. Its evolution mirrors the field of emergency medicine itself. The book began as a series of monthly columns in *Hospital Physician* magazine. While he was still serving as director of emergency services at Morrisania City Hospital and later at North Central Bronx–Montefiore, he and a single collaborator prepared pieces under such breezy titles as "A Rash Decision," and "The Telltale Heart." They were gathered into a paperback handbook of 180 pages in 1978. Such was its success that four years later, after Goldfrank had moved to Bellevue, a 432-page cloth-covered second edition appeared. By that time, others were following Goldfrank's lead in presenting up-to-date clinical case histories on modern poisonings. Goldfrank then had four co-authors, including his principal deputies and the clinical pharmacist and the director of the Poison Control Center.

Soon there was a competing volume with the identical title, *Toxicologic Emergencies.* So the third edition is to be called *Goldfrank's Toxicologic Emergencies*, an imposing tome that will be nearly a thousand pages and have five co-editors. Like all its predecessors, it has been inspired by the memory of Sir William Osler, and its royalty income will be donated to the New York City Poison Control Center. For a book that is expected to generate over $320,000 in revenue for its publisher—it now has a list price of $80—that will represent a substantial gift from its staff—some $32,000—to the Poison Control Center.

At about the time Goldfrank was finishing up his lunch, a 911 call came in to police across the river in Brooklyn. The caller said that a known fugitive was standing with two other men at the intersection of Windsor Place and Prospect Park Southwest.

When the police got there, they found three men, one of whom matched the telephoned description. This suspect bolted from the group. The police gave chase. They caught the man—identified as Nigel Crenshaw—and frisked him. They discovered a nine-millimeter semi-automatic pistol, which they confiscated. Then they wrestled Crenshaw's arms behind him and handcuffed him.

When they brought their suspect back to the street corner, several plainclothes officers were holding the other two men. At that point, police officers Charles Roman and John Franco—both in their late twenties—who had answered a radioed call for help, pushed Crenshaw into their car. The other officers took the remaining suspects and started out toward a stationhouse about two miles away.

Officers Roman and Franco made a mistake. They let their prisoner ride alone in the rear seat of their squad car while they sat in front. Their route took them behind the lead car down crowded streets before they made a left on Parkside Avenue.

After they had gone about a block on Parkside, the prisoner in the back seat somehow bent himself around and withdrew yet another pistol, a .45-caliber semi-automatic, which he had kept hidden in the small of his back, and which had escaped detection in the earlier search. He blasted away with this weapon. A bullet hit Roman in the back, another struck Franco. Roman, the driver, fell out of the car, which was still moving, and took another slug in the shoulder.

Franco had been shot in the face and stumbled out the front passenger's door. The bullet went in one cheek and out the other. By now, Roman had drawn his .38-caliber service pistol and emptied it in Crenshaw's direction. Meanwhile, Roman had suffered two additional wounds—one in the upper arm and one in the hand.

Other members of the police convoy, hearing the shots, closed in quickly on the scene, while the prisoner, his hands still cinched in the cuffs behind his back, attempted to flee.

A flurry of shots from the lead car caught Crenshaw as he sprinted down Parkside Avenue for a few feet. Then he pitched forward, dead, on the sidewalk near the corner of Ocean Avenue.

Officer Roman was rushed by his comrades to Kings County Hospital and within minutes police commanders had made plans to transport him by air across the river to the Bellevue Emergency Department.

At the triage desk the loudspeakers crackled to life. "Multiple gunshot trauma incoming in seven minutes," said the matter-of-fact voice from Maspeth. Nurse Liz Reynolds relayed the message swiftly to the trauma service, which began to assemble a team to meet the incoming ambulance at curbside as it covered the half-dozen blocks from the heliport.

Officer Franco, meanwhile, was admitted to Kings County Hospital, where he was soon listed in good condition.

EMS paramedics accompanied Officer Roman on the four-mile helicopter trip across the East River to the Thirty-fourth Street heliport near Bellevue. From there, a waiting EMS ambulance rushed Roman down to the Bellevue Emergency Department entrance, where doctors, headed by chief surgical resident Glenn Laub, were already starting their examination as the stretcher came out of the ambulance.

As Laub bent over Charles Roman, he saw the breathing tube that the paramedics had inserted, and asked him, "Do you understand what I'm saying?" Roman nodded slightly. "Can you move your legs?" Another nod. At that point, Laub later said, he knew Roman was going to survive. "If you have enough blood pressure to think, then you're going to be okay."

One of the things that Goldfrank brings to this department with particular effectiveness is a sense of order and calm, even though sudden appearances, like that of Officer Roman, threaten to upset the equanimity upon which quality care depends. "You always worry a bit about such transfers. You wonder about the patient and has the other institution really assessed him adequately for transport? And there are a number of other questions—Are large-bore IVs in place? Has he been given blood? Are the x-rays coming with the patient? Will there be enough support of the patient so all the fanfare of an air transfer doesn't adversely affect him?"

Everywhere else in the hospital, Goldfrank has said, a patient is already a disease, has a diagnosis attached to him, already begins to be abstracted and in a sense somewhat dehumanized. But in the Emergency Department you defer all conclusions. "We don't know if a patient is alive or dead when we first see him. And we're never sure what we're going to find, or what kind of emergency medicine we may be called upon to practice—surgery, neurology, pediatrics, psychiatry, cardiology, obstetrics. We have to be comfortable with a broad range of specialties."

During the time the helicopter was vaulting up from Parkside Avenue, a truck door was opening on East Eighteenth Street onto the left arm of bicycle messenger Charles Higgs, a twenty-three-year-old black, who was swept right off his bike into the street by the door. In falling he suffered a lengthy cut as his hand raked over a wire-mesh curbside refuse basket. He wound up in the Adult Emergency Services, where Dr. Stephen Waxman oversaw a surgeon who took six tidy stitches in Higgs's hand with nylon sutures. The messenger also got a diphtheria-tetanus shot, to protect him from a wound infection from the tetanus organisms commonly found on the streets.

Just a few minutes before, Waxman had examined an eighty-five-year-old woman who was not overtly sick, but was brought in by ambulance because the shelter in which she is staying was alarmed by her disorientation. Her blood pressure was 200 over 100—high but in itself not life-threatening in one of her age. "But why has she been living in a city shelter for four days? Who could displace an eighty-five-year-old intact woman from her home without making suitable arrangements? She can care for herself, keep her own home. But for now there is nowhere but the shelter. We'll certainly ask our social worker to look into it, but it's distressing, all the same."

Waxman reflects on his profession: "Emergency medicine demands the most intense involvement personally and intellectually. Every area of clinical medicine is practiced, every emotion is taxed. The challenge is in managing an unlimited variety of disease or trauma at a level of immediacy that is rarely approached in any other specialty."

As he is speaking a male patient slowly shuffles toward an examination room, accompanied by one of the attending physicians. The young man coughs and is short of breath. He is very thin and appears to have a chronic wasting disease. "Possibly AIDS," Waxman suggests. He continues, "The AIDS epidemic is just beginning and it's already the number one cause of death in young men in New York City. The impact on Bellevue will be overwhelming in the years to come because the group in which the attack rate is increasing the fastest—namely black and Hispanic intravenous drug users and their sexual partners—are people for whom our department is the primary source of care. We better be prepared."

$\wedge\!\!\!\!\!\!\!\!\!\diagdown\!\!\sim$

The TV crews are coming one after the other and the print journalists are in evidence also—drawn by the shooting of the Brooklyn policemen. The director fields any number of requests for interviews, and systematically parcels them out among the staff members who have been involved in the care of Officer Roman—which now is giving every evidence of succeeding swiftly and completely.

Meanwhile, it has come to light over in Brooklyn that the 911 caller had deliberately said that Nigel Crenshaw was a fugitive, even though he wasn't, because the caller knew that was one way of getting a swift response from the police. Investigating officers told the press later that they had retrieved about $950 from Crenshaw's body and as much as $4,100 from one of the other men. Seasoned police department observers, putting those facts together with the sophisticated handguns of these detainees, concluded that the trio had doubtless been involved in drug dealing.

# 12 An Alkaloid Plague

Dinah Honeycutt won her first beauty contest in third grade. Wherever she went, the eyes of young men followed. Discovering she was too short to model, she called on another of her gifts: she was an expert typist. She became a paralegal staff member for one of the large law firms in Wall Street. Members of her family revealed these things about her when they came to the Emergency Department from Memphis. By then, Dinah was no longer able to speak for herself.

Somewhere along the line she started using cocaine. She told a girlfriend that one of the bachelor partners of the firm was doing free base, and was having really fine parties out in the Hamptons on weekends. Dinah became part of his weekend circle. She boasted of witnessing a $12,000 coke purchase the partner made from a dealer on a lonely beach road. Then she told her friends how her beau began to take her to the opera, coming around to pick her up in a limousine and, on the way to Lincoln Center, doing lines on the back of a platinum-framed, beveled mirror. "She was really something," her brother-in-law said, "never really happy unless she was flirting around the edge of the abyss."

One Tuesday morning in the Xerox room at work, she suffered a seizure. With a wild cry, she lost consciousness, her breathing stopped and her legs and body stiffened as she fell to the deeply carpeted floor. Her face became blue as her features assumed bizarre contortions. It looked as if she would die on the spot. Her alarmed co-workers called 911.

Shortly thereafter, she revived enough to begin gasping for

breath. In the intervals between these tonic-clonic seizures—in which the muscles wage war against each other in continual spasms of relaxation and contraction at an exhausting rate—her body seemed to relax and she breathed rapidly and deeply. But she did not respond to the questions of her co-workers, who protected her from banging her head and inserted a pencil in her mouth in an effort to keep her from biting her tongue. Volunteers trained in CPR arrived shortly as did members of the firm's medical department. When emergency medical personnel got to her thirty-second-floor location, Dinah's temperature was already well above one hundred and rising steadily.

An ambulance crew arrived within five minutes and began to follow the status epilepticus regimen in their *Treatment Protocols* booklet. "Status epilepticus" is defined as two or more seizures without a lucid interval. They gave oxygen, dextrose, thiamine and diazepam (the generic term for Valium) and rushed their patient by express elevator downstairs and then as fast as they could go up the FDR Drive to Bellevue.

"By the time we saw her," Goldfrank later recalled, "she was in a wildly agitated state, with an exceedingly high temperature—108 degrees Fahrenheit. That's not uncommon for victims of prolonged seizures or cocaine-related agitation. We weren't sure what had caused it when we first saw her.

"I examined her nose and arms, as I had a strong intuition that she had overdosed on cocaine, but couldn't find any powder remnants or signs of a needle injection. But then, on the back of her right calf, we found small needle puncture wounds—unmistakably fresh. And there were several older sites.

"It was there in the saphenous vein of her leg that she had injected the drug. (That was her way of preserving her arms from evident 'tracking' of the sort that marks almost every addict.) We immersed her immediately in an ice bath and gave her anti-seizure medications such as Valium. Within twenty minutes we had her temperature down to normal. But still we were very concerned about the effects of prolonged high temperature. The liver is naturally a hot spot in the body as so much metabolic activity takes place there. A rectal temperature of 108 means that the liver is probably close to 110 or 112. No bodily organ can endure much of that. And, as it turned out, the damage was already irreversible. But we still hoped for the best, and admitted her for intensive care in our emergency ward."

The young woman's sister and brother-in-law arrived the next day, but Dinah never regained consciousness. Two days later she died of what was noted as a "bleeding diathesis"—a disposition to bleed that resulted in hemorrhage from every orifice of her body and into her brain. Liver failure had caused her blood to lose its ability to clot. Then, almost simultaneously, she suffered general breakdown of muscle tissue, which led to acute kidney failure—all traced to the catastrophic overheating of her body. Even had she not succumbed to this cocaine-induced heat stroke, her use of a needle made her vulnerable to another sort of danger.

"To me," Goldfrank later said, "the frightening thing about the needle injection of cocaine or any other drug is that shared needles, which are so much a part of the drug culture, virtually guarantee that such diseases as AIDS and hepatitis will proliferate rapidly. In fact, recent statistics indicate that in New York 40 percent of AIDS is now occurring among drug abusers. Beyond that, stubborn bacterial infections of the heart valves—endocarditis—plus osteomyelitis of the bone and meningitis of the lining of the brain are directly related to the bacterial and fungal contamination of various drugs and paraphernalia."

Those that make it to the hospital are the lucky ones. Others who snort or inject or smoke cocaine sometimes die within the first three to five minutes, of cardiac arrest, before help has time to get there.

The emergency staff at Bellevue has probably seen more of the effects of cocaine than any comparable group of medical professionals anywhere. They know this powerful central-nervous-system stimulant to be a natural alkaloid that comes from the leaves of *Erythroxylon coca*, and they know that this "amino-alcohol" can sweep through the body like a jolt of rogue electricity, shorting out everything in its path.

Goldfrank and his colleagues have done intensive research on the history and lore of this white powder. This is what they have found:

For nearly fifteen hundred years the substance has been known to Peruvian Indians, who have sucked on the coca leaf to relieve their boredom or for ritual observances. As early as the 1100s the Incas used cocaine-filled saliva for local anesthesia for trephinations—that is, drilling into the skull to relieve pressure on the brain. Then in the 1880s, some twenty years after a German chemist had successfully isolated the pure cocaine from the coca leaves, others began applying the refined alkaloid to various disorders. Freud's experiments with cocaine as an antidepressant led him to try the drug in treatment of various other diseases, among them alcoholism, morphine addiction and asthma.

Under cocaine therapy for his tuberculosis, Robert Lewis Stevenson wrote *The Strange Case of Dr. Jekyll and Mr. Hyde* in seventy-two hours—in 1885. The next year, John Styth Pemberton, an Atlanta druggist, mixed up an elixir of cocaine and caffeine plus some vanilla, fruit and sugar flavoring, then carbonated the potion and called it Coca-Cola.

In 1914, the Harrison Narcotic Act incorrectly labeled cocaine as a narcotic (something that soothes, relieves or dulls sensation), whereas it is instead a powerful stimulant—sharply activating the body "almost as if you had stuck your finger in an electric light socket," Goldfrank has said.

Finally, in 1970, the Comprehensive Drug Abuse Prevention and Control Act classified the substance as a member of Schedule II—"high abuse potential with small recognized medical use." In fact, about the only orthodox medical use for more than a century has been to numb the gums for periodontal surgery and to apply local anesthesia to the mucous membranes of nose, throat and respiratory passages of the upper airway, where a surface-administered local anesthetic is desirable.

In that same decade two parallel attempts by United States governmental authorities to suppress Mexican marijuana shipments into the U.S. resulted in the emergence of Colombian drug growers and refiners, who seized the main marketing opportunities. And cocaine was far more profitable. As availability of cocaine increased, prices dropped and the once-rare drug became much more common. A kilo of cocaine cost $65,000 in 1982, but only $30,000 four years later. Each drop in price increased the number of patients at Bellevue with cocaine-related medical problems. At the same time that prices were dropping, purity of the drug was rising, further exacerbating the impact of the substance.

Now, in the late eighties, cocaine entrepreneurs have developed a relatively inexpensive variant of their basic product—crack, a smokable kind of cocaine, which is having a devastating impact on young users especially.

"Perhaps as many as ten million Americans are using cocaine today—more than twenty times the number of heroin users by our best current estimates," Goldfrank has said.

He goes on: "I recently treated a twenty-four-year-old woman and her four-year-old daughter. The major complaint was that her daughter itched all over. The woman was extremely agitated, speaking in a forced and pressured way and gyrating about constantly. Every now and then she reached down to scratch at her daughter's arms. 'Don't you see these small bugs?' she asked me impatiently.

"I looked closely. It could have been scabies—parasitic mites that burrow under the skin. I studied the girl's arms, but there were no mites. Then I checked her hair, where nits, the egg form of lice, can be attached if there were lice infesting her. But I found no nits and no other signs of lice.

"The mother was scratching away at her own arms, meanwhile, and talking in a disconnected way. By that time I had a pretty good idea we were dealing with 'cocaine bugs'—a tactile hallucination often associated with cocaine use."

The staff watched the mother until the drug's effects had worn off. Then Randolph Brown, a social worker whose soft-spoken manner seemed especially well suited to this patient's problem, entered the case to counsel the woman, with particular stress on protecting the child. "What we worry about in these drug cases," Brown said later, "is the child. The children of drug-abusing parents have the highest risk of any category to be killed or become victims of child abuse. In any case like that one, I do everything I can to see that the dangers to the child are minimized."

Wave after wave of public revelation about the cocaine plague has begun to change attitudes toward what was once lightly called "nose candy," "snow" or "gold dust." Each new headline provides a point of departure for the morning staff meetings of Goldfrank and his team.

In the course of a year there were revelations of widespread cocaine abuse in organized baseball, so great that it was acknowledged that the drug was ending careers prematurely and affecting pennant races. Cocaine has become a major factor in discussions of trades between teams, and arouses suspicions whenever a manager sees a skilled player suddenly commit an error on the field. There were also reports of sudden death by cardiac arrest of young star athletes in basketball and football.

At one of their staff meetings, Goldfrank and Lewin recapitulated what is now known about the effects of the drug. This is how they summed up the current state of knowledge:

Cocaine constricts the blood vessels in the extremities so dramatically that blood pressure usually rises sharply after the drug is snorted or injected. The effect is to take a heart pumping blood over a wide network of arteries and to cut that field of circulation, confining the blood to a smaller network. The result is that heart attack or stroke sometimes ensues.

One additional prominent effect of cocaine ingestion is central-nervous-system stimulation "in a rostral-to-caudal" fashion—literally, a "beak to tail," or top-to-bottom sequence. First, the cortex in the brain is sharply jolted, which can bring on a rush of talkativeness, jittery activity, keen alertness. As the toxin affects the medulla—the pyramid-like segment of the brain sitting atop the spinal cord—breathing quickens. As the drug affects the motor cortex and then spreads through the lower centers, such as the spinal cord itself, the user can suffer clonic-tonic seizures. Meanwhile, the drug acts to turn the switch at the axonal membranes near where it is snorted or injected so that no messages pass along these nerve junctions, and they become numb.

When cocaine gets to the heart, it makes that organ doubly sensitive to the body's own hormones, specifically adrenaline. So ingesting cocaine can be like taking a huge dosage of intravenous epinephrine. Under such a strong stimulus, the heart's major chambers can go into the frantic motion of fibrillation, which must be reversed by the timely use of the defibrillator's electrical paddles. The cocaine jolt may result in tissue destruction—muscle injury that can mimic the behavior of a spasm or a clot in the coronary artery. The ventricles no longer perform normally, no longer correctly govern, on the right, the flow of blue blood into the lungs to pick up fresh oxygen; or, on the left, the surge of the renewed, red blood to all the other bodily organs.

"I've seen a number of young men with sudden cardiac complaints," Goldfrank has written, "as a direct result of cocaine use. In fact, cocaine is a common precipitating event for heart attacks in men between the ages of twenty and forty. Many of the coke users who get this kind of ventricular fibrillation are dead within two or three minutes of snorting or smoking or injecting." They never arrive in the Emergency Department. "If we see any young man in cardiac distress, we always examine the septum of the nose—the dividing wall between one nostril and the other. If we find fine white powder, that's virtual confirmation that the patient is a cocaine user. If there is a small hole or other septal damage, he's probably a heavy user. Those indications of damage come from the vasoconstriction and tissue destruction that follows."

At every opportunity to spread the word, Goldfrank and his staff repeat the warnings, hoping to change public attitudes enough to reduce the plague that nearly overwhelms them. Goldfrank tells visiting TV interviewers the same story, hoping that the message will finally begin to penetrate:

"It's as if you are suddenly living life in a speeded-up mode. Like a 45-rpm record played at 78 rpm. We treat these patients by rapidly calming them, slowing the heartbeat and lowering the blood pressure and temperature so as to prevent life-threatening damage.

"People have asked me what makes cocaine so especially dangerous. In part, it's one of the few anesthetics that has a triple effect—on nerves, vessels and hormones. Snorting or injecting cocaine is similar in its results to electrocution. Fibrillation of the heart muscle can take place. Seizures of the brain may set off disorganized movements of the whole body.

"In addition, because blood pressure rises sharply, should there be a weakness in any small artery within the brain, it might well burst. That would be like an explosion of blood into the brain's inner chambers or ventricles. Serious neurological problems follow—seizures, strokes, possibly death.

"But whether there is such a catastrophic incident or not, there will doubtless be a pins-and-needles sensation at the extremities. Frequently thought wanders in a disjointed fashion. A part of you seems to unravel and float away—dissociating from the main stream of consciousness. It is this 'rush' that users seek. But it is their invariable experience that the first such euphoric episode is greater than any that follow. The 'overwhelming goodness,' or 'pleasure beyond description' that such users as Aldous Huxley or Sigmund Freud wrote about just cannot be recaptured. Larger and larger doses produce less vivid feelings of ecstasy.

"Two to four weeks of use lead to addiction, whereas heroin or a painkiller like Percodan may take months to establish a dependency.

"The 'cocaine blues' always follow ingestion, and leave the user in a devastated psychological condition. He feels worse off. So he enters the unending cycle: further use, further risk, further dependency, further depths of the blues.

"Physically, the drug and its contaminants simply destroy the fine veins and arteries of the nose. Then users, unable to get as high as they used to from snorting, up the ante again. They begin to inject the drug, first in the skin, then ultimately in a vein. Here the result is to destroy the vessels that would otherwise keep nutrients circulating and tissues healthy. And always the confirmed user is seeking a new route or higher dose to achieve the unachievable.

"The local effect of the cocaine is detrimental even if the individual injects in the skin—'skin-popping'—with resultant infection, tissue destruction, abcesses and ulcers. In addition, it blocks the small lymphatic vessels that try to clear the body of poisons. Once blocked, these vessels no longer work.

"Crack is equivalent in its effect to injecting cocaine. When you snort, the cocaine must be absorbed into the veins then goes to the heart and then into the lungs and brain. It takes some time to make itself felt. But if you smoke crack or inject cocaine the way the paralegal did, there is immediate impact—in the smoke, right to the lungs, then to the left side of the heart and to the brain. In the injection, the progress is slightly slower—to the right side of the heart, the lungs, the left side of the heart and then to the brain."

Neal Lewin has written, "The onset of action of cocaine depends on the dose and the route of administration. When used IV [intravenously], its peak action is in 3 to 5 minutes. When applied topically to the nasal mucosa—that is, snorted—it peaks in 20 to 60 minutes. When ingested orally, it peaks within 60 to 90 minutes." He adds that free base or crack "can be smoked since it vaporizes rather than burns, and because it is fat soluble it can readily cross the blood-lung barrier."

Lewis Goldfrank looks in on a room with a patient suffering kidney failure. Carlos Mantolo, a thirty-six-year-old Hispanic, is shivering and throwing up. His temperature is over 104 degrees and his legs are swollen. He is pale and there are blood and albumin in his urine and the outlook for recovery is not favorable. Hematocrit—the percent of the blood volume that red blood cells constitute—is 10 percent (as against a normal reading of about 40 percent). He is also suffering nausea, weight loss in spite of swelling, muscle cramps—all signs of nephropathy, or kidney sickness. All the indications are that Mantolo's problem began with substance abuse, probably cocaine. It is enough to give the next morning's meeting its major topic.

"More than half the kidney dialysis going on at Bellevue today is as a result of drug abuse," Goldfrank tells the group, and many other rampant disorders also can be ascribed to the white alkaloid powder.

"In the last three years, the only three cases of wound botulism poisoning that occurred in New York City have been traced to use of this drug. When the coca leaf is picked, handled by the Peruvian, Colombian or Bolivian growers and processed en route to the United States, bacteria naturally adhere to it. Many of these bacteria are *Clostridium botulinum*. The botulinus toxin is the most poisonous substance known to man.

"To control this bacterium, which is everywhere, when you're putting up preserves, you boil everything before canning begins. But the cocaine buyer in America doesn't risk destroying the drug he's spent perhaps $100 a gram* for. So he risks unknown bacterial, fungal and viral contamination. The *Clostridium botulinum* bacteria prefer to grow in the absence of oxygen. So injecting cocaine, which limits oxygen flow to the injected site because it constricts the small blood vessels, increases the chances that such a virulent bacterial organism will grow in the tissue and produce this lethal toxin in a user's body.

"There are many other impurities, too. A small-scale courier, a 'body packer,' is given Lomotil to constipate him, then may swallow up to one hundred condoms or rubber containers filled with cocaine. He travels to the U.S. and then uses a laxative to recover his 'cargo.' In the process, new impurities are introduced, such as the common intestinal bacteria.

"Then, in cutting or diluting highly concentrated cocaine, new materials such as starch, sugars, salts and local anesthetics like lidocaine or procaine are added to stretch the drug. Each of these magnifies the dangers of injecting impurities.

"When the drug abuser injects cocaine, he is likely to be injecting a little bit of feces, a little bit of sugar, salt, starches and fillers (such as talcum powder) and a little bit of cocaine. The user tries to hit a vein, which will carry the drug directly to the heart, lungs and then into the brain and give him the high that he craves. Infection frequently follows. Endocarditis, which attacks the heart valves, is often the result. It leads to abnormal valve functioning and atypical heart sounds—murmurs, clicks, honks, and strange vibrations from the infected and damaged valves called 'thrills.' I have seen the resulting mutilated, deformed heart valves of drug abusers and it is not a sight one soon forgets. Nor is the appearance of the infected pericardial tissue—the lining of the heart.

*A gram is about one-thirtieth of an ounce.

"Next, the drug is pumped through the lungs, where bacteria settle out, causing local infection leading to wheezes, rhonchi [a rough snoring], rales [a rattling or bubbling]. All of these are signs of pulmonary complications and compromise, of decreased ability for red blood cells to take an adequate amount of oxygen from these lungs."

In the autopsy room, Goldfrank has seen "brilliant shining flecks of talc surrounded by the body's phagocytes trying to clear the lungs of these foreign particles," he tells his students and staff.

As the meeting concludes, the noisy door adjoining their conference area opens to admit a stretcher, moving quickly down toward Room 6. Aboard is a young black man glistening with sweat and throwing up.

It is the end of the line for a few milligrams of cocaine among the many kilos that began their journey in the mountains northeast of Bogotá, and continued in a familiar trail, down the Magdalena River valley to Barranquilla, and from there by boat and airplane to their ultimate distribution into the New York streets, and finally into the veins of this superheated young man.

# 13

**The Case of the Crazed Executives**

G. Bradley Tremaine was senior trust officer of a large Kansas City bank, a man large both in reputation and in body. He had come to town for an intensive course on new financial accounting rules. But that was tomorrow. Tonight he would do what he always did when he came to New York—behave like a perfect stranger.

With a stunning woman companion named Judy, he went to a trendy Tribeca disco and gave himself and everyone in the place more than a little amusement by the surprisingly agile gyrations he managed to sustain on the floor as he and Judy danced. After the evening waned, he and his companion got into a cab. The next thing he recalled was the Emergency Medical Services team picking him up in a scoop stretcher. Something was missing. And then he noticed he had no pants and no shoes. He came to his senses gazing up into the worried face of Dr. Dana Gage, one of Bellevue's attending physicians. He was unable to tell Dr. Gage who he was, what day it was or what city he was in.

Furthermore, his great girth was flushed to a nearly cherry-red glow. His heart was pounding, his eyes ranged wildly around the periphery of the Emergency Department examination room, and his hands brushed away what he thought were insects tormenting him. Also, Tremaine's mouth was exceedingly parched. He repeatedly asked for drinks of water, which Dana Gage and nurse Sharda McGuire brought and which he gulped down greedily. But then he couldn't seem to urinate when he felt the need, and that caused him great anxiety.

Something or someone had driven Brad Tremaine to a state of peculiar mania. So, as Unknown White Male No. 2, the second unidentified white man of the evening, he was admitted to the Emergency Services, given 100 milliliters of 50 percent dextrose, 100 milligrams of thiamine and 2 milligrams of naloxone.

Dana Gage leaned over so as to put herself in the center of whatever visual field the patient was sustaining at this point.

"We're going to check your heart, sir," Dr. Gage said. She had been in the Emergency Department only a few weeks now. And, on a normal rotation cycle, she spent half her time in the Adult Emergency Service, and a quarter each in the Walk-in Clinic and at University Hospital's Emergency Department, several blocks up First Avenue.

Dana Gage assured herself that there was nothing really abnormal beyond the sinus tachycardia—unusually fast heartbeats (more than one hundred beats per minute). That ruled out the possibility of an imminent threat to life. Then she examined Tremaine's eyes. She found the pupils widely dilated: 8 millimeters in diameter. They seemed to be stuck open, despite the bright overhead lights of the ward, which would normally set off a sharp opposite reaction in the brainstem, reducing the pupils to mere pinpricks of at most one to two millimeters in diameter.

Before any definite diagnosis could be established, the patient slowly regained his memory and his poise, and by the time the young doctor went off duty the next morning, Tremaine was almost back to normal. Neither urine nor blood tests revealed anything as to the possible cause of his psychotic interlude. Now, considerably chastened, he quietly checked out of the hospital. To Tremaine's great relief, social worker Randolph Brown came up with a pair of trousers big enough to fit him. Brown had walked over to the social work office and gone through the stacks of donated clothing kept there. The size 11D shoes were another matter. "Large-size shoes are always in short supply," Brown says, "as are men's underwear, socks and hats." For shoes, Brown provided Tremaine with a pair of cast boots—plastic slippers that are designed to fit around a leg cast. These items of apparel were enough to see Tremaine back to his hotel. From there he went on home to Kansas City, minus his Rolex watch, a diamond finger ring, his wallet, his credit cards, his personal papers, one pair of pants and one pair of shoes. But he was grateful nevertheless for still being alive and for the care he had received.

He was to become just one of a series of similar men "in two-thirds of a three-piece suit" brought into the Emergency Department over the months that followed. Each aroused the keen instincts of the detective in Lewis Goldfrank.

There was a school superintendent from Maine who was attending a convention, brought in with his clothes in tatters and with his eyes darting around the examination room, unable to tell the doctors anything whatsoever about his misadventure. All the conventional therapy was given without result. Yet his condition cleared up and he, too, regained his composure after twenty-four hours of observation.

Then a real-estate lawyer from Ohio was accosted in the theater district by a statuesque blonde, young enough, he admitted to himself, to be his daughter and pretty enough to make his breath come in little gulps. He ruefully recited his memories, fragmentary though they were, the next day when he came to his senses at Bellevue, without his wallet, two finger rings—one Masonic and one wedding—credit cards and Rolex watch. But at least he survived with his trousers intact.

Goldfrank added the details of the Ohio lawyer's experience to the accumulation that now was beginning to form itself into a pattern: prosperous middle-aged men, usually wearing Rolex watches and finger rings, were being victimized by alluring young women.

Other victims were a leather-shop owner from Sausalito and a high-ranking official with a visiting trade delegation from Eastern Europe. The symptoms were surprisingly uniform: widely dilated eyes, manic behavior, flushed coloring, fast heartbeats, urinary retention, extreme dryness of the mouth and mental confusion. Those who could remember their own names could not tell you what day it was or even what city they were in. Others could recall only the day, but not the city or their own names.

"Clearly we had an ongoing phenomenon," Goldfrank recalls, "and so far were powerless to do anything because we still didn't have a clear etiology, no obvious cause. Also, the hotels were not cooperative. No hotel wants to admit that a guest has been rolled in its rooms. But we began to suspect that was what was happening. The victims were unwilling to complain. The hotels—among them some of the most luxurious in town—wanted to deny that there was any criminal activity on their premises.

"But then there was a banker from Buffalo found dead at the Vista Hotel in the World Trade Center downtown, and another man in midtown who died after cardiac arrhythmias and seizures. The outer signs were so similar to these other cases we had been seeing that the hotels began to be more forthcoming with the police."

Meanwhile, the Poison Control Center's team of clinical pharmacists, headed by Drs. Richard Weisman and Mary Ann Howland, were also on the trail, hoping to gain some understanding of the mysterious toxin. "After we analyzed all the data from the various patients we had seen," Goldfrank recalls, "we figured that whatever the knockout drop was, it had to be something widely available, as there was by that time a total of two hundred cases. So we figured that several or maybe scores of good-looking women were enticing, then doping these men and then robbing them. But how? We were stumped."

The doctor had been nursing a strong intuition after the first few victims came in: the cause of these incidents was doubtless an anticholinergic agent. Such a substance is antagonistic to acetylcholine, the natural substance that is the backbone of the process whereby the hormones at the nerve endings pass the message across the gap to other nerves, thus activating them. The organs affected by such blockage of acetylcholines were the salivary and sweat glands and the smooth and cardiac muscle systems—all of which were obviously involved in the bizarre symptoms these unwary victims had been presenting.

Goldfrank decided to make himself a guinea pig. He went to his office, alerting his secretary to what he was planning. Should he suffer a sharp reaction, would she please call Dr. Flomenbaum? Yes, she would. But how would she know? If he started getting a reaction, he would buzz her. He closed the door between their adjoining offices, then experimented with a series of anticholinergic substances, starting with Elavil, a widely available antidepressant. "The idea was to dissolve small enough quantities not to be toxic, looking for a tasteless toxin." On the roster of possibilities were a number of antihistamines, other antidepressants and even over-the-counter medications such as Allerest and Sominex. Besides that, there were antipsychotic and antispasmodic and eyedrop preparations.

But he never had to buzz his secretary, as none of his experiments produced the needed result—an easily dissolved, colorless, tasteless substance. Most didn't dissolve easily and many, once you licked through their sugar coating, were bitter.

"For some months we were stymied," he later recalled. "Whenever another victim came in, which was about every week or so, I would raise the question anew at our staff meetings. And we spread the word to the ambulance teams, the paramedics and emergency medical technicians, hoping that they would hear something or see something that would help solve the mystery.

"Then I read about a merchant, years ago, who had had a grand opening party at his new jewelry store. Punch was served, and quite suddenly everyone went sort of crazy. There was shrieking and jumping around and wild gesticulations, and by the time the shopkeeper came to his senses, his stock had been pretty well depleted—stolen right under his eyes. Which of course he wasn't able to focus during his delirious state. They thought to analyze the punch remaining in the bowl. The lab report showed scopolamine."

But none of the Poison Control Center lab tests so far indicated scopolamine. There was a possible explanation that occurred to the doctor. Normally, tests are run by a process called TLC—thin-layer chromatography. A very fine film of a substance is applied to a glass plate as a "sorbent"—a stationary "island" toward which, after a drop of the test solution is placed at the bottom of the plate, a second sample moves, almost like a ship sailing toward an island. The amount of motion toward the island is one of the principal idiosyncrasies that aids in identification, as each substance moves a characteristic distance and at a distinctive rate. But this system doesn't pick up scopolamine in blood or urine samples. The far slower and more expensive HPLC (high-pressure liquid chromatography) method of testing is the only reliable way to detect scopolamine in the blood or urine. But it takes more than four hours once tests are under way, and is not available in most hospitals.

During the months when these executives were regularly showing up at the Bellevue Emergency Department, Goldfrank was gathering information for a revision of his discussion of scopolamine poisoning for the second edition of his *Toxicologic Emergencies*. The commonest plant that yields the toxin is a member of the nightshade family—and therefore kin to tomatoes, eggplants, potatoes and tobacco. It's known by various names, but has a large and showy appearance and a notable history. Goldfrank describes it as a "large, erect plant with funnel-shaped white or purple flowers with spreading branches and hard, prickly, ovate—egg shaped—many-seeded fruit." Of its history, he has written:

Locoweed, *Datura stramonium*, thorn apple, or Jimson weed, has been known through the ages. In the *Odyssey*, Homer talks of this plant as a poison.

Cleopatra used *D. stramonium* to woo Caesar. Marc Antony's retreating troops ate the plant as they left Parthia in 38 A.D.; stupor, confusion and mortality resulted.

In 1676 the name Jimson weed came into use. British troops sent to halt Bacon's Rebellion invaded Jamestown, Virginia. They prepared a meal of *D. stramonium* and became acutely poisoned. The plant became known as 'Jamestown Weed.' . . .

Over the years Jamestown weed was abbreviated to Jimson weed. Its use has been reported by Thoreau and Omar Khayyam, among others.

Today, a number of prescription eyedrops contain powerful concentrations of the toxic ingredient of Jimson weed—scopolamine. "I became convinced," Goldfrank later said, "that most of these cases of business and professional men showing up in a berserk condition resulted from a few drops of one of these eyedrop medications. That's all it would take. But we were stuck. Had a thesis but no proof. All we could do was wait and hope."

Just as emergency medicine consists of constant interruption, so this search was less a sustained quest than a maddeningly discontinuous chase—now furiously intense, now in abeyance.

Months went by, and any number of equally bizarre cases came and went, most having nothing to do with the search for the elusive knockout drops. And each of these new dramas had what Sir William Osler likened to the endless fascination of the Arabian Nights: each medical story like a précis of an opera or a novel. Among the myriad tales that unfolded during the weeks of waiting, Dr. Hedva Shamir handled one of the more intriguing cases.

It began when a large, well-muscled black man was brought into the Emergency Department with an obviously altered mental state. He was accompanied by two companions who did not stay around to identify themselves. The patient was reacting to a television football game, which only he could see. But from his descriptions it was clear that he was visualizing a Lilliputian exhibition game on a very compact TV monitor, which seemed to be poised about six inches in front of his nose.

Suddenly the man let out a yelp and lost consciousness. He was swiftly rushed into the largest treatment room, where Shamir and others got his vital signs and—based on his behavior and the alcohol on his breath—gave him 100 cubic centimeters of 50 percent dextrose, 2 milligrams of naloxone and 100 milligrams of thiamine, thereby covering three likely contingencies: hypoglycemia, opioid overdose and vitamin deficiency. The naloxone is something of a miracle substance itself—able to block the effects of any opiumlike drug without any bad side effects. It gets itself lodged in the opiate receptor sites in such a way that it reverses the overdose symptoms dramatically. It also blocks the body's natural opioids—the endorphins that give us a sense of elation and well-being. So it "steals the high"—either chemically induced or naturally achieved. At the same time it reverses the sometimes lethal effects of the opiumlike drugs (such as severe depression in breathing), which can swiftly lead to death.

Shamir gave him oxygen, also, and proceeded to perform the various other tests that might shed light on his condition.

When Unknown Black Male, as he was called on the charts, regained consciousness, he could not recall his name, despite repeated attempts by Shamir and other staff members to jolt his memory. On rounds that morning, he was added to the teaching roster as an example of organic brain syndrome of an acute nature, probably owing to alcohol and an unknown additional toxin. His manner was such that the director used the term *la belle indifférence* to describe the serene detachment that the patient was displaying.

He was admitted to the Medical Service, where Shamir and Goldfrank both made inquiries about him, as something in his behavior seemed too odd to be ignored. Finally, on the third day, he was able to recall his name. It was Sam Mahoney. And he asked if he could call his wife in Park Slope.

"What we finally learned," Shamir recalled, "was that his employer had been calling all over for him. And after Mahoney called home we got a call for him from his employer. By now it was clear that his 'amnesia' wasn't genuine."

Shamir learned that Sam Mahoney was a truckdriver, who, when last seen by his employer, had been behind the wheel of a rig with more than $120,000 worth of clothing in it. Neither truck nor clothing nor driver—until today—had been located. And now Sam Mahoney didn't know anything about the truck or the clothing. As he was about to be discharged, a police officer with a summons arrived to take Mahoney into custody.

How it all turned out for him never was known in the Emergency Department. "That's one of the great frustrations with emergency medicine," Shamir says, "a drama that you never get to see the end of! Or almost never."

\* \* \*

Allegra Morales is sitting stolidly in the examination room as Dr.
Dana Gage bends over her talking quietly. The discussion is short, as
she will be admitted to Surgery for examination of a large mass in her
right breast. The chart is a model of concision: "53 yr. H ♀ mass R
breast × 4 weeks. Also pain R shoulder × 4 weeks. ⊖ discharge or
discoloration or pain in breast. PMH: ↑ BP, DM, CHF." In translation:
she has been aware of the mass for 4 weeks, and has had pain for the
same period of time. There has been no discharge or discoloration.
And her past medical history includes high blood pressure (hypertension), diabetes (diabetes mellitus) and congestive heart failure, suggesting that her heart is having difficulty in preventing her lungs from
filling slowly with fluid.

Gage is pensive as she sets down the spare symbols on Mrs.
Morales's chart, and says a few quiet words to the patient. Then the
surgical resident appears in the doorway and quietly joins them.

A thirty-two-year-old white male is sputtering in rage. He fell
from a cab, which pulled away while he was getting out, tumbling him
along the street. That was several hours ago. He went on to work.
Then his legs began to feel odd. Now Dr. Toni Field listens carefully
as McKenzie Miles, a sales representative for a knitting machine company, makes his recitation.

He struck his head on the curbstone, and now complains of lower
back pain and a heaviness in both legs.

"Any numbness?" Field asks, as she auscultates his chest with her
stethoscope, listening for breath sounds from both lungs to exclude
the possibility of a collapsed lung, while at the same time listening for
regularity of heartbeat, to rule out cardiac trouble.

"No."

"No burning or prickling sensation in the fingers or toes?"

"Nothin' like that."

"Any blood in your urine?"

"Thank God, no!"

"Any nausea or upset stomach?"

"No."

"Good. Okay. Now I have some other questions for you."

"Fire away."

"So you never lost consciousness?"

"No. Just saw stars."

"Okay . . ."

Dr. Field briskly continues her examination, testing and testing again for neurologic "deficits," any abnormalities in cranial nerve function. Just asking such plain questions as Can you hear this? (thumbs rubbed near the ear), smell this? (oil of wintergreen), see here? (finger placed off at one side) or here? (on the other); testing for pain sensation in the face, for the ability to make a grimace like a Greek mask, for the swiftness with which he can stick out his tongue or say "ah" so the uvula (the little punching bag dependent from the roof of the throat) retracts—all can help confirm injury to the brain.

The succession of questions may well be to the tune of an old couplet that many senior doctors—including Lewis Goldfrank—carry with them still: Standing for the twelve cranial nerves, successively the *O*ptic, *O*lfactory, *O*culomotor, *T*rochlear, *T*rigeminal, *A*bducens, *F*acial, *A*uditory, *G*lossopharyngeal, *V*agus, *S*pinal accessory and *H*ypoglossal, is the mnemonic:

> On Old Olympia's Towering Tops
> A Finn And German Viewed Some Hops.

There's a sudden breakthrough in the Case of the Crazed Executives, provided by Pamela Jacobi, a young woman in account management at a large Seventh Avenue design house. To celebrate her thirtieth birthday, she and some girlfriends went from disco to singles bar to disco—a dance, a drink, another dance. At the last disco she encountered a handsome young man who could dance as well as he could talk—which was enough for her to invite him to join her for a birthday nightcap in her apartment. If her girlfriends were concerned, they never let on, and she and the young gallant made their exit.

The last thing she remembers is that her guest poured her a glass of her own wine, she sipped at it and shortly thereafter felt very dizzy.

Her roommate discovered Pamela bound, gagged and lashed to her bed. The apartment had been ransacked. Their furs and money had been taken. As soon as the gag was pulled from her mouth, Pamela went so crazy that the roommate dialed 911.

The paramedics who arrived within minutes had heard Goldfrank's scopolamine alert, so when they took Pamela, they also took the wineglass from which she had been drinking. They deposited her at the Bellevue Emergency Department and took the wineglass to the Poison Control Center.

When she got to Bellevue, Pamela was able only to give her own name to Dr. Kathy Delaney, who examined her. She couldn't tell Delaney what day it was or where she was. As expected, Delaney found her blood pressure to be slightly elevated and her heartbeat rapid. Also, she had a markedly flushed appearance with a low-grade temperature of 100.6 F. Pamela's pupils, like all the others with the mysterious syndrome, were widely dilated: 8 millimeters.

As the victim's agitation was so intense, Delaney decided to give intravenous Valium, which brought rapid improvement. She also administered activated charcoal, for adsorption of the unknown toxin. The young woman's agitation eventually lessened. Routine tests of blood showed only a bit of alcohol—not enough to constitute drunkenness. And urine tests showed nothing notable. "We knew it had to be something more than that trace of alcohol," Delaney later said, "but the tests showed nothing else."

Early the next morning, Dr. Babington Quame, toxicologist for the New York City Department of Health, entered his large, airy laboratory on the fifth floor of the Department of Health Building, just a few floors above the Poison Control Center.

Quame is a native of Ghana, a father of three, and a respected expert in sophisticated qualitative analysis, of the sort that oftentimes provides critical evidence in court trials. As he later described what happened, "I got the wineglass and a small amount of urine sample." Then in a series of steps he used a Quame applicator to place drops of refined samples at various points across the face of an eight-inch-square white metallic plate, which had been coated with a microscopic layer of silica gel. This plate was next put into a tank that contained developing solvent. Then it was air dried and heated in the oven at 75 degrees C. (about 167 degrees F.) for five minutes. This removed all the organic solvent. Next Quame sprayed the plate with a variety of reagents.

"Then under ultraviolet light," Quame continued, "I watched the pattern that emerged in each of the spots—comparing it with my central spots, which were my controls—in this case, scopolamine." There, at last, was the clear evidence: scopolamine in both the wineglass and the urine sample.

Quame called the results of his tests downstairs to Mary Ann Howland in the Poison Control Center, and she relayed the word to Goldfrank and to the New York police department. The long-standing mystery of the cause of the crazed executives was finally solved.

Within a few weeks, police in San Juan made an arrest of a

thirty-two-year-old woman wanted in connection with the death by scopolamine of the Buffalo banker in the downtown hotel. There were other arrests, and suddenly the small group of perpetrators seemed to disappear. At any rate, the incidence of the mysterious plague dropped off dramatically.

"We thought we had it licked," Goldfrank later recalled. "We hoped that the word had gone out so that people from out of town would think twice before having a drink with a strange beauty in a luxury hotel lounge. But that's asking a lot, I know." Too much, apparently, even for the very high-ranking to avoid.

On the weekend of the Fourth of July celebration of the Statue of Liberty's refurbishing, members of the entourage of one of the major leaders in town for the occasion stopped to have a drink with some comely strangers in a midtown hotel lounge. They were discovered later that evening in various states of disarray. One of their number was brought to Bellevue, the others to various emergency rooms at several city hospitals (scattered so as to avoid publicity). Blood samples for all came to the Department of Health laboratories. Given the strong suspicion of scopolamine, high-pressure chromatography tests the next day established that the high-ranking visitors had been felled by the same substance that, in 1676, turned British troops, according to the old histories, into "natural fools" who spent eleven days in addled perplexity.

# 14

# Creepie Crawlies

Will Jensen is back. He bursts into the Walk-in Clinic waiting room with an audible grunt. He's the one whose sharp gaze and shoulder-length hair give him a striking resemblance to Franz Liszt. He's in a state of extreme agitation. "They're driving me crazy," Jensen says, scratching himself vigorously.

"All right, Mr. Jensen, just have a seat here," head nurse Mary Dwyer tells him. She takes his vital signs and most of a history. This time Dwyer thinks Jensen's case merits a rapid evaluation by Stephen Waxman. There's still a medical history on Jensen in the active records, Dwyer thinks, as she recalls sending him on to Dermatology within the past few weeks. She finds his record in a corridor rack and hands a copy of the chart to Dr. Waxman, who leads Jensen down the hall to an examination room.

"I couldn't sleep. Not a wink. Itch all over. I wish I was dead. I really mean it! And I'd like to take the Mayor with me! Wish he could itch like this!"

"All right, Mr. Jensen," Waxman says calmly. Jensen is about forty, looks sixty. Waxman's preliminary examination confirms the medical history already listed—chronic alcoholism, malnutrition, probably cirrhosis of the liver.

At the doctor's direction, the patient strips off his heavy wool shirt and his gray undershirt and places them on the chair next to the table. As the doctor examines the man's torso, he sees further confirmation of the diagnosis. And he sees something else—a rash over most of his upper chest as well as prominent excoriations made by intensive scratching. As Waxman turns to give back Jensen's shirt to him, he sees, at the collar, a tight grouping of grayish-white crawling things ranging from about one-thirty-second to perhaps one-eighth inch in length.

*Pediculus humanus corporis*—body lice. By an act of will that permits the physician to maintain an imperturbable façade, he withdraws the garments, dropping them into a plastic-lined basket. After a hurried consultation in the corridor, Ana Torres gets a bottle of gamma benzene hexachloride (GBHC) and then, clad in a hospital gown, Jensen is quickly guided through Radiology and into the special bathroom where Leon Fields awaits. It's a narrow chamber with a shower stall and tub laid out to facilitate the swift cleaning of patients.

Fields, with a practiced rhythm, already has the shower running, and disposes of the rest of the lice-ridden clothes in a plastic bag so easily that the clean-up is under way almost before the patient is aware of what is happening. As Fields applies the special shampoo to Jensen, he achieves a kill rate of as high as 98 percent of both lice and nits (the tenacious egg sacs of the lice). But he takes care to apply the GBHC with restraint. In excess quantities, it can cause anxiety, agitation, even convulsions or other central-nervous-system disorders.

For now, a thorough application of GBHC will kill the lice and their nits, which are anchored by their parents' spittle to the hairs of Jensen's body. To effect a thorough cure, Jensen's clothes will have to be replaced by a new set, which will soon materialize from the social work office. For him to remain clear of lice, he'll have to improve his personal hygiene and avoid others infested with lice. But as he lives in a shelter for the homeless—when he's not on the street—the chances of achieving either of those conditions seem slight. The best hope of success would be to treat all those who are infested—a practical impossibility.

Case worker Randolph Brown is asked to help get a change of clothing for Jensen. The patient's old clothing is gingerly taken to the incinerator slot. When the replacement trousers, shirt and underwear arrive, Leon Fields helps Jensen into his new things, and reminds him that the itching may continue for a time, that it is a sign that the injured skin is working to heal itself. After a few days the residual itching will go away—provided that Jensen does not get himself reinfested.

On rounds that morning, pediculosis—infestation by lice—becomes the main subject, as the Jensen case is discussed in detail. In a kind of round-table discussion, to which Waxman and Lewin contribute, they tell their attentive audience of the three kinds of pediculi—the head louse, the body louse and the pubic louse.

All kinds of lice are small, bottom-heavy, wingless insects with six little grasping claws that encircle their mouths. "They're used to grab onto the skin during feeding," Waxman explains. "In front of the mouth there's a lancelike proboscis that has a central channel down which they dribble saliva into the wound." The louse injects a bit of this fluid to get the blood flowing. Then he feeds. Researchers have timed him. He averages at least a half hour per meal. And the louse may dine as often as five times a day. Its saliva, besides fostering the free flow of blood, causes an intensely itchy rash. But, unlike the tick, the louse does not swell up. It simply drinks its fill. After its meal, it spends about four hours digesting the hemoglobin and other proteins that it has drawn out of its host's blood stream.

Besides the pincers around the mouth, there are three pairs of stubby legs that end in sharp, scimitar-like claws, which give the louse its sure grasp on shafts of body hair or filaments of clothing fabric.

The louse spends all of its days on or very near its host. Unlike the head louse or the pubic louse, the body louse passes most of its time in the fold of some seam of nearby clothing. When fully mature, the gravid female deposits eggs every four hours or so—about six times a day. She cements the egg to body hair or to clothing fiber with a daub of spittle. Daub plus egg (or ova) make up a "nit," which is hardly larger than the period at the end of this sentence.

Five or ten days later, these nits hatch out larvae, which swiftly grow—keen to ply eager claws and piercing nose lances.

Lewin continues: "Body and head lice travel by grasping hairs and clothing fibers. Some say they swing like Tarzan." The entire life cycle of the louse is lived out on or close to the human body. Without warmth and blood close by, the small insect perishes.

As the descriptions continue, more than one student can be seen scratching surreptitiously—such is the power of suggestion.

Dr. Kenneth Weinberg has just seen another AIDS case, and is in a pensive mood. Weinberg, in his late thirties, did his residency training at Los Angeles County Medical Center, and it was there in the early eighties that he saw his first AIDS victim. "It was before there was even an agreed-upon term for the disease. I was doing rotation in the ICU when they brought in this young man, in his twenties, with Kaposi's sarcoma and *Pneumocystis carinii* pneumonia. He was on a respirator. He was a dancer, from the Midwest, and it was clear he wasn't going to make it. We had him intubated and all the dermatologists were coming in to look at him—that's how unusual Kaposi's was in those days.

"His first week we got to know each other. He was such a nice guy. But then in the second week he lapsed into a coma and had a terrible fluid imbalance. He was swelling up. His family came in from the Midwest to be with him, and his artistic friends, other dancers and musicians, came in, and it was clear that our high-tech medicine wasn't accomplishing much. So they asked permission to bring in a healer, and the head of our ICU said 'okay.' So they did, and then the day after that, his mother arrived, and something in his brain rallied, and he came out of his coma enough to make his farewells. It was very sad, but beautiful, too."

Now, Weinberg is looking exasperated. He's been at Bellevue for several years. He has seen almost everything, but finds himself getting impatient with IV drug abusers, particularly. Down the corridor in Room No. 8 he is listening to the markedly distorted, squeaking heartbeat of Hector Rodriquez, a thirty-year-old Hispanic whose wife has just left him. She took his three sons with her. Now he is obviously short of breath, infected and critically ill—sweating heavily, thrashing about, breathing rapidly and complaining of pains in his chest and lower back. This is Rodriquez's second admission for the same infection in as many months.

Weinberg draws blood samples to help establish the extent of the infection. During a severe infection, the body may increase the number of white cells 500 percent. A complete blood count (CBC) will establish how many white blood cells there are and of what type—whether of normal nucleus or, in the case of severe infection, misshapen with an immature development of the nucleus. Such immaturity suggests that the cells have been liberated earlier than normal—young recruits thrown into combat against bacteria before they are full-grown. These "bands" define the body's response to invasion. Beyond that, Weinberg looks for signs of kidney and liver problems in the various other tests he makes. Among the more important will be a measurement of how much sodium, potassium, chloride and bicarbonate circulates in the blood. Each of these electrolytes plays a key role in maintaining the balance upon which life depends. He will have an hour's wait for the results of the blood tests to come back downstairs from the labs. So, in the meantime, he continues his examination.

The doctor sees what he thinks are signs of septic emboli—little flecks of filth sent out from the aortic valve on the left side of the heart. They move like small pieces of flotsam pushed up against a beaver dam in a river by the rushing waters. To clear out such filth will take at least six weeks on antibiotics, the doctor estimates. Meanwhile, any of these emboli could also float to the brain and induce a stroke. They could also infect a blood vessel, causing a mycotic aneurysm—an infected, deformed vessel with a weakened wall that is liable to burst at any time. Weinberg prepares additional blood tests, to establish what the organism is—whether aerobic or anaerobic bacteria (whether it prefers to grow in the presence or absence of oxygen).

"Why'd your wife leave you?"

"Los drogas, creo."—Drugs, I suppose.

Weinberg sees small hemorrhages on the inside of the eyelids and small, painful purplish pustules on the forearms. These little mounds on the skin contain the products—pus—of the fierce battle between white cells and the invading bacteria. Then under the fingernails in the nail bed he sees multiple splinterlike marks radiating out from the half-moons at the base of the nails. Those are all additional clues suggesting bacterial endocarditis.

It's the tenth case of its kind that he has seen in the last forty-eight hours. He asks Rodriquez a few more questions, and learns that he has a small export business in San Juan.

"What are you doing with your life?" Weinberg asks sharply.

Rodriquez looks offended. In English, now, the patient responds, "What are you, prejudice?"

"I don't care what color or race or nationality you are," Weinberg says heatedly. "Don't you understand what you're doing to yourself?"

Rodriquez shrugs lazily.

"You've got a business of your own. You're young. Why are you doing this to yourself?"

"Why you fussin' at me, man?"

"You've got potential. Why do you want to destroy yourself?"

Another shrug. "All I want is to be *maduro como un queso*"—ripe as a cheese.

Now Weinberg resigns himself to attending to the surface symptoms, abandoning his attempt to confront the patient and awaken his self-defense, even though he feels that he has gotten Rodriquez to recognize what he is doing to himself.

The doctor is hearing the characteristic murmurs associated with valvular heart disease and the sounds typical of pulmonary emboli, a virtual confirmation of a bacterial infection. The affected areas of the heart probably include the tricuspid and aortic valves—infection in the former leading to the rubs heard in the lungs and in the latter to the small hemorrhages in fingernails, skin and eyelids.

Every doctor in the Emergency Department has heard the same dreary stories: How these addicts live out a kind of communal liturgy, using a needle repeatedly until it grows dull, sharing the needle with friends, as a sign of camaraderie. How they use up the easily hit veins, so they are motivated to go to "shooting galleries," where for a "technical fee" someone else will inject them in the neck or the groin.

And then, in a kind of ghostly reversal of medical practice, they will buy antibiotics along with their drugs—such substances as ampicillin or Keflex—and take them, too, as an imagined preventive. But what happens is that the common staphylococcus becomes a rampant and drug-resistant strain of bacteria, making the addict's prospects of shaking such infections as Rodriquez suffers far more unfavorable. Only very expensive antibiotic therapy can contend with these resistant strains of infection.

Finally, there is a tidal wave of evidence that such needle sharing is spreading AIDS among intravenous drug abusers. In some studies nearly 70 percent of the addict population test positive for the HIV viral antibody.

Kenneth Weinberg finishes his treatment of Rodriquez and sees to his admission to the Emergency Ward, across the hall behind the trauma area. He goes on to his next patient, in Room 5—a twenty-eight-year-old black woman named Jocasta Royce, who is conspicuously pregnant and also feverish.

As Weinberg listens to her heartbeat with his stethoscope, at first he disbelieves that the high-pitched murmurs, squeaks and honks are what they seem to be. But all the evidence seems to confirm endocarditis once again.

Her arms and legs show the scars of many needle puncture marks. Her purse, which lies open at her side, has any number of pill bottles visible inside. Finally Weinberg's displeasure bursts forth again. Maybe what didn't work with Rodriquez will work with Jocasta Royce. "What the hell are you doing to yourself?" he asks.

The woman looks at him without apparent comprehension, seeming to be annoyed that her reverie is being interrupted by his intruding voice.

Weinberg repeats his question, more loudly this time.

"I don't know and I don't care" is the response.

The doctor goes to a phone set on the wall and dials Obstetrics. "This is Kennie Weinberg," he says. "Joe? Oh, Lou. I have a patient here eight and half months pregnant with endocarditis . . . Yes, I agree that's usually a problem for Medicine. . . . Yes, I'll be calling them, too. . . . Will you look in, please? We may be admitting her to Obstetrics. . . . I'll have more data shortly—arterial blood gases and the rest of it. . . ."

He dials another number and gets a pledge of immediate consultation from one of the Medical Department house officers. Within moments, both doctors join Weinberg. It will be for him to determine whether to admit the patient to Medicine or Obstetrics.

Obstetrics argues that it is an inappropriate admission for a woman with infection and the risk of withdrawal from whatever she's on to come into their department. Obviously she is infected, and that is a problem for Medicine rather than for Obstetrics. The house officer from Medicine has sharply different views. "She is clearly at term, and that seems to me to make the only plausible admission Obstetrics."

While this discussion unfolds in the corridor, there is a moan from within the examination room. Weinberg glances inside and then comes back outside. "Gentlemen, her waters just broke." The woman is admitted to Obstetrics.

Assistant head nurse Carey Le Sieur is a lively young blonde who moves with the restless energy of a gymnast. Her pet annoyance now is that there is no longer a reliable supply of ether from hospital stores. Now the Emergency Department has to make do with pure alcohol, instead, for maggot control.

"When the warm weather comes, we see maggots. And when you're knuckle deep in maggots you really need ether." She laughs easily at the remembrance of one maggot episode when she had just finished her orientation as a new staff member several years before.

"They brought in this really emaciated man. He was maybe fifty. He looked a hundred and ten. You couldn't tell. Hair all matted, and he was gaunt, filthy and raving. Altered mental state, we call it. Freaking out, you might call it. His legs were all bandaged and we got him into Room 6 and several of us were working on him, trying to find out what the problem was. And as we cut off the bandages here were these grayish, jumping-bean-size bugs—maggots, I learned that day. It was a terrible sight, a wriggling mass. And underneath, we could see his legs with sores on them, which the bandages weren't doing anything to help at that point, they were so dirty and crusted. Anyway, we get the bandages cut off and then Rhonda, the senior nurse, pours ether on the maggots and those horrid bandages.

"Well, usually that's all it takes to knock the maggots dead. I suppose it freezes them to death, as it evaporates so suddenly that it's almost like a quick freeze. So the maggots begin to drop dead, but our patient develops a seizure right then. I tell you I was scared to death. The patient starts to fall off the stretcher, so three of us immediately lower him to the floor—that's how severe the seizures are. Like continuous epilepsy—status epilepticus.

"So, as we are getting him to the floor, he shits all over the place— this watery diarrhea—and Rhonda slips in it and I began to ask myself what am I doing in this profession. But that was the worst of it, really. We got him stabilized and cleaned up, and the maggots were dead, and we called Housekeeping, once he was all right and we had moved him into the Emergency Ward, and Housekeeping came in and they really hated the mess, but they got it cleaned up and we went on about our daily routine. That's pretty much what you have to do in emergency work."

And then, when Nurse Le Sieur had risen to her new post, years later, there was another memorable day, just before the "Liberty" July 4th weekend in 1986, when Bellevue and much of the rest of New York had been spruced up for the arrival of a world of superstars, including the President. In fact, Bellevue was the designated hospital to be on alert should there be any crisis involving the President's health, and this very department would be the receiving spot in case of such a mishap. So through the corridors moved painters and work-men, and an air of excitement and pride permeated the place at this warrant of its preeminence in a great city. It all seemed to be fresh recognition of the hospital's two and a half centuries of leadership in health care; a new public acknowledgment of a record unsurpassed in American medicine.

"It was impressive, but I still was wondering why more effort and funds couldn't be found for patient care rather than such window dressing," Le Sieur recalls. "Anyway, it was the end of the day on July 3, and the paramedics brought in this huge, obese homeless woman in a wheelchair. She was probably drunk; you could smell something on her breath. And who knows what else she may have taken? She had on a hat and a long-sleeve shirt and a brown corduroy skirt and there was long streaming hair under the hat, and the nursing supervisor says something like 'Carey, I think this woman has maggots.'

"I was just finishing up with a man who had fallen off the stoop of a brownstone on Seventeenth Street, another ETOH abuser, as we say, and we had got him fixed up when the woman came in. I noticed that no one stayed close to her in the waiting room. They moved her wheelchair over toward the doors near the bathroom. And where her wheelchair had been there was a little pile of maggots writhing on the floor. And there was this little trail of maggots on the floor and the people there were really upset. So I go get a nurse's aide and tell her, 'Come on, we have to give this woman a bath.' We got her vital signs and could see that there was no life-threatening condition. So our job was to get her cleaned up. It's one of the hardest things. But it's part of your calling.

"So we wheeled her down to the bathroom, but there was no ether available, so we used this 190-proof alcohol, which isn't as good. It just stuns the maggots; it doesn't freeze them the way ether does. Could we put her in the tub? No, she was too fat. So we would have to use the showers, which meant that we were going to get ourselves half drenched, too.

"She had on these enormous Ace bandages over her legs and they were soaked with ooze and we got them off. And got the maggots, finally, under some kind of control, and she's worried, the whole while, that the maggots are going to crawl up inside her, up through her vagina, but I told her she didn't have to worry about that—that I never heard of such a thing.

"So we get her cleaned up, and she says, as we wash her, 'Make sure you wash my back,' and it was at about that point—I was biting my tongue—that along the corridor comes this delegation of the press, being shown our department by the hospital administration. Making a tour of the place that stood ready to receive the President!"

Goldfrank has his lice stories, too. But his principal concern is that someone could die while getting deloused. "It's a hard thing to make the point that critical care comes before cleaning up." He tells the staff about the unnamed hospital in which, recently, as they were delousing an elderly homeless patient, the old man had a cardiac arrest and died. "They hadn't done the medical care first. They had a clean patient. But they had a dead patient."

To make the point most forcibly, Goldfrank occasionally wades in himself, as in the case of a patient brought in by the EMS after getting a call from project HELP—the Homeless Emergency Liaison Project. This is the Mayor's force of people rendering care to people living on the street, deciding whether field treatment or hospital care is best. As Goldfrank recalls: "The woman was in her sixties and had extremely low blood pressure and a rapid heartbeat. I pulled down her lower eyelids to look at the conjunctivae. They were absolutely white. And the stench was terrible, and there were lice—all over. There was a virtual caput Medusae or Medusa's head of lice. I had never seen a case quite like it. But her condition was such that I had to keep on with the assessment. We started an IV after copiously cleaning her skin with Betadine and alcohol. I had on examination gloves, but still I had the sense that *Pediculus humanus corporis* was about to colonize me.

"Her hemoglobin figures were very low. Typically, for every 100 milliliters of blood—a fifth of a pint—normal people have 15 grams of hemoglobin. That's about a half an ounce. This patient had only three grams. The normal hematocrit is about 40 percent. She showed about 8 percent. So she didn't have enough red blood cells, enough hemoglobin, to get the oxygen her tissues needed for normal strength. She couldn't do anything.

"This was really about the lower limit of viability. There were thousands and thousands of lice and her skin was lichenlike, thickened with endless excoriations. She was too impaired mentally and too weak physically to fight them off. And they were eating her hemoglobin. The lice were actually consuming the remaining hemoglobin.

"She was admitted and cared for after that by the Medical house staff. They gave her blood transfusions and saw an improved blood pressure right away. They shaved her head and cleaned her with GBHC several times."

Maggots, Goldfrank points out, perform a useful function, repulsive though they may be: they contribute to the debridement (removal) of dead tissue, and in that sense actually work to keep a wound clean. Small solace, still, for the novice nurse facing the swarming reality for the first time.

There's another negotiation under way: a thirty-year-old black male is swiftly wheeled into the trauma slot with two, perhaps three, bullets lodged in his body near his spine. Dr. Weinberg accompanies the stretcher. The man is moaning and looking panic-stricken despite Weinberg's attempts to soothe him with a calming word. "Just take it easy. It's going to be all right." The patient apparently drew a pistol on two policemen at a Lexington Avenue coffeeshop and they shot him in a wild chase along Thirty-first Street.

As the police wait in the corridor for the hospital to make its determination, there is an intense conversation among Weinberg and house officers from Trauma Surgery and Neurosurgery as to who will care for this man. As the patient is stabilized—owing to the fact that no vital organs are in immediate peril—the Emergency Department has fulfilled its mission. So it remains to be decided whether Trauma Surgery or Neurosurgery. And that will await the momentary arrival of x-rays from the attending radiologist. (An attending radiologist or a senior Radiology resident is always on hand to read x-rays and consult with the ED staff.)

If Radiology says the bullets are actually at the spine, the case becomes one for Neurosurgery. If simply near the spine, Surgery. The radiologist arrives with the x-rays. The quartet of doctors studies the pictures for just a moment and then agrees that Surgery is the appropriate admission. The police fall in behind the stretcher as it is wheeled toward the elevator that is standing open to whisk the patient to the eleventh-floor operating rooms.

Everyone expects a respite now. The loudspeakers overhead click and a voice says: "Trauma: ETA five minutes . . ."

# 15

# A Blonde
# and a Severed Leg

A tall blonde named Melissa Redfield is brought in by ambulance. The paramedics say that a delivery boy on a bicycle, going the wrong way on a one-way street in the East Village, smashed into her as she stepped off the sidewalk with a male companion, then kept right on going.

The companion called 911 and rode with her to Bellevue, and now seems overwrought in the waiting room. As he passes through, Dr. Robert Hessler says a few words in an attempt to calm him down as others work on the patient in Room 6. Then Hessler looks in on the group in the examination room. There was loss of consciousness at the scene of the accident. But now the patient is answering questions, even though her breathing is labored, suggesting the possibility of cracked ribs. Otherwise, her vital signs are not alarming. The medics report a Glasgow Coma Scale of 14—out of a possible 15 for full responsiveness. (The scale reflects three different numerical measurements of verbal, muscular and conscious activity. So the person unable to make any noise, to open her eyes or to respond to a painful stimulus with movement of any part of the body is given a 3. If you babble inappropriately, open your eyes only to a shouted request and withdraw from any muscular stimulus, you might have a 10 on the scale.)

Dr. Rob Pfeffer, a resident in Internal Medicine, on a tour of duty in the Emergency Department, is the examining physician.

He sees a large bruise on Redfield's thigh. She has already re-

ceived glucose, thiamine and naloxone by an IV line, established by
the paramedics on the way to the hospital. As Pfeffer examines her
more closely, he sees a heavy stubble on her legs, which puzzles him
somewhat as she is prepared for further examination.

"Is Toxicology sent?" Dr. Hessler asks Dr. Pfeffer, referring to the
blood sample that the medics have drawn.

"Just went up."

"Good. And she got glucose, thiamine and naloxone?"

"Yes."

Pfeffer and nurse Liz Reynolds continue removing Redfield's garments, in the process of surveying all the bodily orifices—a fixed part
of the protocol in all trauma cases.

"Is the Neurology consult coming?"

"They're on the way."

"Okay. Sounds good." Hessler continues to observe from a few
feet back.

Melissa Redfield had regained consciousness, now, but still is not
"alert and oriented times three" (aware of where she is, what day it
is and who she is). She knows her own name, but not the day or her
whereabouts. And she does not give reassuring responses to the various preliminary neurological tests that the young resident is making.
Her bright chartreuse dress, long opalescent fingernails and elaborate
blond hairdo—which proves to be a wig—give her the appearance of
a young lady destined for a grand ballroom somewhere. But now her
dress is torn and one of her high heels has been broken in two by the
impact with the bicycle.

As a nurse's aide and Pfeffer shift Redfield on the stretcher, to
help her become more comfortable, they remove her dress. A large
penis lies athwart two sizable and hairy testicles. There is a moment
of astonishment, when the air seems momentarily to be drawn out
of the room by a communal gasp. Then there is an immediate return to the matter-of-fact mode with which even the most amazing
discovery is greeted. "As Dr. Goldfrank says," Dr. Hessler offers,
"in emergency medicine you make no assumptions." Dr. Pfeffer
smiles ruefully.

The Neurology resident arrives and joins the group around the
stretcher. He repeats a complete neurologic exam, confirming what
Pfeffer has already picked up—that Redfield probably suffered a con-
cussion when he struck his head on the curbstone.

"Probably should admit the patient for observation," Pfeffer suggests, "and perform a CAT scan?" Hessler nods his agreement and the neurologist concurs.

Hessler goes back out to the waiting room and tells the still-apprehensive companion that Redfield is going to be admitted, and that he can see his friend now, if he wants.

In the first evaluation of any traumatized patient, there's a mnemonic, a memory aid, that governs the order of the doctors' examination—not as colorful a memory device as some of the others, but useful nevertheless. It's "ABCDE." A stands for airway—make sure it is open. B for breathing. C for circulation. D for disability, specifically neurological. And E, expose the patient completely to avoid overlooking any hidden injury.

The loudspeaker comes to life. "Trauma: ETA six minutes."

Upstairs, in an eleventh-floor operating room, where an appendectomy is in progress, a beeper on the belt of a second-assistant surgeon sounds. He gazes momentarily at the head surgeon, who nods his assent. The assistant goes to the wall telephone and dials his number, then leaves the room, still in his green garb and face mask, heading for the elevator bank that will take him downstairs to the Emergency Department.

Meanwhile, on the ground floor, Hessler takes a swing through the Adult Emergency Service, and sees Pfeffer, nurse Liz Reynolds and social worker Randolph Brown all talking to a young black girl, who is with her mother in Room 4. The girl has a cardiac monitor attached to her chest. She is suffering heart palpitations from crack use. Both mother and daughter live in one of the welfare hotels, the Roger Williams, and Brown is about to enter a Bureau of Child Welfare report detailing the story that has come to light:

Both came from Jackson, Mississippi, a few years before, the mother to join her common-law husband, who never materialized. They soon ran out of money and became wards of the welfare system. Both are obese, suffer from problems of self-esteem, and have no interests, really, beyond merest survival. The girl is a dropout and neither mother nor daughter seems to have been caught by the net that supposedly helps such people get themselves on track. Brown tells Hessler that the pair apparently never got linked up with the crisis intervention worker that the city posts in each of the hotels in which its welfare clients are housed.

Pfeffer looks at the monitor again, and sees that the arrhythmia—atrial fibrillation—is now slowing down. This time the girl has been lucky. Another such incident, he tells her, could be fatal. She nods impassively, and the mother looks irritated with her daughter and annoyed at being dragged out in the middle of the night.

Age seventeen is a "nebulous age," Brown says, almost adult but still within the province of Child Protective Services. He has watched any number of homeless adolescents coming through Bellevue over the past few months and he worries about it. "I've never seen so many young homeless with little babies," he says. "You sometimes feel they're already out of reach. But I still try to tell 'em that life *is* rough, but if you just stick in there, don't allow others to rule your mind . . ."

He will fill out a "green sheet," indicating that this case is urgent, and hope that the Social Work Department can get mother and daughter connected to counseling that will help them cope more successfully with their problems.

Hessler moves on, to check the rest of the area, to see that no patients in urgent need remain unattended. He passes Room 2, and recalls, as he does so, the woman who was a friend of one of the nurse's aides, whom he examined in that room a few days ago. The aide told her friend that Dr. Hessler was easy to talk to, she could tell him her problem. His open, Western friendliness seems to encourage confidences. "She had a lump in her breast. Obviously carcinoma," Hessler recalls. "Cancer. She said she first noticed it two years ago. 'I'm going to admit you for some tests right now,' I told her. I was afraid she wouldn't come back of her own accord for a mammogram. But she said no. Refused to be admitted. Said she had her car parked out on First Avenue. Was afraid she was going to get a ticket. I've heard others say they had to go pay their rent, or do some other urgent thing, rather than face up to the fact that it might be cancer. So she went out of here. I couldn't persuade her to stay and be admitted. I tried to hook her up with the Outpatient Department Breast Clinic, but I doubt she'll be back any time soon." He shrugs and gives a wan smile. "Unfortunately we have to let her choose her treatment. . . . We can't very well force people to accept the therapy we feel is best."

Hessler checks his watch. Six minutes since the trauma message. He waits now close to the ambulance entrance, reviewing some charts at the triage desk. Pfeffer comes up with a report on Melissa—actually Michael—Redfield: He has been admitted and has just left with his lover for the seventh floor.

There's a clamor outside, and a quickening, as everyone seems to go into a kind of dance of activity, moving the wheeled gurneys out of the way so the incoming trauma case can have immediate access to the trauma slots. The siren's wail suddenly becomes very loud and then growls to abrupt silence.

On the stretcher is an ashen city policeman, his features obscured by a breathing mask. From the edges of the mask flows his large handle-bar mustache. Neither of his legs seems correctly aligned on the stretcher. Hessler and others move forward to check for problems that might be even more life-threatening than the severely damaged legs, which are bleeding profusely, despite a pair of tourniquets that the paramedics have applied. Now the ABCDE mnemonic helps guide each quick step of the examination. Is his airway clear? Does he have good breath sounds? How is the blood pressure? Does it suggest massive internal bleeding of a sort that might jeopardize life? Are there other injuries?

Because every trauma case presents a distinctive set of symptoms and requires its own array of specialists, the Emergency Department staff physicians and Emergency Ward nurses begin the evaluation as soon as the stretcher emerges from the ambulance. They are immediately joined by the surgical trauma team, who have been summoned from all over. As with the surgeon from the appendectomy upstairs, the signals alerting various members of the team have gone out even before the injured policeman arrived.

As he examines the patient, Hessler detects a very rapid, weak pulse, suggesting severe hypotension—virtually no blood pressure. He oversees the insertion of a third large-bore intravenous line into the officer's body to allow for the replacement of the lost blood by giving "Ringer's lactate" (a solution of various salts in purified water, similar to natural body fluids), which supplies the volume of fluid, and O-negative blood in highly concentrated form, known as packed red blood cells. These cells will assure continued oxygenation and are essential as well as more rapidly available than type-specific or cross-matched blood. The IVs have raised the policeman's blood pressure, but not substantially.

The police officer remains in deep shock and in imminent danger of suffering kidney failure if his blood pressure is not quickly raised further. His heart and brain are also vulnerable and cannot withstand such profound shock for any length of time. Dr. Hessler nods to the chief surgical resident of the trauma service as they continue their assessment; they agree that there is no apparent internal bleeding and no evident chest or abdominal trauma. Still, to be sure, an abdominal tap—in this case an open paracentesis—is agreed on.

The surgeon quickly prepares the abdomen just below the navel for a small incision so that a catheter can be inserted. Nearly a liter of Ringer's lactate is slowly introduced into the abdomen. Then, after gently rocking the officer's abdomen, they siphon the fluid out again, and watch carefully. It comes out as clear as it went in—a sign that there is probably no life-threatening internal bleeding. It seems clear that the injuries to the lower extremities adequately explain both the blood loss and shock. The officer has now received two units of universal O-type blood. And now type-specific blood is available after intensive cross-matching and testing in the blood bank upstairs. Meanwhile, Radiology has taken a series of chest and abdominal x-rays to look for other fractures prior to the surgery that will certainly follow. The catheter is withdrawn and a few sutures are placed to close the wound that has been made.

Now that it is certain that blood pressure is high enough, that the patient has been stabilized, that there is no further immediate threat to his life, Hessler wishes them well as the surgeons move in, to highball the stretcher down the hall. A waiting elevator there speeds them to the eleventh floor, where the microsurgery team awaits.

It is only when the patient passes into other hands that Hessler reviews the details with the pair of paramedics and three police officers from the scene, who are still seated with their coffee cups in the holding area trying to get their own racing heartbeats back to normal and complete the paperwork of their ambulance call reports.

The paramedics, attached to Harlem Hospital, got their summons from Maspeth at 2 A.M. and arrived at the scene three minutes later. There they saw officer Patrick Riordan lying unconscious and his less-severely-injured partner, officer James Giotto, sitting in another patrol car that had come up to stand by.

Both Riordan and Giotto had been driving north on Amsterdam Avenue when they spotted an unmarked Consolidated Edison excavation in the street, about one hundred feet south of 120th Street. Its barrier had fallen, making it a hazard to passing traffic. Officer Riordan, who had been driving, stopped the patrol car, and, with Giotto, got out to replace the barrier. They then had gone around to the trunk of their car, which they had opened to remove several "witch-hat" traffic cones.

At that moment a van slammed into them, knocking Giotto aside but catching Riordan directly, and pinching his legs between the meshing bumpers of the van and the patrol car. Riordan lost consciousness almost immediately. The van backed up a few feet, and Riordan's limp form collapsed to the street.

Later it would be established that the driver had just had his license restored to him by Connecticut. It had been suspended for drunken driving.

Giotto, although himself suffering whiplash and major bruises, immediately went to Riordan's aid and then called on the patrol car's radio for help. Then he made Riordan as comfortable as he could and handcuffed the driver, charging him with drunk and reckless driving and second-degree assault.

Within minutes the paramedics arrived and they applied pressure tourniquets to Riordan's upper legs. He was bleeding copiously from the femoral arteries in both thighs, which, as the microsurgeons later described them, were "hanging by shreds of skin and small strands of tendons."

At the urgent request of Giotto and the other officers who had quickly gathered there from the nearby precinct house, the ambulance bypassed all the intervening hospitals to make the twenty-three-minute run to Bellevue, because of the reputation of its Emergency and Microsurgery departments.

As the ambulance was making its way toward the hospital, calls went out to the trauma team and the plastic-surgery team and to Dr. William Shaw, a Chinese-American who is chief of Bellevue's microsurgery-replantation unit. The doctor was asleep in his apartment several blocks from Bellevue. Ten other surgical house officers, all of whom were already on duty at the time, were also summoned.

$$\sim$$

Now, while Drs. Hessler, Pfeffer and all the ED regulars go about their work through the night, a battle rages upstairs to save Officer

Riordan's legs. As the Emergency Department would later learn, Dr. Shaw and his team of specialists in orthopedic, plastic, vascular and general surgery reviewed their best strategy, while they studied the shattered legs of Patrick Riordan. As the officer's blood pressure improved, he rose just to the edge of awareness. The hurried conversation around him sounded at most like a faint murmuring. Then he submerged again into unconsciousness as the anesthesia that was administered in preparation for surgery took effect.

It was clear that the right leg was simply too badly crushed to be saved. Trying to save both legs might result in saving neither. It might even subject Riordan to strain too great for him to bear. Before he left the ED, Hessler and the trauma team had got Riordan's blood pressure up to 80 over 40—not robust, but considered high enough to justify a start on a long operation. The figures, as Dr. Goldfrank would explain the next day to an inquiring journalist, represent the maximal force of the heart, and the response, like a kind of echo, of the vascular system. So 80 millimeters of mercury would be driven up a glass pipette by the outgoing pressure (systole), and 40 millimeters by the returning pulse (diastole). A normal figure for systole is usually said to be roughly your age plus 100. So, for a thirty-six-year-old man like Riordan, 136 over 70 might have been the expected blood pressure. When he first got to the ED, it was 50 over 0—catastrophic—allowing only moments for correction. That was the condition from which the medics' rapid treatment and the Emergency Department fluids and transfusions had rescued the policeman.

Upstairs, they decided against amputating both legs, but considered, momentarily, attempting to attach the severed right leg in place of the left, which had a hopelessly shattered kneecap. Then they asked themselves if they could possibly save the left leg. It would have no functioning kneecap, and therefore could not bend. Yet there remained a small but significant shred of flesh in which a tiny nerve—the posterior tibial—was still intact. From this small core of normality they could possibly restore the leg to some sort of function. That was the plan they adopted.

The right leg was cut off directly, but it was not sent with other such amputated tissues to the Pathology Department, where materials of this sort will be studied in hopes of developing useful data for future treatment of similar injuries. They saved it, instead, as a source of skin and blood vessels that could later be "harvested" in the fight to save the other leg. It was wrapped in saline gauze, enclosed in a plastic bag, wrapped in a second bag, which was then filled with ice and a small amount of water. The saline gauze helps prevent drying and destruction of tissue. And the two plastic bags helped minimize direct contact of tissue and ice and the possibility of frostbite developing.

(On several occasions, detached fingers or limbs have arrived at Bellevue incorrectly packed. Goldfrank recalls one instance when a pair of fingers arrived packed directly on a block of dry ice, and a leg, on another occasion, came packed in a bag of chilled water. The fingers were frostbitten, and the leg had developed trenchfoot. And neither was any longer usable.)

Shaw's surgical team then set about gauging just how badly damaged were the bones, arteries, veins and nerves of the remaining leg. "We had to make sure we could match them up on the bottom and top," Shaw later said. At the same time they had to take the momentous step of seeing if they could revive the nearly moribund leg. It had gone more than two hours, now, without a normal flow of oxygen and the other critical nutrients that normal blood circulation would have supplied. They attached a catheter that linked the severed femoral artery with the lower leg, and waited to see how the blood would flow. After some minutes of watching and waiting, they knew that they still had a viable leg, as it grew pink and warm.

It was now nearly dawn and the wives of the two policemen had arrived at the hospital under police escort, to join the vigil.

Dr. Shaw and the chief resident in orthopedics sawed off uneven edges of the femur (the thighbone) and tibia (the shinbone). Then parts of the smaller bone in the lower leg—the fibula—were cut away, as were some cartilage, tendons and muscle tissue. Ultimately, four pins would secure these repairs, and they in turn would be supported by an exterior brace. Meanwhile, working with tiny instruments and microscopes, Shaw and his team were making the meticulous attachments upon which the daily life of the restored leg would depend—the arteries, the veins and the nerves.

Officer Riordan was to keep his left leg, eternally stiff, oftentimes painful, but able to support him, with a prosthesis for the other leg, and two canes as outriggers. He was to be retired from the police force on full disability and to spend much of his time thereafter in a wheelchair.

Dr. Rob Pfeffer is smiling broadly and accepts congratulations from a young cardiology attending physician. Pfeffer has made a difficult diagnosis. Shortly after 7 A.M., a fifty-two-year-old man with chest pain was brought in by his alarmed wife, who sat nervously smoking cigarette after cigarette while Pfeffer watched the tracings of the ECG, and saw a subtle elevation in the "ST segment" of the tracing. Normally, there is a hummock, a short plateau, a great peak, a valley, another plateau and then a final hummock. In this case, instead of a valley after the major peak of the heartbeat, there was a high plateau. "I kept looking at it and thinking, This is an anterior-wall heart attack," and he alerted a group of young cardiologists who had been having good results with a new experimental drug, TPA—tissue plasminogen activator—which opens up a clot in a thrombosed coronary vessel within three minutes.

"I was pretty sure we had a clot here with this patient, and called the TIMI Team (shorthand for TPA In Myocardial Infarction Team). Then I had a few minutes of suspense, as to whether it was an appropriate call. And I waited a bit, taking further steps—applying nasal oxygen and giving morphine to relieve pain—to keep the patient comfortable. Then the cardiologists got here and they took one look and told me I was right on. It was a really good feeling." The patient was admitted to the intensive cardiac care unit and was treated with medication by his house officer—to spend seven to fourteen days in the hospital, if all goes well.

For Pfeffer, who is specializing in internal medicine, his tour of duty in the Emergency Department has been a revelation. "Goldfrank was the attending physician when I was an intern on Medicine and I was struck by his sense of mission—how respectfully he treats everyone . . . the same for a bum off the street or some big executive. I came here from NYU Medical School and before that I went to Brown, and people I see, people I grew up with, my old classmates and friends, they think oftentimes they are insulated from what we see here—the homeless, the drug abusers, the alcoholics. I tell 'em they ought to go back and read 'Ozymandias.' "

The allusion doubtless gets past most people, but Pfeffer recalls the impact the Shelley poem had on him when he first encountered it at Brown—the shattered remains of a statue in the desert, which used to be a green and flowering land. Of the old civilization, which thought itself insulated too, only this vainglorious claim remains:

> "My name is Ozymandias, king of kings:
> Look on my works, ye Mighty, and despair!"

But nothing else survives.

> Round the decay
> Of that colossal wreck, boundless and bare,
> The lone and level sands stretch far away.

It's half past seven and Goldfrank has arrived. He comes by the doctors' station, where Hessler and Pfeffer are updating their charts and getting ready to take their leave. But they decide to stay for the 8 A.M. meeting, so Pfeffer's TIMI case can be presented. And Hessler's trauma case. And perhaps the statuesque blonde, as well.

Nurse Jeanne Delaney comes by to convey a bit of news: Toby Wilts is on his way in this morning. In fact, he's wheeling himself down the main corridor toward the triage desk now.

# 16 A Lesson and a Crash

Leon Fields is showing a new employee around the Emergency Department as he moves through the various examination rooms distributing gloves—green for normal use, white for sterile procedures.

Some patients have objected to being examined by doctors with green gloves on. The doctors, they say, don't want to touch flesh. Fields has heard the replies that silence the patients' objections. There are new strains of staphylococcus bacteria, highly resistant to penicillin. The green gloves keep these nosocomial (hospital-related) organisms from proliferating in the mere act of the laying on of hands. Fields explains that to his young charge this morning in his own less formal manner.

Members of the housekeeping staff are also moving through the corridors, swabbing the dark three-foot-square masonry slabs that make up the floors. They're durable, those slabs. Unfortunately, they seem to act as amplifiers for the sounds generated in the corridors by the throngs of passing humanity. These, plus the thin walls, make silence virtually unattainable in the Emergency Department. Even as the housekeeping staff sweeps and dusts in the office of Neal Lewin, the voice of a vociferous prisoner from the holding room, some distance away, can be plainly heard.

This morning, everyone is especially pleased to see a clipping from yesterday's New York *Post*, an article by the City Council President, praising Fields as one of New York's unsung heroes—one who performs outstandingly without seeking to vaunt himself in any way.

Fields tells his young companion, a beginning nurse, what he is doing, and why he is doing it, step by step. He hardly wastes a motion, moving with the same grace that attracted the major-league baseball scouts years ago when, as a nineteen-year-old, Fields was a regular outfielder for the New York Indians, a semi-pro team based in Freeport, Long Island. "We were having a great season—this was in 1960—and I was looking good, maybe because I was the youngest member of the team. One scout told me I had a shot at a professional career. I don't know. Maybe I did. But then the team fell apart, started playing lousy, and the scouts didn't come around so much anymore.

"Then I played windmill softball—that's where the ball comes up to the plate maybe eighty, eighty-five miles an hour—and then I started working in the municipal hospital system." He continues to distribute bandages and gowns and paper towels to the various examination rooms.

Fields can hear a disturbance down the main corridor near the doctors' station as he finishes setting up. He gazes down the long passageway. Just past the double doors, he sees a nurse motioning to him with a familiar gesture. "Come on, Billy," he says to his young apprentice. Fields stops for a moment to get a bottle of disinfecting shampoo and a water-repellent gown and cap out of a supply cabinet. As they go through the double doors, the familiar form of Toby Wilts presents itself. Wilts is sputtering mad and more than a bit drunk. Goldfrank is listening to Wilts's labored breathing with his stethoscope. He nods to Leon, with a sorrowful gaze, signaling that Wilts's condition is one of personal filth more than urgent disease. Wilts's decubitus ulcer still torments him, but does not seem to be much worse. "We'll get you cleaned up and then have a look at that bedsore," Goldfrank says.

Wilts greets this remark with a stream of highly audible abuse, directed not specifically at Goldfrank but at the world in general—motherfucking city, nobody cares, gonna get some revenge, hate this place too . . .

Fields approaches the wheelchair. "Morning, Toby. Come on, let's get you cleaned up." Wilts continues to rant, but Fields ignores his outburst and seems to have so completely mastered his own emotions that he doesn't give even a sign of the extreme offensiveness of Wilts's person. His young companion, however, is keeping his distance, trying not to vomit.

Offering a key in his outstretched hand, Fields says, "Here, Billy, get around the corner, there, and unlock the second door on the right. You can get the water started in the tub in there. And get a plastic bag ready, too." The young man nods emphatically, apparently relieved to be able to put some distance between himself and the malodorous Toby Wilts.

As they get around the corner and reach the bathroom door, Fields pushes Wilts on through. There is another stream of abuse from Wilts, but Fields's calm rejoinder seems already to be having a soothing effect on the irascible man in the wheelchair. "I know your stump's hurtin' and I know your backside's givin' you a fit," Fields says equably. "We're gonna have you cleaned up in a moment."

The refurbishing goes with surprising rapidity. "First of all," Fields explained later, "we get the water going and the liquid soap suds up, and that already begins to kill some of the smell. And then we get his clothes off quickly into that plastic bag and as much of the waste matter he has all over him in at the same time, and then you tie up the bags and the smell is already half gone." Young Billy hasn't learned this part of the routine. He is throwing up in a toilet bowl on the far side of the bathroom, while Fields is helping Wilts out of his wheelchair into the tub.

"It's a fact," Fields later says, quoting Goldfrank, that "after the first few moments, you can't even smell it anymore." He has the gist of the passage in Goldfrank's book on the phenomenon of extinction—the highly offensive odor that paralyzes the olfactory nerve and shortly ceases to be so acutely annoying.

"As long as the patient doesn't hassle me, I can stay cool," Fields says. "But when someone gets really loud . . ." Even then, however, Fields does his task, if a bit less steadily than he is doing it this morning.

When the chore is done and Wilts is taken around to Room 5 for examination, Fields can tell his young charge that it won't be so bad hereafter. "The first time is the worst. Then you begin to learn that your mind can control your body."

"Even that smell?"

"Even that."

The young man shakes his head in evident disbelief.

∿

The subject of the morning staff meeting is how to get reliable medical information from a patient, fast. "The point is," Goldfrank tells his students, "you have to learn more about a person in three minutes than most people learn in weeks of acquaintance."

The inspiration for this morning's discussion is a missed diagnosis. Two days ago a patient with atypical chest pain was sent home after his condition was said to be esophagal pain. He returned that same evening with an acute myocardial infarction. The question this morning is: How do you accomplish in a few minutes what the in-hospital doctor achieves in several hours? In the Emergency Department you have to take the vital signs, then get a history that will somehow convey the most essential information in the midst of noise, constant interruption and lack of privacy. You may have to break off with your patient in the middle of taking a history, because of a more urgent case that suddenly comes in.

Goldfrank and Neal Flomenbaum have written extensively on this very technique. One of the sources they have cited in their own work is an article, "Clinical Hypocompetence: The Interview," by the Denver General Hospital team of Drs. Frederic W. Platt and Jonathan C. McMath.

In that article, Platt and McMath tell how they watched their own colleagues and students during a series of three hundred interviews with patients, and discovered these five common shortcomings:

- The failure of the doctor to give emotional support and healing to the patient during the interview;
- Neglecting to learn much about the patient's life style;
- Not taking into account "the sort of person the patient is";
- Not asking for basic symptoms, but being content with secondary information such as lab data told to the patient and the interpretations of other doctors as relayed by the patient; and
- Adopting a "high-control style" in which the doctor talks more and more, the patient is reduced to "yes" or "no" responses—none of which is likely to be very illuminating.

What can the young doctor do? First, try to develop sensitivity to the "loss of integrity" that the patient feels in an emergency department. "And if you call someone 'hon' or 'dear,' or by first name from the very start of the interview, that only makes the patient feel like a child. The recommended approach: shaking hands and introducing yourself in a friendly, relaxed manner. Also," Goldfrank says, "you can simultaneously grasp the epitrochlear nodes—bracketing the point of the elbow—and get a pulse at the same time, and save a bit of time." Osler himself used that technique, Goldfrank says, to make a quick check for syphilis, which might cause the lymph nodes flanking the elbow to swell.

"The key," Goldfrank says in summary, "is that you have to be interested in people and actually like people. If you like people, you'll be able to hear things they haven't told you and to pick up patterns that they themselves may not be aware of."

Neal Flomenbaum offers some practical aids to young physicians and nurses in sharpening their diagnostic perceptions. Before anything else, you have to get five things: blood pressure, pulse, depth and rate of respiration, temperature, and level of consciousness. Then the person taking the history has to add two other "vital signs": (1) the color of the patient—whether white (possible shock), blue (not enough oxygen), bright red (too many red blood cells) or yellow (jaundiced) and (2) his "motor status"—whether writhing, hunched, immobile or uncoordinated.

The case of a twenty-three-year-old black female is brought up by one of the medical students. He saw the woman last night, after she was stabbed by her boyfriend. She was sewn up by Dr. Wayne Longmore. The student and Longmore both noted a whole pattern of past keloids—or excess scar tissue—on the woman's arms and neck, as if she had been repeatedly slashed and stitched up. Goldfrank suggests that Dr. Longmore respond to that observation. Wayne Longmore is board certified in emergency medicine, an expert on wound care, who started his training in Halifax, Nova Scotia, and continued it in Birmingham, Alabama, before coming to Bellevue.

Longmore, a self-assured Canadian, says flatly, "Sterile technique is a myth in emergency medicine. In essence, we are providing 'clean wound care,' and it's a struggle to keep from contaminating a wound with various noxious agents, germs, microscopic flecks of dirt, found in the Emergency Department."

Oftentimes, says Longmore, keloid formation in blacks is thought to be unavoidable. The patient in question has had bad suturing in the past—and that may account for her keloids. There's also a possibility that her scars reflect her own psychiatric instability—that some may be the result of self-mutilation.

Longmore goes on to say: "Before examining a wound, you have to infiltrate it with 2 percent Xylocaine"—a local anesthetic. "But you don't want to apply it right in the wound, as the needle will unwittingly track dirt into the opening." Better, he says, diagramming his point on a large pad of paper, to infiltrate outside the perimeter of the wound to keep from contaminating it. So then, when you've examined, rinsed and scrubbed the anesthetized wound and have decided there are no injuries to underlying deep structures and no foreign bodies caught in the opening, you get on with the suturing.

Pick the wrong material, however, and you'll create a new keloid in your patient. "Silk should never be used to close wounds in the Emergency Department," says Longmore. "The infection rate with silk is 10 to 15 percent versus 2 to 3 percent with nylon." Apparently the strands of silk trap debris, whereas monofilament nylon—very similar to fishing line—is far less likely to cause infection, even if it is harder to tie.

Then there's the matter of matching up one side of a wound with the other. Epidermis, the outer layer of skin, is dead and does not heal, Longmore reminds the group. The under part, the dermis, is where healing takes place. But if you don't take pains to match up the lower, healing layer with like tissue, there'll be poor healing, and perhaps a tendency to infection. But match up dermis to dermis, you maximize the potential for good results.

And finally, "dark-skinned patients, because of their tendency to form keloids, deserve careful attention to detail to maximize cosmetic wound healing."

That afternoon, shortly after lunch, Goldfrank was walking back from his daily rounds at the University Hospital Emergency Department four blocks up First Avenue from Bellevue, where he is also director. He took his customary route—a walkway that parallels the FDR Drive. He was enjoying the sparkling clear day, thinking about a gardening task he has set himself at home: restoring an old grape arbor.

Suddenly he heard the percussive cough of a helicopter behind him, taking off from the busy heliport some five blocks to the north, at Thirty-fourth Street and the river. Then came an eccentric sound instead of the normal rhythmic popping of the engine. He turned around in time to see the aircraft angling over on its side and dropping rapidly toward the river.

He thought for an instant that he ought to run to the site to help.

Then he realized that he was at Twenty-ninth Street. To get across the FDR Drive so that he could actually reach the river, near Thirty-first Street, where the craft had apparently foundered, he would have to go several blocks to one of the walkways spanning the drive, and then retrace his steps to get to the wrecked airship. Better to hurry on to Bellevue, where the ED or Bellevue's mobile ambulances would probably be hearing about the mishap already or could be swiftly mobilized. Once there, he could help organize the effort to respond.

As he and the Emergency Department staff would later learn, a parking valet at the Water Club riverfront restaurant at Thirtieth Street, just three blocks north of Bellevue's Emergency Department ambulance entrance, saw exactly what had happened: A red-and-white helicopter, bound for JFK Airport from the Thirty-fourth Street heliport, gunned its motor and leapt up into the air. Suddenly the valet heard something that he later said sounded to him like "a loud clunk." The time was 2:17 P.M.

The flying machine rose a bit more, until it reached fifty feet or so above the East River, then tilted to its left and began to fall toward the water. The craft, a ten-passenger French-made Dauphin 360C, flipped over, flailing the water with its giant rotors, which churned up a giant cloud of spray. Then followed a huge splash and a hissing cloud of steam from the hot engine. The downed helicopter slowly drifted toward shore and gradually began to sink.

Meanwhile, as Goldfrank ran the remaining distance to Bellevue, entering via the ambulance portal, the department was already mobilizing. Kay O'Boyle, Associate Director of Emergency Services, was coordinating activities. O'Boyle knew exactly what moves to make in response to what seemed to be a sizable disaster.

At the same time, she was alerting the crew of the MERVan, also. This well-equipped mobile emergency room would shortly be pulling up to the ambulance entrance to pick up an Emergency Department team that Dr. Kathy Delaney, the senior attending physician on duty, was now recruiting.

Team members were getting the "tackle box" from the nursing station. It's a large traveling case with all the medications and supplies needed to supplement equipment aboard the MERVan.

Kay O'Boyle had just received the first notification via police radio and conveyed the preliminary estimates to Delaney and her team—a possibility of ten gravely injured passengers and crew members. Anticipating that relays of casualties might be ferried to Belle-

vue, Kathy Delaney led the way to the scene in an ambulance, while
the MERVan with the rest of the team followed.

A West German passenger later said that he heard a loud pop as
the craft was about one hundred feet from shore. "Then we dropped
right into the water," he said, "touching down in a gentle manner. The
helicopter went on its left side. I was on the right side. The pilot got
up and opened the door. I unfastened my seatbelt before I went into
the water." The machine submerged—but did not sink entirely. There
was no panic, no screams, despite the fact that some of the passengers
had trouble getting their seatbelts unfastened.

Almost immediately a jogger leapt into the river to help the survivors, and an inflatable raft was launched from the heliport, also making for the wreck. Moments after that, a police lieutenant, hearing a
short-wave radio report about the mishap, pulled his car off the FDR
Drive near the heliport, got out, and also dived into the river to help.

Meanwhile, the pilot and co-pilot emerged to cling to the landing
gear of the upside-down machine, helping passengers—who included
a number of West German business travelers as well as tourists from
California and Kentucky—out of the cabin. One of the passengers was
still trapped within. Apparently he couldn't get his seatbelt undone.

All of the others safely got out the door or an open window of the
overturned helicopter. They were soon met by the inflatable rescue
craft and ferried to the heliport ashore. Within minutes, police officers
and, later, a Fire Department diver, arrived at the wallowing craft.
They tried to enter the cabin to rescue the remaining passenger, who
remained immobilized by his seatbelt. None of them—until the Fire
Department expert got there—had scuba gear. For want of such gear,
the crew members of the helicopter had also been frustrated in their
attempts to enter the cabin.

Finally the Fire Department diver, dodging floating kapok cush-
ions, reached the last passenger and cut away the restraining belt to
free him. He was not breathing. But, as the water temperature remained in the low fifties, there was a hope that this last passenger's
metabolism might have been slowed down enough by the chilly water
to give him a chance.

Ashore, Delaney and another ED attending physician, Henry Sen-
dyk, treated the drenched passengers as they were lifted back up to
solid land. Then the MERVan rushed the victims and their chilled

rescuers to Bellevue. They were a forlorn group, dripping wet and thoroughly chilled. A team of attending physicians and nurses gave them immediate attention, but were relieved to discover that there was not much needed beyond restoration of normal body temperatures and securing dry clothes. As the river water was so chilly, there was a danger of hypothermia. And, as Goldfrank insisted, there would be shots for hepatitis and tetanus—two microorganisms that the passengers could well have encountered as they struggled in the East River. Hepatitis they might have swallowed in a drop of river water; tetanus might have been scraped into the skin by any rusty object. Beyond that, the staff was primed for a complete check of their physical condition.

Back at the heliport, the last passenger does not look good. As he is lifted out of the inflatable, his body is limp and frigid. He's quickly loaded into the remaining ambulance, which races off toward the hospital. His papers say his name is Rolf Schneider, age thirty-five.

By the time they get to Bellevue an oxygen tube has been inserted as have several IV lines, but still there is no pulse and no respiration. He is rushed immediately into Trauma Slot 1. His skin is pale and cold. As Delaney bends over him to peer into his eyes with her ophthalmoscope, she can see the pupils are widely dilated and fixed. And also she can see in the vessels radiating out from the disk or optic nerve that there are "box cars," minute blocks of blood separated by small spaces, faintly resembling a string of freight cars sitting on a track. It's a sign of a grave prognosis.

She resumes cardiopulmonary resuscitation, using the oxygen-powered "thumper," a mechanical device with a pad, about the size of a shoe's heel, set about four or five inches above the sternum, which rhythmically presses on the chest, tirelessly emulating the natural motion of the heel of the hand during extended resuscitation efforts. Through IV lines, Schneider gets sodium bicarbonate, to counteract any acidosis that lack of oxygen could have caused. He has also been given epinephrine, a solution of 0.9 percent saline and also 100 cubic centimeters of 50 percent dextrose. Still there is no sign of heartbeat or independent breathing. The electrocardiogram shows a bleak, flat pattern.

They cling to hope, nevertheless. There is a chance that the chilly salt water has acted to slow down the effects of oxygen deprivation on Rolf Schneider's brain—which is the key organ involved. So long as his brain maintains its integrity, there is reason to keep trying to save the patient. Furthermore, there is the firm belief that Goldfrank has inculcated in his entire staff: the patient isn't dead until he's warm and dead. And Schneider is still very cold indeed. His rectal temperature is 90 degrees, low enough to account for the flat ECG reading. The warming continues by means of a surrounding gray blanket and heated oxygen and fluids being given.

The blood gases taken before intubation showed a pH of 6.8—as against an ideal of 7.4. Six point eight meant acidic, a bad sign. And now, after a few minutes of resuscitation, Schneider's pH has barely moved upward to 6.9.

An hour passes. His temperature is up to 92 degrees, now, and the ECG is still flat.

The oxygen in his blood measures 30 millimeters of mercury as against 100 normally. And his carbon dioxide is at 90 millimeters, whereas 40 is a typical reading in a healthy person. That's clear evidence that the river water is sitting in Schneider's alveoli. These are the little air cells in his lungs through which oxygen moves toward his blood and by which carbon dioxide is withdrawn from his blood capillaries. From these alveoli, carbon dioxide should be drawn into his bronchioles, then his bronchi, and then exhaled up his trachea and out his nose and mouth. But now the gases cannot get past this liquid blockade. And there is no good way to remove the barrier. Even now, in fact, the walls of his alveoli are collapsing, going into atelectasis (a Greek word literally meaning "imperfect extension"). As Schneider's lungs lose these essential surface areas through which the gases normally move, the end becomes inevitable.

Dr. Delaney is looking defeated, and gazes over at Goldfrank, who nods his head in agreement. They must abandon hope. They have been bested by the East River and the balky seatbelt.

Now an unwelcome ritual. Each of the IV lines (going into the left arm, right arm, femoral vein) and the Foley catheter in the penis are clipped off, leaving some six inches of tubing attached to each of the still-embedded needles as well as the endotracheal tube in the trachea. Each of these vestigial tubes is tied in an overhand knot, so that no fluids will leak out. All the tubes in place will give the coroners a clear idea of what steps have been taken in the vain attempt to save this last passenger's life. Now nurses Elizabeth Swanson and Patricia Kunka move in to clean up the body, which has been soiled from the involuntary evacuation of the bowels—a normal occurrence in such an extremity.

When that is done, each wrist is wrapped in a loosely woven sterile gauze, Kerlix, and then crossed and gently tied together. The face is swathed in a sterile, more tightly woven dressing called Surgipad, and this is covered with a wrapping of the soft Kerlix also, as a protection against rough handling in the morgue or in transit.

Two mortuary tags are being prepared by Swanson and Kunka with the patient's name, age, sex, color and religion, plus a listing of all rings, dentures and other personal effects left on the body. One of these tags is tied to Rolf Schneider's right forefinger, another to his left great toe—insurance against loss of identity or personal property. The body is covered, and Delaney calls the Medical Examiner's office to tell them what happened and to ask them if they want to take the case. Yes, they do want to take the case. They give Delaney a four-digit number, which the doctor enters on the pink death notice. An Emergency Ward clerk has already partially filled out the death notice with chart number, name, age . . . Now Kathy Delaney fills in the last of the information on this notice.

Kunka, who moves with swift assurance at such times, goes to the wall telephone and dials a number. "Transport? This is Pat Kunka in Emergency. We have a body for the morgue in the trauma slot."

Meanwhile, outside, Kay O'Boyle is handling queries from the press, asking Social Work to help find dry clothing for the survivors, taking calls from the West German embassy, and making sure that the information desk, out in the lobby, which has the responsibility of notifying next of kin, is informed of Rolf Schneider's death. A transatlantic call will soon be made to convey the bad news.

Within minutes, an elevator door slides open and a stretcher-sized cart topped by a boxlike green metal housing is pushed by an orderly down the crowded corridors toward the Emergency Department. It slips through a narrow door, past another bank of elevators, goes by the Pediatric Emergency Service and Goldfrank's office, makes a right at the triage desk and then pushes up against the double doors of the trauma slot. It pops them open and traverses the gray tile floor with a soft rattling sound. It moves up next to the stretcher bearing Rolf Schneider's shrouded remains. His body is gently lifted into the cart. Its open top is covered with a green cloth, to disguise its true nature. Almost no one is fooled. The cart glides back across the tiles and retraces its course down the corridor to the bank of elevators, which will carry it below to the morgue. Everyone quickly makes way for it.

There is hearty gratitude from the survivors as they are discharged and take their leave. They express unalloyed admiration for the dedicated care they have received from everyone—jogger and crew members, police and firemen, doctors and nurses and social workers, who radiated an entirely unexpected measure of warmth in rendering aid.

Later that afternoon nurse Elizabeth Swanson shows Goldfrank a page from Albert Einstein's works, which could almost stand as a credo for what goes on in these 22,000 square feet of intense activity:

A hundred times every day I remind myself that my inner and outer life depend on the labors of other men, living and dead, and that I must exert myself in order to give in the same measure as I have received and am still receiving.*

*The World As I See It, by Albert Einstein, translated by Alan Harris, Philosophical Library, New York, 1949, p. 1.

# 17

# Human Warmth and a Drink of Gasoline

It's a costly service to run a large Emergency Department, says Susan Callaghan. Only 15 percent of the patients are truly "emergent"—in imminent danger of dying. Another 35 percent are "urgent"—which means they should be seen within an hour or two, at most. And the remaining 50 percent are non-urgent. Callaghan used to be assistant director of nurses for the department. Now, she has a new job upstairs with the Emergency Care Institute. But her heart remains in the examination and resuscitation rooms on the first floor.

"Toby Wilts," she says, "doesn't fit into an emergent category, but he demands talking to—to get him calmed down, with time to bathe him and then three people to get him up on the examination table, because he's an amputee and is drunk just about all the time. So he's in the second category—'urgent.'

"But today, whatever the category, there's all the new pressure to get the patient in and out fast—the 'DRG' mentality. It's hard to preserve a feeling of human warmth when all the forces seem to be pushing against compassion and concern."

"DRG" is the acronym for "diagnostic-related groups," and is the shorthand for federal regulations developed since 1980 that require that each disorder or disease be exactly labeled, and then treated within a fixed average length of time. The intent is to help control medical costs but it also gives doctors and nurses strong motivation to move the patient along as fast as possible—even though, say some critics, it can mean low-quality care. DRG also gives strong motivation to avoid the tough cases.

The budget makers rely on patient census figures and then they decide how much care each patient needs. So in intensive care units there are two nurses to every patient, whereas in emergency medicine the ratio is one nurse to every ten or fifteen patients. "But how do you explain to an accountant that you need time to tell a little old lady that her husband has a brain tumor and must wait maybe four hours for a bed? And therefore the budget figures are always defective because they can't take account of such things.

"Or take this case that I happened to be involved with. A young widow and her three children come to New York from Boston on Amtrak with her father. Grandpa, the boys—all under ten—and the young widow. They are touring the city and the grandfather has a cardiac arrest on the Staten Island Ferry as it's coming back to Manhattan. They rush him up here but there is nothing we can do. We work for well over an hour, but he's gone.

"Now, here I am, the nurse on duty on the case. And it's my responsibility to accompany the doctor and approach this young woman, surrounded by three wide-eyed little ones, and tell her that her father, their grandpa, is dead. So first we get someone from Social Service to take the children for ice cream so the mother can stop being brave and have a chance to grieve when we tell her the news. Then we have to be available to support her when she tells these three children what they at some level already fear. And then we want to be on hand to help answer their questions also. After that, you're supposed to go back on automatic pilot and continue business as usual—saving lives. How do you describe that to accountants?"

Or there's another case that Callaghan remembers with particular vividness, illustrating the kind of dedication that no budget-making entity can comprehend, really, because there is no "line" for it, no way the field of vision of the hard discipline of cost accounting can see such a thing:

"A woman in the East Forties was in the laundry room of her high-rise, doing her sheets, and somehow she got her arm in the machine and set the thing in motion at the same time. Caught her hand above the wrist and nearly tore it off. The police were called and they took the machine apart to get her free and they brought her in here and her severed hand—it was cut off—amidst all the bloody laundry. Of course they rushed her right upstairs to eleven and Microsurgery, but there I was cleaning up after we had completed our evaluation in the Emergency Department, with this pile of bloody laundry—still wet. 'What shall we do with this?' I asked the supervisor.

" 'Just put it in the property closet,' she said, 'along with the woman's other belongings.' Well, I knew that wouldn't really be very pleasant, when the woman later gets her things back. So I got a plastic bag and put the laundry in it and took it home with me that afternoon, and washed it and dried it. Not that what I did was all that unusual. It's just that you can't really provide for that sort of thing if you let people with stopwatches and long sheets of numbers define what is good health care. Intravenous drug abuse, alcoholism, homelessness, prisoners—things we see—all raise costs of emergency medicine but due notice is not taken of what we try to do here."

Callaghan is now working in an educational capacity as Program Consultant and Education Coordinator for the Emergency Care Institute. She has responsibility for the education of health workers from throughout the metropolitan area. She also helps in the organization of conferences that draw participants from all over the country.

The woman's hand was successfully reattached by the microsurgery team, and she was most grateful for her clean laundry and her wedding ring—which Callaghan had found among that laundry— both returned to her when she was discharged.

As Callaghan concludes: "We have an unusual kind of leadership that everyone in the department responds to very positively. But still we know that Emergency Medicine is also mundane and it's vomit and maggots and addicts who hate you for saving their lives because the naloxone you gave them just blew their $100 high. . . . Nobody in business management or cost control really sees any of that or thinks it's especially important. Or so it seems, most of the time."

The siren at the ambulance portal dies away abruptly and there's a jangling and thumping as the doors open and a tormented-looking little woman steps to the sidewalk behind the paramedics, who quickly lift out a middle-aged man who is coughing violently, writhing as if he's drunk and blubbering as if he's on a crying jag.

They rush him swiftly down the entry corridor and through the double doors, then left to the triage area. His wife runs to keep up, wringing her hands and chewing her lips in distress. Assistant head nurse Elisabeth Weber makes a quick triage decision—category one, emergent.

The patient himself is unable to answer any questions at all as they wheel him into Room 3. His pulse is 120—high—and his temperature 99.9, which seems unremarkable to Lewis Goldfrank, who is conducting the examination himself. Blood pressure is 140 over 80—again within a normal range. The man is obviously in respiratory trouble, coughing and retching. From his distraught wife Goldfrank learns the salient facts:

Niels Zimmerman is a forty-eight-year-old cabinetmaker. His wife heard him coughing in the small garage next to their apartment, where he keeps his prized possession, a 1939 Packard automobile, which he was working on. When she had looked in an hour earlier, she had noticed that he was pouring several solvents, which he kept stored under his workbench, into canisters in which he had placed small engine parts of the Packard.

His sputum appears frothy and the emesis—as the chart will call the vomit that flecks his worksuit—has a distinctive odor. Goldfrank is acutely attuned to odor, and stresses to his colleagues and students how helpful it can be in diagnosis. In this instance, Beatrice Zimmerman tells the doctor that the smell is similar to that of the industrial solvents her husband uses on his Packard.

The patient is sweating heavily and his nose and mouth bear quantities of frothy secretions that pervade the entire upper airway. As the doctor moves his stethoscope down over the chest he hears a marked series of wheezes and rhonchi—rattling sounds, somewhat like snoring—in both lungs. Because of Zimmerman's irrational behavior, there's no assurance that the results of the neurological test that Goldfrank gives are reliable.

Yet, as the physical exam progresses, Zimmerman begins to come out of his confused state. "Hi, Bea," he says to his wife, with a little smile.

"You're feeling better then?" she asks, and he nods his head a bit. Soon his mental status is normal.

"I'm Dr. Goldfrank, Mr. Zimmerman."

"How do you do?"

"You had us a bit alarmed, there."

"After this I'll watch what I drink!"

"What happened?"

"Bea probably told you already."

"No, I don't know what happened, honey. What happened?"

"Well, doc, I was siphoning gas out of the Packard's tank. Wanted to get some more to clean the plugs. And I don't know, the fumes sort of got to me, and I coughed and the next thing I know, I swallowed I don't know maybe a couple of ounces of gas. I can still taste it."

"You will taste it for a while, probably. . . . And then what?"

"Then I'm feeling like looped. I mean really crocked . . . sizzled. Whatever you want to call it."

"Drunk?"

"You know it! And I couldn't seem to get my body working. I mean I couldn't hardly stand or even walk and that's when I guess Bea called the ambulance."

His wife is holding his hand and squeezing it in thanksgiving that he has come back to himself after his accident. She looks at the doctor.

"We're going to want to watch you at least overnight, Mr. Zimmerman, so we'll want you to stay in the hospital just to make sure your lungs haven't suffered from the hydrocarbon."

"Okay by me, doc. I wasn't planning to go to the race track or anything!"

Zimmerman's mental status improved rapidly and his chest x-rays showed nothing to arouse apprehensions; and even the wheezing, rhonchi and rapid breathing cleared away in several hours. His mouth was rinsed out repeatedly with water and he was given humidified oxygen over the next twenty-four hours, on the sixteenth floor in the Internal Medicine Ward. And his recovery was virtually complete by the time he walked out of Bellevue the next day.

Almost at the moment Zimmerman was being discharged, two-and-a-half-year-old Dionisio Evergaru was brought in with his hysterical mother, who had discovered the little boy lying near the sink, coughing and nearly blue and breathing with difficulty. On the floor nearby was the can the boy had mistakenly drunk from. His precocious fingers had popped open the snap top of a gasoline additive, packaged by its makers in a can indistinguishable in size or general design—for one who cannot read—from a soft drink. He had taken a sip, coughed, and thus breathed in some of the additive—a naphtha-based hydrocarbon that also contained some benzene. Rushed into the pediatric section of the Emergency Department, the child was carefully examined to assess the amount of the substance he had ingested. If just a few cc's, the severe effects of a more massive dose would be avoided. But those possibilities were enough to cause apprehension, as such hydrocarbons can have a harsh impact. They can induce leukemia or the related disorder, aplastic anemia, which kills red blood cells in their infancy. In this instance, the child made a rapid recovery without any apparent complications.

A third, related case involved deliberate paint sniffing by a young man who arrived at the Emergency Department in a gravely impaired state. He carried a hip flask filled with toluene and a washcloth. From time to time, he would drench the cloth with the liquid from his flask, then cover his nose and mouth with it and breathe deeply. This classified him as a "huffer." Not quite the same as the "bagger," who inhales from a plastic bag containing toluene. Toluene, too, is a petroleum distillate. Originally it came from a Caribbean balsam tree, but it is now extracted from coal tar or petroleum. As Goldfrank explains, such volatile hydrocarbons actually attack the white sheath of the nerves—the myelin, which has a large lipid, or fat, component. Such hydrocarbons can dissolve this outer wrapping of the nerves, causing a sort of short-circuiting of the nervous system that is often indistinguishable from drunken behavior. Fortunately, that was not the explanation for Niels Zimmerman's demeanor. He had not suffered irreversible effects but rather the passing intoxication that comes from momentary oxygen starvation and inhalation of the strong hydrocarbon fumes.

"There are both immediate and delayed effects from breathing in gasoline or other volatile petroleum distillates. First there is the irritation of the membranes inside the nose and mouth. There's a burning sensation and usually coughing, choking and gasping. And what turns victims blue right then is that the exchange of carbon dioxide for oxygen in the lungs is upset and these petroleum fumes—vaporized hydrocarbons—actually replace oxygen in the lungs, starting at the bronchi, the large major branches of the pulmonary system, and moving to smaller and smaller branches until they reach the alveoli.

"In addition, as the substance spreads down the respiratory tree, spasms of the bronchi may take place, making oxygen starvation that much worse and even causing depression of the central nervous system." As he diagrams the process, the trachea—the tree trunk—is affected first. Then the branches (or the bronchi) react. They lead to the smaller branches (the bronchioles) and finally to the leaves (the air sacs), which are affected last of all. There can be actual destruction of the function of these organs, because the vaporized gasoline or other light hydrocarbon destroys the cells that make up the alveoli and irritates the bronchial and tracheal mucosa, causing excess secretions. When this happens, there is no further basis for air exchange.

"In extreme cases," Goldfrank continues, "the lung may begin to fill with fluid—hemorrhagic pulmonary edema—with pink, frothy sputumlike fluid. Hypoxia (oxygen starvation) can lead quickly to shock and heart failure—or to cardio-respiratory arrest.

"Some are lucky and suffer just a transitory cough. Others develop such destruction of the alveoli and bronchi that severe complications ensue."

Later that week, Teresa Rizzo, Goldfrank's secretary, inserted a cassette into her Dictaphone and began transcribing the doctor's notes on these cases.

Rizzo is a stylish young woman who listens closely. She came to Bellevue after a semester of college. "I got the job," she says, "but it was a tremendous shock when I typed at first. I had no experience with medical terminology." At first she resorted to phonetic spelling, and a patient Goldfrank bore with the early efforts. "It was really like learning a new language. Half the terms didn't appear in the dictionary, and others I couldn't even begin to spell well enough to look up!"

She persisted, and made rapid progress, aided by steady recourse to two large books, *Stedman's Medical Dictionary* and *Physicians' Desk Reference*, and a small one, *Webster's Medical Speller*, which lists 35,000 medical words spelled and divided. Today, she breezes through the passage that Goldfrank dictates on the complications of aspirating hydrocarbons, of the sort that felled Niels Zimmerman:

"There may also be direct destruction of airway epithelium, alveolar septae, and pulmonary capillaries resulting in atelectasis, interstitial inflammation, necrotizing broncho-pneumonia with polymorphonuclear exudate, and formation of hyaline membranes lining the alveoli." She is unaffectedly delighted with her mastery of such material.

Rizzo is thinking about her boyfriend, who is on the West Coast, trying out for a place on the line with the Los Angeles Rams professional football team. They're to be married in a few months, but their plans are still somewhat up in the air—depending on how his tryout goes.

A patient sticks her head in the door as Teresa Rizzo is typing. "Where's the doctor? I want my shot!"

"Good morning, Deborah. He's in a meeting."

"Well, I want my shot!"

"I know. The best thing is just have a seat in the waiting room. They'll get to you in a minute."

The door closes and Rizzo confides to a visitor, "She's got sickle-cell anemia. I never heard of it before, when I was growing up. But Deborah comes in—I don't know—several times a month. I used to, you know, sort of shrink back from homeless people or sickle-cell or things that I now see every day. We don't have any of those things in my neighborhood in Queens. Not that it was all that fancy a neighborhood. But it is a nice, solid middle-class community. But you see it here, and you see how the doctors react and how now and then things get better, or there's some sort of warmth or something. And your own attitude can't help but change; you become more understanding. I mean I even see some of these people on the street on my way from the subway and I don't have a big reaction like I might have years ago."

There's one recent case of a homeless person with a happy ending that has elated everyone.

"There was this old guy in his sixties, some kind of programming expert for one of the banks. Got laid off and didn't have any savings. The shock of being laid off at I think it was sixty-one or sixty-two was too much for him. He was divorced. Didn't have a family. Children were in the Midwest or somewhere, grown and out of touch. He sort of went to pieces. Began drinking. Fell behind in his rent. Got evicted. Was living on the streets when we saw him. He was in here a couple of times that I heard about. It looked like he would, you know, end up like a lot of the homeless—staying in shelters, or hanging out in Grand Central or Penn Station and getting chased or fighting for a spot on a heating grate. I mean I never knew any of that sort of thing before. Didn't want to know. I guess you are afraid it will be catching.

"Then we heard that he got himself together. Got detoxed and cleaned up and the social work people were counseling with him and getting him back on the track, and we heard last week he got a new job. His old skill was still in demand. I forget what it was, something like programming in C language. So he's back on a private payroll and is supposed to be moving into his own place this week sometime. It's great when you hear someone succeeds after being down and out for a time."

Outside, Deborah Morales, a thirty-three-year-old black female with sickle-cell anemia, waits to be seen. This makes her third visit to the Bellevue Emergency Department so far this month, and it's only the tenth of the month. She complains of pain in her stomach and back and vaginal bleeding. Dr. Susan Davies, a willowy young woman who was born in Britain and took her medical training at Syracuse, greets Morales and leads her to an examination room, where she listens anew to her complaints before administering oxygen, fluids, Dilaudid and folic acid—four medications that may do something to ameliorate the pain of the chronic disease.

Among the most frequent visitors to the ED are patients with sickle-cell disease and chronic asthmatics with no financial means to speak of. Both conditions require frequent medication—one for pain, the other for bronchospasm—breathing distress. By default, the only place where they can get help is in such places as Bellevue's Emergency Department. Few private physicians are interested in such poor patients, and many hospital clinics refuse the pain medication for the sickle-cell victims, because of their organization and philosophy of care. Clinics dislike anarchy and cherish order. People like Deborah Morales have unpredictable diseases. They cannot be treated on the basis of a distant clinic appointment for a follow-up visit, if symptoms grow acute at 3 A.M. Furthermore, any patient with a complaint that doesn't fit a standard clinical category also faces rejection. So to the private physician or hospital clinic such frequent visitors are trouble. They often require complicated care. "I'm sick," the patient says. "You're not on our schedule for today," the private hospital clinics respond, in effect. "You should have your symptoms in a way more compatible with regular office hours," seems to be the effective message delivered.

Because of the strong desire to control health costs that has marked the Reagan era, hospitals increasingly have come to be regarded as businesses. In that spirit, there is a "GOMER" attitude in many private institutions: "Get Out of My Emergency Room," they say, to those who do not have health insurance or the ability to pay for service. But for 250 years Bellevue has consistently rejected the "GOMER" viewpoint, which is exactly why Goldfrank wanted to practice medicine there in the first place.

Goldfrank quotes Hippocrates: "Discussing money before caring for the sick lacks propriety." Yet he fears that "financial triage" is already eroding medical ethics and leading some to forget that "the health providers' responsibility is above all the health of the patient."

⌇⌇

Deborah Morales begins to feel better now. Ironically, as Dr. Davies explains it, sickle-cell anemia was at one time a defense against malaria. As the cells become half-moon-shaped, they no longer provide habitat for the malarial parasite, which needs a round, full-bodied host for survival. Hence, over the eons, those Africans with the sickle-cell trait proved more resistant to malaria, even though the oddly shaped cell imposed severe costs on its carrier. When enough cells assume the sickle shape, the blood becomes thickened and reluctant to flow into the tiny arterioles and capillaries that radiate throughout the body. When circulation slows, then breaks down, it causes oxygen deprivation and then acidosis, both of which encourage further sickling, leading to a painful crisis of the sort that has brought Deborah Morales into the hospital. "One of the principal dangers," Davies says, "is dehydration, which increases the tendency for cells to sickle."

Morales is looking much more comfortable as the medications take their effect. She thanks Davies and makes her way out past the triage desk. She stops by to say goodbye to Teresa Rizzo. She opens the door of Goldfrank's and Rizzo's offices, and there is a congregation of staff members just coming out of the morning administrative meeting. Deborah Morales is unfazed. "Goodbye, honey," she calls through the group to Rizzo. "Goodbye, Deborah," Rizzo answers.

Near the ambulance entrance, assistant head nurse Elisabeth Weber gets the salient details on the thirty-eight-year-old patient lying on a stretcher. The condensed account that she sets down: "Lawrence Scranton, 38, W ♂. AIDS. Collapsed on street today and could not walk. Patient found sitting on street. Was discharged yesterday from another city hospital with pneumonia. IVDA [intravenous drug abuser]; A + 0 × 3 . . ."

This will be another DRG "dumping," a patient without resources whose disease will take up more time than the budgetary limit set for his illness. Weber has seen three such patients in the last few days. Each had been discharged by a private hospital with prescriptions, which the patients could not afford to have filled. In the case of the pneumonia victim Scranton, he was told to get bronchodilators and antibiotics. Retail cost of his prescriptions: $150. A second patient, recovering from a myocardial infarction, was discharged with three prescriptions: beta blockers, calcium channel blockers and anti-hypertensive medication. Retail cost: $110. A third, an abscess victim who needed codeine, four-by-four pads, hydrogen peroxide, tape and antibiotics, would have to find $70 for her medications. All came to Bellevue because there they get free medicine. The ideals, embattled though they be, remain: No one gets turned away. No one is denied care or medication. And each time, Goldfrank sends a report to the

Health and Hospitals Corporation, so that the record will clearly show the pattern of abandonment of patients in this fashion by private institutions. It is a new kind of cost control that shifts rather than actually controls the costs. And Bellevue, disproportionately, winds up saddled with the expense.

Back in her office, Rizzo has just received a phone call from Los Angeles. Her fiancé got cut from the team. At six feet three and 250 pounds, he was a bit small, they told him.

# 18

# A Leap
# and an Inspection

The identification they found later on the bearded young man said his name was Cornelius Waddell and he was twenty-nine years old. Witnesses said they saw him standing on the wall on the north side of the Brooklyn Bridge, studying the traffic coming up the ramp from the FDR Drive that leads to the bridge. They said he seemed to be waiting for the right moment to jump.

A cab bearing Donna Leslie, a forty-year-old legal secretary, who had been taken ill at work and was now on her way home to Brooklyn Heights, was making its way past City Hall and up the ramp to the bridge when suddenly she heard a tremendous thumping sound. She was driven sharply lower in her seat by the collapsing roof of the cab. She felt sharp pain in both her neck and back as the rear window of the cab suddenly burst and exploded outward, showering the car behind and the pavement below. Both she and the cabby thought that someone had thrown a bomb at them.

Witnesses told police that the bearded man seemed to be timing his leap, and that he fell the fifty feet from the wall to the ramp below in such a way that his head buried itself in the middle of the cab's roof.

When the police got there, they cordoned off the ramp and re-routed traffic so that two EMS ambulances could get right up next to the stricken cab. When they got to the scene, the medical technicians strapped the man onto a backboard, so as to immobilize him and prevent any further injury as they moved him. He was still breathing, even though his back appeared to be shattered. To try to insert a tube into his throat would risk doing further injury because that would mean moving the neck. Doubtless that would reduce whatever slight chance for survival he had. But it was clear from the gross injuries to his cervical spine that it would be impossible to make an adequate assessment of his airway in the field. About all that could be done was to place an oxygen mask over his mouth and nose and rush him to Bellevue.

The second ambulance crew slipped a backboard under Donna Leslie, who was lying on the back seat of the cab, disoriented and still unaware that it was a would-be suicide, not a bomb, that had crushed the roof and injured her neck and spine. Her speech was confused, but she opened her eyes and her motor response was such that she was given a 13 on the Glasgow Coma Scale—out of a normal 15 for maximum responsiveness. So her condition was not bad. In the other ambulance, however, the patient made incomprehensible sounds, would open his eyes only when a medic shouted in his ear, and could not move, even in response to painful stimuli—6 on the Glasgow Coma Scale. The lowest possible score is 3.

The cabby was left on the bridge with his crumpled cab and the press, able only to say as he awaited the tow truck, "This city! You can't even drive a cab without somebody landing on top of you!"

As the ambulances race uptown to Bellevue, the white telephone at the doctors' station rings.

"Dr. Delaney."

"We have a request for a standby."

"What is it?"

"A multiple trauma: skull and spinal injuries."

"How soon?"

"ETA three minutes."

Kathy Delaney alerts the trauma team, Neurosurgery and the orthopedic house officer upstairs and she meets the first ambulance, with Waddell in it. Clearly there is not much chance for the man. She follows the stretcher down the entryway. Donna Leslie's ambulance arrives just a moment later, to be met by other members of the staff. Her condition is far more favorable. Both have been treated under the "altered mental state" protocol in the paramedics' Emergency Medical Services procedure book, which means that both have intravenous lines established and have received dextrose and injections of naloxone and thiamine. Because of Waddell's low Glasgow Coma Scale score, he is receiving 100 percent oxygen at the rate of twenty-five breaths per minute, about twice the normal rate. The hope is to use this hyperventilation to induce alkalosis, which will reduce brain swelling.

The two go right into the trauma slots, one of which has just been vacated by a stabbing victim, who has been stabilized and rushed on to surgery upstairs. Leslie is almost immediately declared out of danger after lateral x-rays of her neck show all seven cervical vertebrae safely in alignment. But tests of her neurological functioning continue on her neck and back injuries, and only now does she grasp that her cab had not been bombed, but had been hit by the man lying across the room, there, surrounded by a half-dozen intent staff members. She is soon moved into the Emergency Ward and admitted to the Neurology service for observation.

Dr. Delaney can immediately see that there are problems so overwhelming in Waddell's case that the only way to get an airway established is by cricothyrotomy—actually cutting into the throat. After making a three-quarter-inch vertical incision just below the Adam's apple, Delaney spreads the wound with her white-gloved thumb and index finger to make room for a short tracheotomy tube topped with an obturator, a metal plate that holds open the wound to assure free passage of oxygen. The breathing is now more regular, but it is an island of normality in a sea of devastation. As she and the Trauma, Neurosurgery and Orthopedic residents scan the thirty-three segments that once described the long S-curve of the spine, they can see little but destruction—the cervical, thoracic, lumbar and sacral vertebrae are grossly misaligned and smashed, denying function to everything from the diaphragm to the anal sphincter, from the flexing of the neck to the wriggling of the toes.

It is clearly hopeless. Even as they stand there, Cornelius Waddell fulfills his last purpose. He dies.

There later was a discussion of the *chronic* versus the *acute* suicide. The former typically slashes a wrist, or drinks to oblivion, or overdoses on sleeping pills—taking halfway measures in some latent hope that there will be a timely rescue from the consequences. There is obvious ambivalence in the attempt. With the other kind of suicide, of the sort represented by Cornelius Waddell, there is no margin for reconsideration. Jumpers and shooters, Goldfrank maintains, are the acute variety, with a high success rate.

He recalls his own first experience with an acute suicide, when he was walking down the street in Brussels as a fourth-year medical student and actually saw an obese woman leap from a fifth-story window. "Of course I rushed forward to help if I could. But, when I got there, it was clear that there was no chance at all. There was nothing left."

Jumpers normally land feet first, fracturing the calcaneus, or heel bone, driving the long bones, the femurs, through the pelvic girdle and wreaking destruction on their lumbar vertebrae, making impossible demands on the skills of physicians and surgeons, which is doubtless the intent of such determined suicides.

A problem that Goldfrank confronts every day in himself and in others is the anger that can well focus on a patient with self-inflicted injuries. More than one attending physician has griped about the extra work suicide attempts cause. Those who have dedicated their lives to saving life don't relate easily to those who would throw it away. The difficult task that all of these doctors are called upon to perform is to give the best possible care to every patient—even those who spit at you, revile you or reject your guiding principles by trying to kill themselves.

Dr. Susan Davies is examining Claude Henrique in Room 4. Henrique is a thirty-six-year-old Québecois who was driving a car at about thirty-five miles per hour when a bus sideswiped him. His chart will read: "36 W ♂. Pt. had seatbelt on—did not hit head or chest on windshield or steering wheel. Denies dyspnea [difficult breathing]. Pt. A + O X3 [alert and oriented to 3 dimensions]; complains of right femoral pain and lower spine pain. Pt. can move R foot /c difficulty. PMH [past medical history]—ETOH [alcohol]."

Two examination rooms down the corridor Dr. Carlos Flores is examining Patrick Toole, forty-six-year-old male, who complains of falling from a step three feet high. His chart: "States + last drink 24 hrs. ago. PERL sluggish [pupils equal; react to light sluggishly]. PMH: ETOH abuse. T & R. [treat and release]."

Flores moves on to the next patient, Confessadora Valladio, age twenty-nine. She wants to be evaluated for diabetes mellitus.

It's inspection time at Bellevue. A team of severe-looking men and women materializes once a year and causes more than a small *frisson* of apprehension, as the areas on which they choose to concentrate can seem entirely unpredictable. This year, the rumor has it, they're looking especially for poorly documented drug administration.

Early this morning, Goldfrank comes out of Grand Central at his usual time, 7:15 A.M., and in a light drizzle strides past the gleaming façade of the Grand Hyatt Hotel. Lying on the grating in the sidewalk is a man whose head is entirely covered with a large black blanket that twitches as he tries to get more comfortable on his pallet of corrugated cardboard. Overhead, the cantilevered greenhouse-like dining rooms where breakfasting guests are enjoying their first meal of the day intercept the rain that would otherwise have fallen on their unseen fellow citizen below. A luxury bus idles at curbside, its great cargo bays open to receive the luggage of a group of Swiss tourists. Two nearby doormen, resplendently liveried in crimson twill, keep their gaze carefully away from the grate dweller. The doctor swiftly moves down Lexington Avenue.

When he gets to First Avenue and Twenty-seventh Street, there is a fresh look to the entrance of Bellevue. The gardens have been recently tended and the litter has been removed from the edges of the small island of green behind the wrought-iron fence at the main entrance. The dogwoods and hedges have had the soil around their bases cultivated, and inside the corridors have been waxed and buffed and some of the furniture is being repainted in the waiting areas along the main thoroughfares.

In the Emergency Department waiting room two women in Moslem garb, fully veiled, have their heads together over a single copy of the Koran, which they are reading intently. Near them are two young Hispanic women on stretchers, each accompanied by her mother, and each ready to give birth, breathing with effort, trying to answer the questions of the triage nurses: "Did your water break? Any past medical problems? Any allergies? What's your date of birth . . . ?"

On the other side of the triage desk, in the doctors' station, the staff meeting starts at 8 A.M. This morning Dr. Eric Alcouloumre, the director of Los Angeles County pre-hospital care, in charge of ambu-

lances, paramedics and technicians, tells of a patient dumped at the
University of Southern California Hospital by another hospital, which
had done nothing at all—not even stitch up the lacerations suffered
in an auto accident. He reviews the new legislation expected in California to cover such dumping of the indigent. Then Goldfrank responds with his own accounts of indigent patients being dumped here
at Bellevue. At this point head nurse Pat Kunka comes to the edge of
the group and, when she gets the director's leave to interrupt, makes
an announcement:

"Okay, doctors, we've got a state inspection going on here today,
as you know. And they're looking for lidocaine and heparin especially. . . . So if you open a vial, put your initials on it, time it and date
it, and don't drop anything on the floor. We're doing the best we can,
but you guys are messy. We need your help."

Paramedics John Duval and Moe Padellan from Bus 13-X come
jauntily through the hall, having just delivered a young man in sunglasses whom they found in an altered mental state among the potted
chrysanthemums on the median strip of greenery on Park Avenue.
The young man and his girlfriend both have on cadmium-yellow
shirts. The girl also has canary-yellow heels and white bobbysox, and
an upswept hairdo cinched about with a pale-yellow bandana. She is
asking around for a cigarette for her boyfriend, who is lying face
down on a stretcher. The triage nurse figures him for an overdose, by
the well-recognized odor of Placidyl, a tranquillizer. Brought in on his
back, he keeps turning over on his stomach.

Duval and Padellan are delivering the blood samples they drew
for analysis and turning over their paperwork to the triage desk.
While Duval still sports his "Don't Panic" button, Padellan has a new
button. "(Trust Me)" it says. The parentheses, he explains, are meant
to suggest that whatever is inside is true. "Saying 'Trust Me' makes
people wonder if they really should," he says. The parentheses act as
a fence to keep out doubt and suspicion.

It's 9 A.M. and Duval and Padellan are ready to go off duty. Since
1 A.M. they've helped at two auto accidents, one self-inflicted gunshot
wound (big toe), several drug overdoses—ETOH and cocaine—and
one myocardial infarction.

Out in the waiting area a blood-pressure reading is taking place:
the sharp sibilance of Velcro being unfastened from the pressure cuff
vies with the swishing sound of a mop moving over the floor—as if in
imminent expectation that the inspectors will materialize. A cigarette

mashed out on the floor by a thoughtless relative of a patient is gone, today, within minutes.

Down the corridor in the entry area next to the Walk-in Clinic ranks of plastic lobby chairs have been repainted and their signs, two-thirds in Spanish, give their warnings: *recien pintado*, just painted, *ojo mancha*.

In the adjoining locker area, Dr. Hedva Shamir is putting her jacket into her locker, ready to go home. She sits for a moment studying her papers, before adding them to the bundle of small paisley-covered notebooks that are her journal, the repository of her most vivid memories. A few minutes later, down the corridor in the coffee shop, she tells a companion of the case that has left her emotionally drained.

They brought in a woman yesterday, eight and a half months pregnant. There was a beautiful little two-year-old boy with her in the ambulance. Child Protective Services took the two-year-old. Mother and son came from one of the dingy old establishments that the city now uses to house welfare mothers and their children. It's a place where children run in packs at age six or seven and where drug abuse is said to affect perhaps half of the residents.

This young mother had overdosed on iron. Her name was Celestial Harnett and she said she didn't want to have the baby. Shamir had a strong reaction. She herself had fought and struggled for each of her three daughters, and could not bear to see the life of this unborn child ended right here before her.

"I had eight pregnancies for my three daughters," Shamir says, "and the last time I stayed in bed for five months, so I really have a kind of fervor for pregnant women and newborn babies. I even started out to be a pediatrician . . . until those leukemia cases during residency changed my mind.

"This woman had a bomb in her stomach. If you take on extra iron, over what the blood can assimilate, and it stays in the system for any time at all, it can destroy the stomach or intestines. It destroys the liver, the heart, the brain. And it would certainly threaten brain damage to the baby, too. She took—she wouldn't say exactly—perhaps sixty iron pills, ferrous sulfate pills. One or two per day would be normal. So, when I see her, here's this twenty-seven-year-old woman who is refusing gastric lavage and all other attempts to help her."

"Your baby will die."

"I don't care."

Shamir became furious. "I just blew my top. Inwardly. She didn't see how agitated I felt. I am pretty good at staying calm outwardly. Such indifference to the life of a child! I got a lavage tube down into her stomach, but she just pulled it right out.

"As I got the tube near her mouth to reestablish it, she bit me!" Shamir displays an angry-looking left index finger, under the fingernail of which is a bright purple hematoma, memorial of the seventy-five pounds per square inch of pressure that the human jaw can develop in the act of biting. "So I gave her a bolus dose of diazepam—5 milligrams of Valium—and that calmed her down."

With help from two nurses, Shamir persisted, and got the tube established. "There was this large number of pills in her stomach, and her stomach was being squeezed by the baby, so when I started to lavage her, I knew it was going to be a mess, but I had to go on with it."

Normally, in such cases, an x-ray can be taken of the stomach and the iron pills will show up radio-opaque on the film, giving an exact idea of how much lavage will be necessary to remove all of the pills. But in this instance Shamir was flying blind, because she did not want to risk additional jeopardy to the fetus with x-rays, to which the unborn are particularly sensitive. "Instead, I decided to do an aggressive lavage—eleven liters instead of the usual three or four."

She used the largest-diameter orogastric tube she could get, which would permit the passage of the sugar-coated green pills—they resemble M&M candies. "So I started with a bottle of saline solution, and almost immediately she started to have gastric upset, and there was this spray of fluid into the air, and I was getting drenched, and thinking, What if this woman has AIDS? What if she's an IV drug abuser? I didn't see the warning signs—abscesses, needle tracks, swollen hands—but the thought still crossed my mind."

After the first liter—a bit more than a quart—of solution had been poured into the funnel leading down the large-diameter tube, Shamir lowered the tube and thus induced a siphon action. "I watched the fluid drain back into the bottle down below the level of the stretcher, and I saw the first of those green pills appear, and I knew we were on the way to saving the baby." But it was to be a long, spattered evening, and it was to require blood test after blood test.

As the lavage continued, Shamir got the toxicology lab on the phone. "Mike? Hedva Shamir. I've got a life-threatening OD of iron and need a serum iron concentration level, stat."

"Okay, doctor, send it up."

A volunteer working at the triage desk called for the messenger, who took the blood samples that Shamir had drawn up to the fourth floor for the first of what would be a series of tests through the evening and on into the late hours of the night.

Shamir had read the chapter in *Toxicologic Emergencies* on iron as a toxin and knew what she was up against. As Goldfrank has written in that work, there are four stages of iron poisoning. "The first stage, lasting about six hours, corresponding to the acute rise in the serum iron level, is characterized by acute gastro-intestinal distress, including vomiting and diarrhea which may be hemorrhagic. Signs of impending cardiovascular collapse, including metabolic acidosis, cyanosis, pallor, lethargy, and tachypnea from metabolic acidosis may develop at this time." Shamir knew that if Celestial Harnett started turning blue, and breathing very slowly, this phase would have been entered. As the third and fourth liters of saline solution went into the tube and came back out again, green pills and parts of green pills showed up, until the doctor had counted nearly twenty pills and innumerable parts, and more appearing with every additional liter.

*Toxicologic Emergencies* lists three other stages, which include a deceptive period of six to twenty-four hours of apparent improvement followed by the violent impact of the iron on the stomach, intestines, kidneys, liver, heart and brain, leading finally to coma and death. At any point, from the first phase on, the iron has the power to bore right through the mucosa of the stomach and produce the equivalent of an instant ulcer. This can result in such massive internal bleeding that only a gastrectomy—stomach removal—can save the patient's life.

The protein molecule that carries iron through the body is called transferrin. If its carrying capacity is loaded to capacity, free iron then roams the body, poisonous in the extreme.

The first lab report came back. The starting value was 76 micrograms of iron per 100 cubic centimeters of serum. (At somewhere between 200 and 300 micrograms, there is mild toxic impact. At 500 micrograms, the iron becomes seriously toxic.) Shamir kept on lavaging. Another liter poured in. Harnett coughed, and sent another geyser aloft, to spatter those working over her.

Two hours later there was another laboratory report: 160 micrograms per 100 cubic centimeters, approaching the danger zone, when the transferrin would be fully loaded. The lavaging continued. More green pills appeared, until the total reached nearly fifty. "I just kept on, even though I was drenched and my arms were really getting tired," Shamir recounted. One option that she did not want to use was an antidote derived from a fungus, called deferoxamine mesylate, which can remove the iron from the blood by chelation, from the Greek word for "claw." The deferoxamine actually grabs iron and forms a compound that makes it possible to excrete the iron readily—turning the urine to a vin rosé color. The more iron chelated, the redder the urine, in fact. But, in consideration of the fetus, Shamir withheld this substance, as she feared it would cause a potentially dangerous drop in Celestial Harnett's blood pressure and could even put her into shock. And it could also lower the placental blood flow to the fetus.

Suppertime passed and Shamir was still at her lavaging task. And then at 8 P.M. the results of the third test arrived. The serum level of iron was 110. It looked as if they were winning the battle. Eight liters had been lavaged by then, and the patient was still furious with Shamir. At midnight the results showed 76, and the eleventh liter of saline had been cycled through Harnett's stomach. There were no more green pills in evidence. The total number of pills collected was fifty-six. Many more in degraded form gave the returning fluid a greenish tinge.

The next morning Shamir went to see the patient in the intensive care unit in Internal Medicine, but Harnett turned her head to the wall and would not accept Shamir's greetings. "It was one of my most frustrating cases," Shamir said later. "An uncooperative patient with an explosive charge, really, in her stomach . . ." But, as she walked out of the ward, two young interns came to shake her hand, to congratulate her on a job well done. The patient was transferred to Psychiatry later that day.

The State of New York's hospital inspection team found nothing of consequence to criticize in the Emergency Department during their audit. On a day with four hundred patients—one hundred more than usual—there was one open heparin vial without a doctor's initials on it. And one of the areas of the holding area near the trauma slots was not as clean as they would have preferred. But otherwise the marks were good and everyone was able to relax a bit until next time.

# 19

# A Doctor on Call

As he struggles to set a large locust-wood post upright into the ground, Lewis Goldfrank has the help of his muscular son, Andrew, now in his early twenties and a student of archeology and historical preservation at New York University. Andrew and his father are both shirtless, and the doctor is wearing wooden sabots on his feet—a favorite item of leisure footwear since his student days in Brussels, and just the thing to wear on a damp day like this, when there's an autumnal nip in the air.

The old post-and-rail fence they are repairing encloses an unusual parcel of land that surrounds a handsome old house built on a hill above the Hudson in the town of Ossining, hardly thirty-five miles north of Bellevue. The surroundings are as wooded and wild as those around the house on Nannyhagen Road in Thornwood where Goldfrank grew up. For his own children, too, his mother's vision of a "golden rural childhood" has been at least a partially realized dream.

The house, built in 1814, needs constant looking after, a task that the Goldfranks share. Lewis maintains the large lawn and plantings, while Susan does most of the vegetable and flower gardening. In addition, Susan has set herself the exacting task of restoring the old moldings and intricate carvings on the main stairway to their original natural mahogany finish.

"Okay, dad, that's it," Andrew says, as his post-hole digger completes a tubular excavation in which the pair will re-erect one of the uprights. They grasp the six-foot pole together and horse it into position, then move it to the edge of the hole Andrew has dug. A moment's further study and they are satisfied that all their preparations are correct, and the post is shoved home.

Inside, Susan and daughter Michelle, now married and a mother herself, are organizing a grape-jelly assembly line—sterilizing jars, melting paraffin and setting the high-spirited younger girls, Jennifer and Rebecca, to work. The girls have finished picking through a large basket of grapes. Then they boiled their harvest and now are watching the juice strain through a cheesecloth pouch into a bowl in a small pantry. They gaze through a window that has bright stained-glass medallions and old bottles dug up by Andy ranged about its sill and mullions. Shuttling between one activity and the other is Michelle's five-year-old son, Benjamin.

As Lewis and Andrew set another upright into a fresh post hole, the piping voice of grandson Benjamin can be heard calling his grandfather as he dashes across the lawn toward the fence. "Telephone, Grandpa Lew, telephone!"

Goldfrank thanks Benjamin for the message, and breaks into a rapid lope—the sabots making an audible report as they slap the ground on the way to the house. Benjamin also runs back toward the house in a playful race with his grandfather. Andrew raises his digger and starts to cut another post hole.

It's a call from a new poison-information specialist, Assumpta Agocha, who has recently won a doctorate in pharmacy from St. John's University. Agocha, an Ibo tribeswoman from Nigeria, holds a part-time postdoctoral fellowship in toxicology at the Emergency Department. In a few years, she will return to her native Nigeria as an expert toxicologist.

Now she has Dr. Royster Parsons on another line from a hospital in western New Jersey, near the Delaware Water Gap. There's a pause as a three-way conference call is established. Dr. Agocha makes the introductions and explains to Goldfrank that the caller is treating a Vietnamese man who has been out gathering mushrooms with his wife. The man is hallucinating, with a glassy, almost imbecilic look about him and the appearance of altered facial features, as if he were making faces and his features froze. He has all the earmarks of advanced drunkenness.

Dr. Parsons lists the physical findings: pulse of 120; blood pressure 120 over 80; respiratory rate, 20 per minute; temperature 100 degrees. The patient's pupils are dilated, lungs are clear and the bowel sounds—a sign of normality—are "present but diminished."

"How long ago were the mushrooms ingested?" Goldfrank asks Parsons.

"Hard to say. Their English isn't the best. I think about two hours."

"Any ideas about the species?"

"Well, you see, that's the trouble. We don't know."

"How's his wife?"

"She's okay. She didn't eat any. He was snacking out of their basket as they collected. She thinks it was shriveled, sort of like a brain in appearance."

That would suggest a false morel, *Gyromitra esculenta*, but the season is wrong, and the symptoms don't sound right. False morel is a spring mushroom and would typically cause vomiting, diarrhea, muscle cramps, acute abdominal pain and seizures.

"Can the wife get the basket with the mushrooms in it?"

"Yeah. She's gone out to the car to get it right now."

"Okay, you can start lavaging."

"We're doing that now."

"What are you seeing in the lavage return?"

Parsons describes a number of fragments of mushroom flesh.

The discussion that follows establishes that the mushroom is very likely a fly agaric, *Amanita muscaria*, of the same variety Goldfrank has used in demonstrations before meetings of toxicologists. If pulverized and put into a little sugar water, it acts as a powerful lure and toxin for houseflies, causing them to sip and then to drop paralyzed to die. Hence its common name.

"Assumpta?"

"Yes, Dr. Goldfrank."

"Can you get the Poisindex data on *Amanita muscaria* for Dr. Parsons here? And after I've finished you can give him additional background material."

"Fine. Just a moment, please."

"Dr. Parsons? Plan for supportive care after lavaging. Give activated charcoal and if he really remains agitated use Valium. It sounds to me like the patient's ingested muscarine—which I'm sure you know affects voluntary muscle movements. . . ."

"Yes."

". . . And atropine—which will dilate the pupils. I've never used physostigmine for that. Its risks outweigh the benefits, I'd say."

"Oh, here's the wife now. She's got the basket but she says she isn't

sure which kind he sampled. There're maybe four different varieties
here. Oh, wait a moment, doctor. She's got a partially eaten specimen
here. Could be the kind he ingested. Yes, he says this is the kind. He's
nodding his head. It's yellow."

"Has it got gills under its cap?"

"Yes."

"And how do they meet the stem?"

"Looks like the top of a parabola where it meets."

"Is there a kind of ring on the stem?"

"About halfway up? Yes."

"And scales of a contrasting color on the cap? Looking sort of like
baked sugar on a sweet roll?"

"Yes."

"Sounds like *Amanita muscaria*."

"Okay. Can I give the Poisindex data now?" Agocha asks.

"Good. And, Assumpta, why don't you go over the basics again,
and, Dr. Parsons, if you aren't satisfied with management, let me
know. Dr. Agocha has my home number, which she can give you if
you need to call me directly."

"Many thanks to you, doctor," the voice from New Jersey says,
and Goldfrank signs off with Agocha, also.

Dinner is over. There have been several other calls, but they're
mostly for daughter Jennifer, who's fifteen. She and her father have
an understanding. She gets a lot of calls, but keeps them short. After
a string of calls for Jennifer, the next one is for him. It's one of the
young attending physicians at Bellevue, Dr. Kevin Smothers.

They have a twenty-three-year-old narcotics abuser, well known
to the ambulance team, who has been found unconscious in an alleyway on Fourteenth Street. Earlier in the day, the same man had been
taken to another local hospital, where he made a quick recovery from
an overdose. From all indications, he was discharged shortly thereafter, only to be discovered hours afterward in the alleyway.

As Goldfrank later explains, there are probably both social and
toxicological reasons for his second overdose. On the one hand, many
hospital doctors have strongly negative feelings toward treating addicts and they sometimes discharge them prematurely. Or they let
these patients sign themselves out against medical advice. On the
other hand, the standard antidote, naloxone, has a short half-life. It

blocks opiumlike drugs so long as it maintains its own effectiveness. But naloxone is metabolized by the body more rapidly than the narcotic itself. So the lethargic patient may become alert under the influence of the naloxone, and the inexperienced or inattentive emergency department might well discharge such a patient, forgetting the differential rates at which the drug and its rival substance are consumed. Then, as the naloxone is metabolized, there's a repetition of symptoms and the patient can become comatose again, as the opiates regain their unchallenged influence.

Now the patient was hardly breathing and the pupils of his eyes were miotic (abnormally contracted). Nothing could arouse him—neither shouts in his ear nor pinpricks on his arms and hands. By the time he arrived at Bellevue, he was inert and blue—"comatose and cyanotic," as his chart would say. Now Smothers is checking the handling of the case with Goldfrank.

Already the patient's airway has been cleared, and he is being ventilated with 100 percent oxygen and is getting intravenous naloxone, but he goes into withdrawal. His pupils become widely dilated and he becomes agitated, spontaneously urinating and defecating and thrashing about at the same time.

Is it a narcotic episode? Yes. Naloxone has provided an answer. It actually erects a barrier at the receptor sites in the brain for opioids (the term for all opiumlike drugs that induce sleep). It can end the narcotic effect within two minutes, without depressing the central nervous system and without causing other unwanted side effects. In fact, so benign is this substance that it is used as a therapeutic and diagnostic medication at one and the same time. If an overdose is suspected, naloxone is given. If there is an obvious improvement in clarity of mind and normality of breathing, or any signs of withdrawal, that's taken as proof that the patient has ingested an opioid of some sort—heroin, methadone or one of the host of other molecularly similar substances (with such trade names as Darvon, Dilaudid, Demerol, Talwin, and Percodan).

As Smothers describes what reactions they are getting, he and Goldfrank are acutely aware that they are trying to chart a course between narcotic withdrawal—an unpleasant but not life-threatening condition—and respiratory depression, which can lead to impaired brain oxygen or immediate death.

Much of the original clinical work on managing this kind of problem was done by Goldfrank and his colleagues at Morrisania, when they experimented with first giving an overdose victim a bolus dose—a spurt—of naloxone and then a continuous infusion by IV drip, adjusting the quantities to the reactions of the patient.

"His pupils are mydriatic now and he's still comatose," Smothers says. He knows as well as Goldfrank does that there are a number of other possible explanations, but it is not clear which one is most likely to apply here, and that is the reason for the phone call.

They review the commonest possibilities again. First, there's oxygen starvation (hypoxia) because of aspiration of saliva. Or there's water building up in the lungs (pulmonary edema). This is normally a sign of an aging heart, unable to pump enough blood. In the case of a drug overdose there might be normal heart function but compromised lung capillaries. This could be owing to central-nervous-system depression, which can account for the symptoms they're seeing, also. But matching the physical evidence with the symptoms makes a few things clear: There is no indication of pulmonary edema. The lungs sound clear to Smothers's stethoscope. Head trauma? "No sign of any trauma at all," Smothers says. Finally, hypoglycemia (low blood sugar, the lack of one of the body's basic nutrient substances) could produce the patient's comatose state. But glucose has been given and the blood tests indicate normal levels of blood sugar. Finally, after eliminating the remaining possibilities—brain damage from oxygen deprivation, epilepsy or a cerebrovascular accident (stroke) of some sort—the doctors are left with one highly probable possibility: a mixed overdose of several substances, some of which don't respond to naloxone.

In *Toxicologic Emergencies* there is a roster of some common combinations, including the familiar (speedball) and the recherché (4's and doors). The first is a combination of heroin and cocaine of the sort that killed comedian John Belushi. The latter is a mixture of Tylenol No. 4 (acetaminophen and codeine) and Doriden (glutethimide). There is also a combination called "loads," which is codeine and glutethimide, and one called "blue velvet," which is tincture of opium (paregoric) and an antihistamine (tripelennamine or pyribenzamine).

Drawing on his experience over the years in such narcotics cases, Goldfrank says flatly that nine out of ten overdoses involve more than one substance—a narcotic plus cocaine, alcohol, a barbiturate, a tricyclic antidepressant (such as Elavil or Tofranil), a benzodiazepine compound (such as Valium or Librium) or a phenothiazine (such as Thorazine).

Eventually, based on the fruity, vinyl odor on the young man's breath—similar to the smell of new linoleum—the doctors deduce what the patient himself later conceded: that he had taken cocaine with Placidyl (a common drug of abuse) and had very nearly done himself in.

There's another call from Dr. Agocha at the Poison Control Center. She has heard from a doctor at Brooklyn Interfaith Hospital that they have just admitted an eighteen-month-old girl who has apparently swallowed a number of iron sulfate tablets from her mother's bottle. These are the same type of tablets that Dr. Shamir lavaged a few weeks ago from her pregnant patient. In this new case, Agocha reports, the mother gave her baby castor oil and the child moved her bowels, but there was a bloody diarrhea and then no further symptoms. Now it's nearly seven hours later, and the baby is drowsy and has developed a metabolic acidosis.

Goldfrank says that in view of the depressed sensorium—the clouded consciousness—of the child, and in light of the blood tests that show a high white-blood-cell count, high glucose levels and clear signs of acidosis, the medication to give is deferoxamine. This substance has the power to cull the free iron out of the blood and promotes its excretion through the kidneys.

It's after eleven and it's time for the day's last chore. So Lewis takes his seat in a heavy wooden armchair, and Susan drapes him with an old towel and begins to give him a haircut. "I'm so good at it," she says, "because he hasn't been to a conventional barber in years."

Whether she intends it or not, the beard that she maintains for him is forever reminding people of Lincoln. As Susan clips away they recall together an incident when he was still at Morrisania, working over a victim of a heroin overdose. When the man revived, he blinked his eyes several times, staring up at Goldfrank's bearded visage in disbelief.

"Good God!" the man said. "I thought I'd gone to my reward but I didn't expect the first person I'd meet would be you, Mr. Lincoln!"

Susan continues clipping away, and is about to finish the job when the phone rings again. It's Dr. Eric Alcouloumre, calling from Los Angeles Medical Center, where he is overseeing a resident's treatment of a comatose patient who has overdosed on some unknown drug. Just as they are beginning the lavage of the patient's stomach, the orogastric tube has become doubled over in the patient's esopha-

gus. The resident remembers reading a case report by Goldfrank and
his Poison Control Center colleagues in a recent number of *Annals of
Emergency Medicine*, but they can't locate the article fast enough to
solve their problem. The patient isn't feeling pain, but still they want
to avoid any harm and get on with the lavage.

Will Goldfrank tell them the correct procedure?

"Sure. You get the gastroenterologist in and then take your esoph-
agoscope and place it on the upper portion of the kinked tube—like
one arm of a tuning fork. Just ease down on it while you gently pull
back your orogastric tube."

"Okay, thanks a lot! If we get in a jam can we call back?"

"Sure."

They ring off and the haircut concludes.

The next morning on the train trip down the Hudson to Manhattan, Goldfrank embarks with a collection of the botanical specimens
that he and his daughter Rebecca have assembled over a long weekend for study by the Poison Control Center staff. A large shopping bag
that he carries contains a varied array of specimens for the instruction of the pharmacists and toxicologists as well as attending physicians and residents in the Emergency Department.

In the bag Goldfrank carries with him, he has samples of *Amanita
muscaria* mushrooms, carefully segregated in their own plastic sandwich bags. He also has specimens of poison ivy, yew berries, rhododendron, lily of the valley, foxglove, ragweed, may apple and nightshade leaves. Finally, he carries a sampling of highly distinctive
peach-colored ginkgo fruits, which in their fully ripe state exude a
pungent odor.

They are a leading cause of contact dermatitis among Chinese
who gather the fruits in New York City. At this time of year, the mere
sight of a rash on a Chinese makes one suspect the ginkgo. The pit at
the core of the fruits has a gelatinous nut within that is a highly prized
addition to Chinese cuisine. Its outer husk is the irritant. In fact, there
are always eager hands in the city to pick these dangling fruits from
the female ginkgo trees that grow along the streets, and in many city
parks, as well. And many of these hands find themselves itching and
inflamed as a result of splitting open the fruits and handling the nut.

On his arrival at his office Goldfrank has a crisis to handle. A
young attending physician, just starting her sixth week at Bellevue,

says she's already burned out. As she sits wringing her handkerchief in his office, she details her reasons:

"I've been here all night. And it's not the transvestites or the patients with AIDS or the overdoses. Not just them. But it's all those things *and* the lice and the abusive patients and the crazies. I mean I'm at the end of my rope, Dr. Goldfrank—I'm afraid I've made a terrible mistake."

He studies her quietly for a moment or two. "Was there a specific case last night?"

"Yes."

"You want to tell me about it?"

She does. There was an overdose last night of an addict with a hand so grossly misshapen from skin popping that it had taken on bizarre dimensions. Its skin was marbled and discolored and the fingers had become swollen and malformed at the same time, skewing out of their normal relationship to each other until they resembled a baseball glove that had been fought over by two dogs, and halfway torn apart.

Yes, he can well understand her aversion to such self-destructive horrors. "Initially you said you found the intellectual challenges exciting." He pauses, waiting for her response.

"That's true. But you don't have a residency program. So, even though the whole thing is very organized and all, just like a residency program, there isn't that official recognition. So it's not really like a recognized specialty. There was so much respect for emergency medicine at my university. There is a special place for teaching residents the way you teach attendings here. Here we teach residents from various other departments. By the time they've learned some of the principles, they're gone. They'll never be secure in many of the basics. People would get better care with more Emergency Department residents. And I feel I should be teaching them."

He concedes her several points. She is right. There is resistance to Goldfrank's long-cherished dream of having a full academic program in emergency medicine at Bellevue-NYU with a residency program equivalent to those prevailing in internal medicine or neurosurgery or obstetrics and gynecology. The young specialty has earned a place, but not commensurate with its contribution to the activity of the hospital and the medical school with which the hospital is affiliated.

Time after time Goldfrank has made proposals, often with support of the Health and Hospitals Corporation. Every time the establishment of a full teaching program similar to that prevailing in some three dozen other leading hospital centers around the country has been advocated it has been opposed. So the funds and support typically associated with a residency program, necessary to develop the educational base and commitment, must come from varied grants and sources such as the royalties from *Toxicologic Emergencies* and from honorariums from speeches and conference appearances that Goldfrank and his senior colleagues earn. These monies form a fund that finances special research projects. It supports the purchase of books, subscriptions to journals and educational trips for staff members. It also pays for acquisition of advanced equipment, like a fiberoptic laryngoscope, recently purchased for experimental use in the department. Other new acquisitions include a sophisticated new computerized electro-cardiograph and an advanced spirometer, to measure lung function and capacity. Some of these devices are as costly as an automobile. Other essential equipment also is financed from these funds, which thus sustain a climate of growth and intellectual ferment among staff members.

But Goldfrank is blocked, for now, in his hopes for a full residency program; and the young attending physician who wants to quit symbolizes one of the frustrating realities. Vast as the field is, and varied as its potential, here at one of the great health-care institutions in the country, responsible for three-fourths of the admissions to the hospital, seeing more than 100,000 people a year for emergency care, the team giving daily expression to the ideals and traditions of Bellevue is denied the chance to fulfill its own destiny.

So it is agreed. The young doctor will work through to the end of the month and then accept an appointment at a hospital with an emergency medicine residency program. She gets up to go.

"Good luck," he says. "I hope your new position is everything you want it to be."

She says a hardly audible "Thank you" and makes her departure.

As there are numerous candidates who have applied for places in this department who had to be turned down, there will probably be no difficulty in finding a qualified replacement quickly. It's just that the complaints about the residency program are justified. This limitation causes problems in hiring people who have trained with and are used to working with Emergency Department residents rather than residents from other services, with no particular emergency expertise, on a one-to-two-month tour of duty there. It is an ongoing difficulty and seems unlikely to reach an early resolution.

Nurse Elizabeth Swanson has had a memorable night. She was the receiving nurse when a police car came screaming through the Queens Midtown Tunnel with a hysterical woman in the back seat supported by her trembling husband, led down the ambulance entrance by two badly shaken policemen. The woman's hand was embedded completely in a giant meat grinder.

She had been grinding tomatoes in her husband's deli in the Laurel Hill section of Queens when her hand got caught in the blades, and now no one knew how badly she had been mangled. Her husband and the police removed the whole grinder from its counter mounting and carried it and her to the police car and came racing through the tunnel and down to Bellevue.

No attempts to free her have succeeded. The Fire Department metal cutters were brought in, but they could not penetrate the case-hardened steel. The woman remained highly vocal—crying, swearing—though nearly passing out as the doctors and Swanson worked on her. Finally, as yet another call went out for a still-heavier-duty metal cutter, Swanson and trauma team members got IV lines established and gave a bolus dose of Valium to calm the woman down.

"We were waiting for the metal cutter to get here and were comforting the woman and trying to tell her husband it was going to be okay, when suddenly the Valium took effect and she calmed down entirely. She shook her hand. It came out of the grinder. It came out looking gory because it was covered with tomato pulp running down her arm. But she had all her fingers and we couldn't find a scratch on her. A couple of bruises but not a scratch!"

Everyone started to laugh—even the woman, whose tomato-stained hand may have looked terrible but was now feeling fine.

They figured out, as the heavy-duty metal cutter arrived, that had the grinder taken an additional quarter of a turn, she would have lost at least one finger, perhaps two.

# 20
# Notes on a Juggernaut

Fitzcarlo Napoles is a twenty-three-year-old Cuban hired to work for a building contractor who refurbishes small apartment rental units on the East Side of Manhattan. Napoles was given a bucket and a large brush and told to mount a mobile scaffold dangling from the front of an Eighteenth Street brownstone. He and two others were to apply a cleansing compound to the façade of the building. In such work, usual practice would be for him to don a raincoat, rubber gloves, a hat and mask. But none of those protective garments was supplied.

He freely swabbed on the stuff, knowing only that he was getting $6 an hour, that the liquid worked remarkably well. Later, when another crew member followed with a high-powered jet of water, each new treated brownstone emerged from a century of sooty accumulations with surprising swiftness.

Fitzcarlo's right index finger began to hurt him some time after he inadvertently dipped it into the cleansing compound on the morning of his second day on the job. He wiped it off with an old cloth and later washed it in the spray from his partner's hose. He thought no more about it and during the rest of the day got some more of the stuff on his finger, and again wiped it and washed it off.

That night his whole arm began to burn, and his wife, Rita, urged him to go right to Bellevue, which he did first thing the next morning.

A new addition to the Bellevue staff is Dr. Robert Nadig, a specialist in occupational medicine, a subspecialty of internal medicine. He's attached to the Poison Control Center and also to the Emergency Department, and was in the ED when he heard about the case.

Nadig, a tall and thoughtful-looking young man, grew up in New Jersey, studied at Lafayette College and NYU School of Medicine, and then spent two years at Johns Hopkins as a resident and later as a fellow in the Division of Occupational Medicine of the School of Hygiene and Public Health, where he learned much about the characteristic diseases that attach themselves to various occupations.

It's a venerable area of medical knowledge, dating from Hippocrates, who collected information in the fourth century B.C. on the distinctive ailments of metalworkers, tailors, handlers of horses, farmworkers, fishermen and cleaners and dyers. Through the ages, others have also noted disorders particularly linked to the different callings—culminating in the work of the great Italian of the seventeenth century, Bernardino Ramazzini, whose *Diseases of Workmen* listed every malady from the "weak eyes, headache, vertigo and dyspnea [difficulty in breathing]" of wet nurses to the "severe flatulence, pallor, arthritis and melancholy" of sedentary professors. Nadig is acquainted with all those illustrious predecessors in occupational medicine, and sees the field as especially relevant in an age of environmental concern.

Now, as he looked on during a senior attending physician's examination of Napoles, he could see, as Nadig later said, that "Napoles's finger was turning white. It was necrosing—dying. The senior physician called the employer, to get some information on the stone cleaner. And I actually went over there to the worksite, to see what I could discover. But there wasn't anybody around when I got there.

"Meanwhile, we were both convinced, based on the clinical signs, that it had to be hydrofluoric acid. That's the stuff that etches glass. In concentrated form, it has to be kept in wax-lined containers." Anyway, the senior physician finally got through to the contractor, who said yes, it was a fluoride compound, but he had never had any trouble in the past. Had Napoles only worn the gloves and other protective gear that he was supposed to, nothing would have happened.

"You can imagine how we felt about that. The patient had told us quite clearly that he had not been given any protective clothing to wear. And here was the employer trying to shift the responsibility to the worker. But we didn't get into that argument. We had a patient with progressive necrosis of the flesh. It's in the literature that if you get a patch of greater extent than five centimeters (about two inches) square on the skin, it's considered a potentially fatal exposure.

"But this was a dilute solution of hydrofluoric acid—not immediately life-threatening, perhaps, but still in many ways more insidious than exposure to a concentrated form of the acid, because you don't get the same kind of immediate warning with the dilute solution." Goldfrank's *Toxicologic Emergencies* describes what fluoride can do. The book says that hydrofluoric acid draws so much calcium and magnesium out of healthy tissue that it precipitates rapid breakdown of flesh.

As the doctors recalled later, "The employer had told us the name of the product and said it was 3 to 4 percent hydrofluoric acid. We checked with Poisindex, and they had three products listed with similar names by the same manufacturer, but none showed hydrofluoric acid as a component. So then we asked for the manufacturer's 800 number, and they referred us to the company chemist who knew the formula. But then they discovered he was on vacation. We told them it was an emergency, and they gave us the man's vacation address. We tracked him down to his lakeside cabin in northern Wisconsin. He was very helpful. Yes, it was 12 percent hydrofluoric acid in concentrated form, which was supposed to be diluted by the end user three to one, which agreed with the employer's percentages."

By now, Napoles had been admitted to Plastic Surgery, and the senior plastic surgeon had his own fund of clinical information on such injuries. Using his personal computer, he presented an extensive bibliography on treatment of such poisoning, which did not accord with the approach that the Emergency Department doctors thought correct. "What you faced in a case like this," Goldfrank later said, "was a hypocalcemic crisis—a radical loss of calcium. Our information was that replacement of that calcium was the best therapy." To resolve the impasse, they turned to a doctor at the University of California Irvine Medical Center, Dr. Philip Edelman, a recognized expert in occupational medicine and toxicology, who has special expertise in fluoride, particularly in its application to etching silicon chips and circuits for microelectronics.

"Edelman agreed with us that perfusion through the radial artery [on the side of the wrist] with calcium gluconate was the best course of treatment." So the ED physicians connected the West Coast expert with the plastic surgeon, who was thus persuaded to perfuse Napoles's hand, which he did, with quick and dramatic results.

"I went around the next morning to the surgical ICU," Nadig recalls. "Napoles was in good spirits, even though he was going to lose the tip of his finger. At least he wasn't going to lose the whole finger, as he well might have. He didn't have any concept of workman's compensation, which I urged him to file for right away. If you don't make timely application, sometimes, your claim can well be delayed. Anyway, he would have the cost of a day in the ICU plus three or four more days in the hospital with follow-up perfusions to make sure we got all the fluoride out of his system. It could go up as high as maybe $8,000 or $10,000. So he won't be getting those bills. His employer will—indirectly, in the form of higher insurance rates. It should be clear to such employers that you can't run cheap and fast all the time and get away with it."

Still, it is an era when buildings are torn down in the middle of the night by rogue developers. (That happened to two single-room-occupancy hotels on Forty-fourth Street in midtown Manhattan.) It is a time when contractors, in their pell-mell desire to maximize their cash flow, disregard normal standards of safety (as in the crane that felled Brigitte Gerney). Some observers have characterized it as an era in which the city itself seems to be for sale, a time when many are being injured and some are being killed by the juggernaut of headlong real-estate development. All the more reason for the urgency with which Goldfrank espouses the view that emergency doctors should act as advocates of the rights of patients.

Flo Botte is a senior nurse who became an administrator of a clinic in the Bellevue Shelter for the homeless, which occupies the old Bellevue Psychiatric Building on Twenty-ninth Street. Dr. Dana Gage is medical director of the shelter clinic and has three physician's assistants to help her and Botte take care of the medical needs in the twelve-hundred-man shelter. It's thought by many to be an especially good shelter because its small rooms permit two or three men a room—in contrast to the great open bays in some shelters, which give them the semblance of giant warehouses for the dispossessed.

Just now Botte is on her way down First Avenue to join morning rounds at the Emergency Department and to check on a half-dozen of her patients who have been sent down to the ED for treatment of various complaints. Flo Botte's full name could well be Florence Botticelli. But an immigration clerk sheared three syllables off her grandfather's last name when the family came to America in the 1880s. Then she herself has shortened her first name, in the interests of simplicity.

Simplicity and directness are two conspicuous traits she brings to bear in her work with the homeless. In frequent consultations with Goldfrank, Gage, O'Boyle and other members of the Emergency Services staff, she tries to soften the institutional realities that confront increasing numbers of citizens who suddenly find themselves without a place to call home.

"Any one of us could end up like this. For example, one woman I have seen over the years if you just heard her talk you'd think she was a duchess. A beautiful, plummy British accent. She was born in India. Had her own ayah—nursemaid. Then came to this country. She taught at a country day school out in Bay Ridge. But that was years ago. When I knew her she had had a drinking problem and had what you would call a borderline personality. Worked for years at a coffee-shop on Ninth Avenue in the Forties. Then the shop went out of business and she lost her apartment at about the same time. For years she lived in front of the old shop.

"Right there in the street. Surrounded by boxes she had with her. Boxes and bags, full of old clippings and her possessions. And then one night someone set fire to her boxes. It's the same mentality that ties a tin can to a dog's tail. And she was admitted to St. Vincent's, and then later to a women's shelter. She was one of those who was just getting by, until her life crumbled. First her job and then her room. And then even her square of sidewalk."

Botte knows what toll such a life exacts. Leg ulcers, trauma and complications of exposure to the inclement environment are three frequent problems she sees. To get a better sense of what life is like for those on the streets, she undertook an experiment. "I tried to see how I could get along in similar circumstances."

She put on old clothes—neat, but obviously worn—and stationed herself on a street in the West Thirties. "I told myself I would just see what the problems were, firsthand. And the first problem is finding a bathroom. If you don't have any money, you can't use the bathroom in a coffeeshop. They throw you out. So you've got the terminals—Grand Central and Penn Station—and the few public urinals (horrors that they are) in Bryant Park, on Forty-second Street, and in Central Park. Here I am a chubby, middle-aged woman and the first thing you are aware of is that nobody wants to look at you on the street.

"I noticed that the shopping bags I carried made my hands puff up. And I didn't have all that much in them. That's why you see people using shopping carts. That's sort of the station wagon of the homeless. The medical consequences of this kind of life are largely unknown to the public—but not to the emergency rooms or to our clinic. If you stand around for long stretches of time, the veins in your legs lose muscle tone, and the valves that act to keep adequate pressure along the length of your legs fail to function satisfactorily and a condition of venous stasis leaves you with leg ulcers. Transudation of fluids— oozing through the blood vessels into the flesh—leads to minor infections, which cause tissue breakdown from any minor injury. In other words, standing for long stretches of time is a real threat to your health. And, when muscle tone is lost as a result, the body is unable to maintain correct circulation and the upshot is cellulitis. Thrombophlebitis, a clotting of a vein with inflammation, can send flecks of coagulated blood floating through your system, resulting in pulmonary emboli. Cellulitis and the ulcers associated with it in turn can lead to osteomyelitis (inflammation of your bones), or can promote sepsis (when bacteria from anywhere, such as your skin or bones, proliferate and are distributed throughout your body) and death."

This Friday morning, Flo Botte sees more than the usual number of people sitting in the main lobby of Bellevue. Many of them have shopping bags with them. A few address invisible companions in middle distance. One appears to be conducting an orchestra. His ears are covered with small headphones plugged into a battered Walkman unit that he has hooked over the beltless waist of his badly stained trousers.

It is Marathon Weekend. Two days from now, Sunday, more than twenty thousand runners will form into a giant phalanx on the Staten Island end of the Verrazano Bridge and begin the twenty-six-mile run that will take them in a giant arc through every borough of the city, ending up in Central Park. Millions will watch, and many of the homeless will be obliged to spend the weekend in the streets because of the Marathon.

"It's the worst day of the year from the standpoint of the homeless," Goldfrank says. "The runners outbid the homeless for the remaining single-room-occupancy space in the city—the last housing that might be said to be economy housing. Young athletes without the wherewithal to pay $85 or $100 a night for a hotel room in a decent hotel wind up at places where they pay perhaps $30 a night. Usually, the regulars pay from $15 to $20 a night. And at other, still more rundown places, they spend $5 or $10. But for the duration of the Marathon all these places fill up with runners and many homeless are back on the streets."

As Flo Botte knows, if the weather is cold and rainy, the problems for her clinic are compounded. Hypothermia—low body temperature—cases will multiply, sometimes to the point of overwhelming available services. On the other hand, if the day is hot—as it has been some years—the problem of heat stroke among competitors may make distinctive demands on Goldfrank and his staff.

It's a drizzly morning, warmer than usual. In fact, the forecast for the weekend is for clearing and much warmer: near eighty with high humidity predicted.

Botte walks into the Emergency Department and joins the meeting just gathering at the doctors' station. Today the discussion is heat stroke. There was a story in the *Times* some weeks back that Lewin mentions. A cautionary tale that everyone should know about before Sunday's race: A thirty-four-year-old city official of Rochester died of heat stroke after running a ten-kilometer race—about a fourth the distance of Sunday's New York Marathon. He collapsed during the race and was taken by ambulance to Highland Hospital in Rochester. After his temperature was taken, he lay for two hours unconscious under a diagnosis of suspected head injury. Then his temperature reached 106.7 and he developed seizures. At that point he was rushed into an intensive care unit, where he died shortly thereafter. It was an unusually busy time at the hospital, and they told the press that the ambulance crew gave the emergency staff insufficient information on the patient's condition. But still the man was dead.

Goldfrank adds his comments: "When you get a hot day and high humidity, sweat runs off the body instead of evaporating, and that's one of the problems. Little cooling occurs." You have to get the patient cooled in a hurry—removing his clothing en route to the hospital and opening ambulance windows to allow breezes to accelerate the cooling process. And oxygen is given en route, also. Then, when the patient gets to the hospital, a rectal thermocouple is inserted, to monitor temperature. And the patient is immersed in an ice bath up to his neck with cold water running or sprayed on the patient and he is massaged with the ice, to hasten cooling. "And you can get a fan blowing across him, too," Goldfrank says.

Lewin tells the group that the Israeli army on maneuvers in the Negev Desert has ice trucks that accompany the troops for this very reason. They have virtually eliminated the incidence of heat-stroke mortality or serious morbidity as a result.

The U.S. Army has had similar success by different means. Dr. Joseph Wilkinson, a U.S. Army captain visiting the Poison Control Center and Emergency Department for a week of training, is taking part in this morning's meeting. He adds his description of how the Army handles the problem at his base at Fort Hood, Texas:

Whenever any soldier falls out from heat stroke, a helicopter is called in immediately. "By the time it arrives, we will have poured water all over the victim and then, once we've loaded him into the helicopter and loosened his clothes, we leave the doors open, for maximum breeze, and the chopper goes right up to four thousand feet altitude and then flies to the field hospital. That is usually enough to effect rapid cooling."

"Yeah," says Lewin, "and the soldier has a myocardial infarction from anxiety!"

There's a round of laughter and, almost simultaneously, a running entrance by two paramedics pushing a stretcher through the doors of the ambulance entrance and then a left turn into the trauma resuscitation room.

The stretcher bears a thirty-five-year-old male who is looking bad, but not nearly as bad as he should look, based on the story that comes with him: Charles Gerry had a disagreement with his roommate after an all-night party. The roommate got annoyed and locked Gerry out of their apartment. Gerry thought he could gain access anyway. All it would take would be to leap from a roof to a terrace outside the apartment—a gap of about three feet. He missed, plunged four floors, wound up on the roof of a parked car, suffering two broken legs and a shattered pelvic girdle, according to the x-rays that have just been developed. Dr. Toni Field is stabilizing the victim.

After a few minutes, she and the trauma team are satisfied with his blood pressure and the surgeons shuttle Gerry down the corridor to the waiting elevator for rapid transfer to the operating room on eleven.

Flo Botte checks several of her shelter's clients. One, "Mr. Green Beret" they call him here, is a psychiatric patient with a long history of alcohol abuse. As Botte looks in on him in the Adult Emergency

Service, he is shaking the IV lines as he slashes the air with stylized karate-like motions of the hands.

She can hear the familiar voice of Eighth Street Eddie, too, who has been found unconscious, again, on East Tenth Street. Now he's coming around, garrulous and highly audible.

Botte can see nurses Lillian Conrad and Pat Sorensen moving a small, apprehensive-looking woman toward the bathroom. Her clothing now hangs in tatters. She seems to be intoxicated and smells strongly of alcohol.

"No, no, dear," she is saying, "I don't want a bath!"

Conrad, a mother of five who is working toward an advanced degree at Hunter College, has been a regular for eleven years and has seen everything. Her husband is one of the hospital police at Montefiore. Between them they have an enormous store of experience in handling just such cases.

She and Sorensen, who has been here for five years, exude a great sense of calm and assurance. The patient is looking acutely worried about her possessions—several shopping bags, with the bright Christmas logotypes of B. Altman and Saks Fifth Avenue among them.

"Come on, now, Miss Norris, we're just going to get you cleaned up."

"No, no! I told you I don't want a bath!" But there is something in her bearing that suggests that her opposition is diminishing. Still, she has adopted an almost fetal position, as if to guard her midriff most of all.

Later Pat Sorensen described what happened. "We got her out of those rags she was wearing, and then we saw why she was resisting. She had her life savings with her. We didn't count it, but it had to have been several hundred dollars in $10 bills in a money belt she had around her middle. That's why she didn't want to be bathed. Thought she'd lose her money. Of course we gave it right back to her and got her some new clothes from Social Work. And got her spruced up, cleaned up, on her feet. And then she wanted her meal. She knew when it was time for lunch!"

Pat Sorensen worries most of all about the prisoners—convicts transferred from various prisons and lockups, and the newly arrested suspects who have not yet come to trial. There are at least five different sources for prisoners in the ED—regular city police, transit police, Port Authority police, housing police and the Department of Corrections. And every new prisoner-patient adds just that additional element of noise and danger.

"Just a few weeks ago we had five Colombians arrested in a drug bust over in Brooklyn. They had swallowed bags of cocaine—which if they burst inside you, you die. So we were examining all of them, and there were ten police in here, and one of the prisoners—no one knows how—wriggled out of his cuffs and escaped. Ran over to the ambulance garage. They got him back, but it's very upsetting to have that sort of thing going on.

"And then a few weeks ago six federal marshals with bulletproof vests and shotguns come parading in here with one single prisoner. He was supposed to have swallowed a bag of cocaine, too. He must have been a very important suspect to rate so much attention. All we knew was his name was Ambrizella, but we never knew what the charge was—mostly we don't want to know. If you know too much about the crime they're charged with, it could interfere with your commitment to give your best effort, no matter what.

"Every time a prisoner comes in here there are two police with him. And, even though mostly it's all right, there are times when things get really tense. Prisoners will do anything to come here for a day or so. They swallow razor blades to get out of the Tombs or Riker's Island and spend a few days here, get attention, good food, comfort. A razor blade can destroy your insides. Mostly, though, they tape it before swallowing, but that doesn't show up on the x-rays. We observe them for bleeding or pain, and then they'll pass it in time. So the prisoner beats the system for a day or two.

"Or another prisoner will say he's suffering diabetes, and until you draw blood you don't know for sure. So he'll get out for a while, too. Of course if the police or Department of Corrections refuses a request like that and the prisoner gets sick or dies, then there would be a real problem. So they mostly seem to give the prisoners the benefit of the doubt. . . .

"The violence is very upsetting. The other night the Port Authority police brought in a man from Penn Station who was acting suspicious and they had him in the prisoner holding area back there by the Walkin. He was cuffed, and somehow he grabbed the pistol off one of the officers, and the other jammed his finger in behind the trigger, so the suspect couldn't pull it, but he was trying, and they were wrestling around and we were yelling 'HP! HP!' (hospital police), who came running and it was very tense until they got the gun away from him.

"So," says Sorensen, in an introspective mood, "you wind up being skeptical but you still try to maintain your compassion. And then you hear yourself starting to bark at everybody. Then it's time to go on vacation." And when is she going? "I just got back!"

Flo Botte is going back up First Avenue toward the shelter. There's the recurring report that the city has declared the whole block a redevelopment site. That means the showcase Bellevue Shelter is living on borrowed time. As she approaches the building it is hard to believe that a massive, well-built structure like that is already doomed. It's a handsome edifice, built in the thirties of red brick with eighteenth-century detailing. Now the land value is so high here on the East River that the redevelopment juggernaut seems destined to roll right over this part of Bellevue.

She recalls an interim redevelopment plan, to build housing on this site for health professionals—interns and residents, nurses and others on the staffs of Bellevue and University hospitals. Now that, too, may be superseded by the inexorable pressures of land values. And there will be room for but a few of the many nurses, aides, technicians and house staff members without whom Bellevue could not operate. Even now, the paucity of affordable housing in the hospital's immediate area makes it increasingly difficult to maintain staff levels.

"It's very hard to comprehend what the city is doing, except yielding to special interests in a way that is just devastating for the homeless. The city owns thousands of dilapidated buildings. But it pays up to $6,000 a month to house a family in these welfare hotels. So around the city you have the homeless warehoused, really, in armories, old schools, rundown hotels. And when you get a place like this—the old Psych Building—it's a structure that makes it possible to maintain a human scale, not great open rooms crowded with rows and rows of cots. Solid construction. Adequate elevators and bathroom facilities. We run a good shelter. No drugs. No alcohol. No obstreperous behavior. No stealing. And some of the men take brooms and bags and clean up the litter in the local parks. So there's even a community sense beginning to develop among these resident homeless.

"It's very difficult. This land has been in the public domain since 1795, and it seems to me it should remain in public service, that it should not be for sale to the highest bidder."

At the Walk-in Clinic, head nurse Mary Dwyer listens to a woman in her fifties with her hair pulled back in a tidy bun. Her eyes are rimmed in red. "I'm worried about the rent and they're putting me out."

"I'm sorry. But this is not a shelter. This is a hospital."

"But I need a place to stay."

"I know. It's very hard. I wish I could help you."

"I can't stay here, then?"

"No, dear. People pay $600 a night to stay here."

The woman turns away. "Thank you very much," she says. "Good luck, dear."

# 21

# About Children

Monique Tittle, a thirty-four-year-old mother of five, tried crack for the first time, and during her high she suffered a stroke that left the right side of her body paralyzed. When the Crisis Intervention worker looked in on the family's nine-by-ten room at a welfare hotel after a neighbor alerted the front desk to the trouble next door, she found the stricken victim and her seven-month-old baby boy, who was having trouble breathing. Social workers located three of the children, sisters, with a group of children playing on the sidewalk outside.

The mother and baby, whose name was Ricky, went to Bellevue. It was discovered in the Pediatric Emergency Service that the baby had pneumonia. The sisters were taken by the city's Child Protective Services and placed in foster care.

Ricky lies in the Pediatric Emergency Service, a tiny IV line dripping antibiotics into his arm. He cries fitfully from time to time, but head nurse Denise McLean says he is doing better than he was when he first came in. He is ready to be admitted to the in-hospital Pediatric service.

The children that Denise McLean sees appear wary, as well they might. They are the offspring of the urban poor, almost without exception. The vast majority of them live with a single parent, usually the mother. "They don't have a fantasy life," Denise McLean says. "Look at little Ricky, here. Isn't he a beautiful baby?" There is no doubt that he is unusually attractive. Plump and active with a dramatic shock of black hair, piercing dark eyes and a finely shaped nose. "But you notice he doesn't smile?" McLean makes a silly face at the baby. There is an impassive response.

She tickles the sole of his foot. He withdraws the foot without amusement, staring with those great eyes, soundlessly studying the grownups at the perimeter of his small crib.

Denise McLean has been at Bellevue for ten years. She graduated from Hunter College and trained here at Bellevue, and she often finds herself upset by the sharply deteriorating conditions in which children of the poor are born and raised in the city. "I think the threat of homelessness haunts the parents—hardens them, when they experience it, so that little children are hardened, too. Joyless and street-wise at very early ages.

"Some of the things that go on just make you want to weep. For instance, we had a three-year-old in here yesterday with an earache. He is calling out, 'Mommy, mommy,' and his mother is outside here in the hall, lighting up a cigarette and she says—it sounded so cold— 'I'll be back,' and then she walks down the hall to have her smoke.

"Or another case. A six-year-old was riding his bike down on Avenue D, got hit by a car and was brought in by ambulance. He was shaken up and had a broken arm and we called his mother. She came in with the child's aunt a while later saying, 'Everything's all right, Tyrone, mommy's here.' But she didn't stay five minutes. She left. Went off with her sister. Three hours later she shows up again. Three hours! With shopping bags! The little boy was almost frantic by then. Had his arm wrapped in a cast, but she wasn't here when he was most frightened. And I kept saying, 'Now, now, Tyrone, mommy will be right back . . . it's going to be all right. . . .'

"Many of the children we see are withdrawn, suspicious. From their earliest years they show signs of damage. They don't talk. Won't be drawn into a conversation. No imagination. It's really heartbreaking."

Within a few days Ricky Tittle is well enough to be discharged, but his mother is not. The prognosis for her has turned from fair to poor. Her mood is one of suicidal depression. When she leaves the hospital, it will be in a wheelchair, and it seems unlikely that she will be able to resume the care of her children. And she repeatedly threatens to end her own life.

She has been an intravenous drug abuser and a prostitute, and Ricky and his three older sisters and older brother were fathered by three different men. The case is a difficult one, and it finally involves

Frances Gautieri, Deputy Director of Social Work for Bellevue, a gentle-mannered woman whose calm style does not hint at the sense of crisis that pervades her every working day.

"There is no foster care available for this baby," Gautieri says, "and the numbers have increased so rapidly that child-welfare agencies cannot recruit enough foster homes to meet the need. Pressures are placed on us to keep children in the hospital beyond the point when there's medical need. But we have to keep hospital beds available for children with acute problems. The hospital cannot become a custodial institution for boarder babies—become what used to be called an orphanage—or it stops being able to function as a hospital."

She goes on to say that she negotiates case by case to work something out for each baby. "For Ricky, the only alternative today is to send him to a temporary nursery in an office building with homemakers who change each shift. Then we move him in the evening to a 'crisis nursery'—an emergency nursery that will keep a child overnight." Such nurseries are run by a foster-care agency. Then, the next morning, the agency sends the baby back to the Child Welfare office, where the cycle will be repeated for days or weeks until an adequate arrangement is finally worked out for the baby. "It's a real dilemma. We will keep Ricky and continue to try to work out an acceptable plan for his future care. But we will not admit the five-year-old developmentally disabled child who was found abandoned today and has been released from the Pediatric Emergency Service. They would like us to keep him for a few days but that could turn into weeks or months."

Gautieri sees an arid desert for families—no neighborhood support of the kind that used to be natural, no networks of mutual encouragement, no grandparents to share their wisdom. Connections seem to be lost as families break apart and neighborhoods are redeveloped and transformed. And, since the era of budgetary stringency that began in the early eighties, case workers who used to have an ongoing relationship with a mother and her children no longer have such long-term connections.

"Homelessness and child abuse and neglect have been rising side by side," Gautieri adds. "There are more than five hundred cases of child abuse a year reported from Bellevue, and many other cases doubtless are missed."

A four-year-old girl, Jessica Gonzales, was brought in for evaluation by Child Protective Services. The child lived with her mother and an older sister, who had earlier been removed from the home on the complaint of the grandmother that her daughter was abusing her grandbaby. The divorced father was a frequent visitor in the home, and he was brutal with the children, beating them with a ruler on occasion. The mother admitted that she had whipped both girls in the recent past.

When a pediatric resident examined the girl, there were scars on her legs, arms, thighs and buttocks. Her stomach was notably protuberant. She had what is called a "flat affect," which means she was generally unresponsive to the doctor's questions, or answered without animation. Radiology's pictures showed old breaks in two ribs. The verdict was that the child had suffered abusive treatment.

There is a large fund of information on such victims. Abused children have different patterns of lesions and fractures from those who have been accidentally injured. Bruises predominate. Cuts are rare. Hemorrhaging in or about the brain—a subdural hematoma—is the most frequent kind of fatal head injury, and many go undetected. The frequent cause of such injury is the strenuous shaking of a child, which can actually slam the brain against the bony vault of the skull, leave no outer mark of the damage, yet do severe harm within. As there are no signs of exterior trauma, the child's injury goes undetected.

Doctors and nurses try to stay alert to such obvious signs as bites, radiator burns, rope marks, hand marks from slapping, and to the various patterns of bruising. New bruises (less than two days old) will be swollen and the child will recoil if you touch the spot, even gently. After five to seven days, bruises grow greenish blue; then, after a week or ten days, yellowish; and after two weeks, brownish.

The staff try to stay alert to the differences between accidental and intentional burns. The accidental kind usually reflect the object that caused them—hot coffee, a chewed extension cord, an oven door. The burns of child abuse are brandlike—as if something hot had been pressed against the child (a cigarette or matches), or the child had been dipped in a scalding bath.

The law clearly states that suspicion of child abuse must be reported, yet parents almost always deny any abuse and frequently, as in the case of an infant suffering a subdural hematoma, there is no outward evidence to support the physician's suspicion.

"Sometimes we just admit a child to get him out of a potentially murderous home situation," Goldfrank has said. "It's one of the more difficult conditions to assess accurately. So we attempt to apply all those diagnostic skills that we try to develop in ourselves and in our colleagues. And, when in doubt, we have to come down on the side of the child."

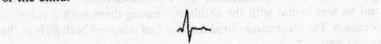

An eighteen-year-old student named Arne Olsson is brought in by his school supervisor and two police officers. He comes right into the Adult Emergency Service. Goldfrank glimpses the passing stretcher and almost immediately has a hunch. He defers it, as he is in the midst of a crisis involving a woman suffering heroin withdrawal in Room 2. But he makes a note to look in on the boy as soon as he can.

Olsson is very blond, a bit small for his age and extremely agitated. His eyes seem to widen, as if he is grasping a moment of clarity from the whirling mass of perceptions within his troubled mind.

As they push him into the resuscitation room, Olsson tries repeatedly to get up from the stretcher. "Easy, Arne!" the school supervisor shouts in his ear, while both police officers and the supervisor add their strength to keeping the boy in place.

Suddenly the boy is quiet. He stares at the bright overhead lights. His right arm twitches involuntarily. The school supervisor, a balding man in his late thirties, describes how the boy went berserk in the school cafeteria during lunch, creating a major disturbance by throwing chairs and shouting out obscenities. Goldfrank enters the room, and puts his hand on the boy's head. Arne Olsson sharply shakes his head, as if to be rid of an unwanted presence.

"What happened?" Goldfrank asks.

"Poo poo!" the boy responds, laughing mindlessly.

"Can you tell us what your name is?"

"Raster blaster shithead asshole," he says, flailing his arms out as if they belonged to a rag doll. Then he seems to lapse into unconsciousness. Goldfrank listens to the boy's breathing with his stethoscope, then to his heartbeat.

A brisk physical examination then goes forward, as the police step back and the school supervisor takes a seat, exhausted, out in the corridor. Olsson is obviously a well-nourished boy with no signs of trauma on his body, which now lies bare on the examination table. His skin is cool and slightly moist—"diaphoretic," as the doctors will record it. They lift his eyelids, and he stirs.

"What's up, doc?" he asks thickly, and then titters in evident amusement.

"Are you awake now?" Goldfrank asks.

The eyes flutter open. The pupils are small, 2 to 3 millimeters. And there's rapid, involuntary movement of the eyeball, from side to side (called "nystagmus," from the Greek word for drowsiness).

"I'm Dr. Goldfrank. Can you tell us your name?"

"Ka-ka, doo-doo, poo-poo," is the response.

Goldfrank sniffs the air around the boy—a technique he urges on his colleagues as a great diagnostic aid. He knows, for example, that the smell of sweet russet apples would suggest diabetes out of metabolic control; that garlic would indicate possible arsenic poisoning; that the musty odor of fish could signal liver failure. But here there is no discernible odor.

It is agreed that they ought to proceed as if this is drug ingestion, which suggests they get the boy into a room with dimmer lights and start administering activated charcoal, which is a good binder for many kinds of drugs. Meanwhile, the rest of the work-up continues.

Altered mental status of the sort the boy is suffering includes coma, confusion, agitation and anxiety. Here, there is a strong presumption that, as the school the boy attends has a reputation for a lot of drug activity, drugs are probably involved. So, in addition to the assessment of his vital signs and a complete physical examination, blood and urine samples are taken. These samples will be screened for toxicology (drugs or poisons); electrolytes (sodium, potassium, chloride, bicarbonate and calcium); glucose (the principal fuel that powers the body's cells); blood urea nitrogen (a measure of nitrogen waste to be cleaned out by the kidneys after protein metabolism); and a complete blood count (which tells about red and white blood cells and platelets—clotting factors).

There's a summons for Goldfrank from the triage desk. A Senegalese with chest pain has come in and no one else on duty just now is fluent enough in French to take a history.

The working hypothesis arrived at before Goldfrank's departure is that the boy has a drug-induced altered mental status, perhaps owing to a trick played on him by some friend or acquaintance.

Goldfrank goes to examine the Senegalese patient, a man in his forties, whose vital signs suggest that he may be suffering some sort of cardiac event. Goldfrank gets the history and says a reassuring word to the patient, before turning the case over to Dr. Robert Hoffman. Goldfrank returns to the case of "altered mental status."

Arne Olsson is much calmer. The police have left and the school supervisor is talking with the boy's mother, an attractive blond woman in her forties, who has just arrived and appears very distressed.

In a rush of revelation, the staff learned from a now-lucid Olsson that while he was at lunch one of his friends had put mustard on his sandwich. "He called it 'hog dust' mustard. It tasted terrific," the boy says, a self-deprecating smile on his lips. "I finished the sandwich and then slowly I started freakin' out. I just didn't know who I was or where I was."

"What was it, doctor?" the mother wants to know.

"Phencyclidine—angel dust," Goldfrank responds.

He went on to describe to mother and son what was known of phencyclidine (or PCP). It was developed in the late fifties as an anesthetic and was used for a time in operating rooms, until patients began reporting weird postoperative states of mind. Then, in the midsixties, the stuff showed up on the streets of San Francisco and was called the PeaCe Pill, whence its acronym, PCP. But it was not popular then as it plunged users into stygian depressions. So, its continuing use is normally as a part of "polydrug" use—this killer weed, as it is called, combined with other more euphoria-inducing substances. Among its nicknames besides "angel dust" are "rocket fuel," "super grass" and "wobble weed."

Because the drug is fat-soluble, it gets into fatty tissue and the brain itself, and lingers—which can mean a half-life of as long as sixteen hours.

"So Arnie's going to have to stay in the hospital?"

"We'd recommend it. At least for several hours, until he's absolutely normal."

Their visitor thanks the doctors warmly.

As Goldfrank takes his leave, the young patient calls after him from the dimly lit room. "Hey, doc?"

"Yes?"

"They tell me I was a little raunchy when I first came in. If I said anything, you know, out of line or anything . . . ?"

"That's all right, Arnie. I hear that sort of thing all the time."

"Well, I'm sorry."

With a smile and wave, Goldfrank goes on down the corridor.

Administrator Lawrence Dugan feels like a stage manager, he says, in the varied duties he performs in the non-medical administration of the Emergency Department. Just now he saw the end of a drama that unfolded this afternoon, and this had nothing to do with his usual responsibilities. A woman fell in the subway and was brought by ambulance. She lay on her stretcher-cart in a distant corner for somewhat longer than she or Goldfrank would have preferred. Had the director known about it, in fact, someone would doubtless have been reproved. For the woman was thought to represent another aggravated case of maggots. Her head appeared to be a writhing mass of the repulsive larvae. But, when nurse Carey Le Sieur approached the woman to tackle the unwelcome but necessary task of getting her cleaned up for further evaluation, the cap of maggots turned out to be the twitching fur of a small black rabbit, virtually molded to the woman's head in fright at its unaccustomed surroundings. That explained the curious box the woman also carried—it said Cutty Sark, but the Scotch whisky was long gone and in its place were lettuce, carrots and a water dish. The patient was promptly treated and, with her rabbit, released.

# 22 An Aneurysm and a Dangerous Diet

Magdalena Suarez is a forty-one-year-old mother of three, a widow, who has just completed a course in ultrasound technology, aided by a church group that helped finance the cost of her training. She is within days of becoming financially independent of public assistance, a cherished goal that she has pursued ever since her husband's premature death in an industrial accident four years ago. His insurance and death benefits have not covered all the family costs, but the children and Magdalena have been hard-working and lucky. Her eldest son, Estaban, has just won admission to the Bronx High School of Science. The younger children, both girls, show signs of academic excellence, also.

Now, as Magdalena prepares for her own final exams, studying far into the night, she does not feel well. There is a sudden tearing pain in her stomach, which alarms her so much that she feels she has to get to the hospital. Her preference would be to take a taxi from the family's crowded Avenue B apartment right up First Avenue to Bellevue. But no cab is likely to be around this neighborhood at this hour. So she calls 911, quietly, so as not to alarm the children. She leaves a note for her eldest, Estaban, telling him where she is going and asking him to take charge in her absence. It is nearly 2 A.M. The pain is getting worse. She goes downstairs to await the ambulance.

Dr. Robert Hoffman is a third-year medical resident working in the Emergency Department at Bellevue. He grew up in Flushing, Queens, went to Brandeis and then NYU Medical School and traces his determination to be a doctor to a scouting experience. The sum-

mer he became an Eagle Scout, he and a partner saved a fellow
camper from drowning in a Catskill lake. After three diving searches
of the area where the boy went under, they found him unconscious
under a float. As Hoffman and his chum brought the boy to shore,
where the camp doctors and a throng of campers awaited, the victim
was already breathing again, weakly but regularly.

Hoffman is just finishing up treating an overdose case. As he
comes by the doctors' station to complete his paperwork, he gets a nod
from Dr. Robert Hessler, the senior attending physician on duty that
night. An emergent case has just come in. Is Hoffman free now to
handle it? He is. He goes across to Room 8 to a stretcher where nurse
Jeanne Delaney has already heard the gist of the patient's chief complaints, has ascertained what, if any, allergies she suffers, and has
taken a medical history in addition to getting the vital signs of Magdalena Suarez. The patient's blood pressure is 180 over 120, markedly
elevated. The young physician introduces himself and, as he checks
her pulse, he ponders her pallid appearance and her complaints of
tearing pain. He takes her blood pressure once again, on the other
arm this time. The readings differ.

Nurse Delaney, meanwhile, is making Mrs. Suarez comfortable
on her stretcher, loosening her garments, helping her into a hospital
gown. But now the patient is saying the pain, which was localized in
the front, seems to have migrated to her back.

As Hoffman said later, "I began to think she had an aneurysm, but
Dr. Hessler, I knew, was busy with a cardiac arrest, so rather than
wait for a consultation with him I called Cardiovascular Surgery and
Radiology to organize an angiogram. The radiology technician
wheeled in the portable machine and got a chest x-ray immediately.
Meanwhile, we started several IV lines, so Mrs. Suarez could have
continuous infusion of nitroprusside to lower her blood pressure as
well as beta blockers, which we hoped would stop the effect of adrenaline and similar neurohormones. We were hoping these would halt
the shearing effect of the elevated blood pressure on her vessels. We
also gave her Valium. And we had an intra-arterial catheter in place
also, for a minute-by-minute blood-pressure reading."

The angiographic examination involves injecting a radio-opaque
liquid dye, which means the usually invisible arteries will now stand
out as sharply as bones normally do on x-ray pictures. The radio-
opaque substance can be introduced through arteries either in the
upper arm or in the leg. In Mrs. Suarez's case, it was decided that a
leg catheterization was safest, to avoid passing the site of a possible

aortic aneurysm and allowing the catheter to further injure an already-impaired vessel—and possibly perforating the outer covering of the artery.

The chest x-ray and angiography results both appear normal, leaving Hoffman somewhat crestfallen. Still, he's convinced that it has to be an aneurysm. As he later said, "I was a bit perplexed, because here we had the gold standard of tests and it was negative. Angiograms are supposed to be nearly infallible and this one indicated no aneurysm. Yet all the symptoms—the pain and the pressure and the sense of tearing, all these things—argued for a dissecting aneurysm, and it was probably enlarging as we stood there studying her."

Such an "aortic dissection" is a bulging of the principal blood conduit in the body, the aorta, much in the manner of a rampaging river overrunning its banks and establishing a new channel outside of the mainstream. It begins with a tear in the inner lining of the aortic wall. The blood splits through this layer of the vessel's wall, then forms a second lumen (the term for a passage within a tubular organ). This dissecting channel forms an aneurysm that is a virtual time bomb, threatening to go off at any moment, with devastating effects.

The patient's blood pressure has been reduced by the medication from 180 over 120 to 165 over 105, a good sign, but Hoffman is still getting unequal pulses and pressures on each arm. "I was thinking, here it is nearly three o'clock in the morning and this woman is in obvious distress and I have the results of the angiogram and I'm a new boy on the block and Radiology is already overworked and I have pushed them and I'm about to push them again." Mrs. Suarez is pursing her lips, clearly in torment, as the minutes tick away.

Jeanne Delaney has already conveyed Mrs. Suarez's worries about her children to social worker Mary Caram, who has the address and phone number. Once the outlook has been clarified, she will make arrangements to notify and care for the children.

Inside Mrs. Suarez's chest, a drama continues to unfold. The best surmise now is that such dissecting aneurysms as she is indeed suffering result from the constant pulsing of high blood pressure, which works to lacerate the innermost layer (the intima) and then weaken the middle coat of the organ (the media). Finally, the outermost layer (the adventitia), like a dam under intolerable pressure from the swollen river upstream, weakens and collapses—sometimes loosing a catastrophic deluge that explodes into the chest cavity, causing nearly instant death.

While a classic aneurysm blows out, like an inner tube, at a single spot, a dissecting aorta can shear off many vessels as it extends, blocking various branch arteries to brain, arms, legs, kidneys. Or, occasionally, a flap of the innermost intima that has been torn free by the torrent of cascading blood will jam itself into a branch vessel and block one of the three chief arteries that are tributary to the "aortic arch," which rises from the heart itself. These three govern the head, the brain and the arms.

Sometimes, to complicate the patient's diagnosis, the victim feels no pain, because mental status may have been altered by the shock of sudden blood loss through internal bleeding, or by the jamming of a flap of the intima into the carotid artery—the main source of the brain's blood supply.

As Dr. Robert Hoffman knows, about 20 percent of untreated victims of a dissecting aneurysm will die within six hours of first pain. Half will die within two days, 90 percent within three months.

Mrs. Suarez is writhing on her stretcher now, so acute is her distress, which continues to migrate from its original location. Despite the morphine that she has been given, she feels increasing discomfort. That is all the additional indication that Hoffman needs. He calls Radiology again, getting a drowsy resident.

"This is Bob Hoffman. We've got a patient here presenting all the signs of a dissecting aneurysm."

"The one we just did an angiogram on?"

"Yes. I want a CAT scan."

"It's the middle of the night. Give us a break!"

Hoffman nevertheless persists, and he and Jeanne Delaney, both making reassuring explanations to Mrs. Suarez, move her down the hall and around the corner to an elevator that will take them upstairs to the third floor, where an unenthusiastic resident stands ready to make the CAT scan. Such a procedure makes a series of x-ray "slices" through the body, which are then linked by computer-aided graphics to present a three-dimensional portrait of the principal organs and structures of the body. The patient is slowly moved through the plane of the x-ray camera, until the entire area in question—even the entire body, if need be—has been scanned.

"We watched the results," Hoffman later recalled, "on the display screen. The first cut—nothing. The second cut—nothing. I admit I was a little uneasy, wondering if I could have been wrong, but still having a strong belief that nothing but a dissecting aneurysm could be causing all the symptoms we were seeing." Three—nothing. Four—nothing.

"Of twenty segments the CAT scan was doing, the first fifteen were negative. No sign of aneurysm. Then, on number 16, we got a positive. And it was positive on all the others, also."

Events moved with extreme rapidity then. Within half an hour Mrs. Suarez was on the eleventh floor undergoing cardiac surgery for the replacement of a length of her descending aorta.

"It was a very good feeling," Hoffman says, "that we caught it before it got out of control."

Mrs. Suarez recovered satisfactorily and was able to return home within two weeks. She will sit for her final exams later, but immediate anxiety has been reduced by a special purse collected by the church group to help tide the family over in the meantime.

Monday morning. Robert Hoffman has just presented his dissecting aneurysm case on morning rounds and has gone home elated at the attainment of a real diagnostic coup. Goldfrank and Lewin have both congratulated him, reminding their colleagues that a past failure—the Italian who died of a dissecting aneurysm after his transatlantic flight—has been an essential ingredient of this fresh success.

As the meeting breaks up, a chubby Chinese girl and a friend with a worried look are talking to Sharda McGuire at the triage desk. Vikki Chong is an eighteen-year-old college freshman preparing for midterm examinations who has passed out three times in the past week. She seems more than a bit embarrassed that she has been brought into Bellevue by a classmate, as by now she has regained consciousness entirely.

Goldfrank decides to take this case himself. She is taken to Room 6, where a cardiac monitor is attached. She tells Goldfrank that she has never before passed out.

He asks her if she uses drugs, alcohol, cigarettes or any medications (even things such as vitamins). To all she answers "no." And she denies any history of fainting—"syncope," as it is called in medical terminology. Nor is there other evidence of a particular problem, except in the last week Vikki has had two episodes of light-headedness. One occurred while she was seated at her desk in her dorm room and the other while walking to class. Both times she lost consciousness.

And this last time, just this morning, was witnessed by her friend Wendy, who has brought her to the hospital. In both instances, Vikki became dizzy and unsteady on her feet, tried to sit down, then collapsed without doing herself any harm.

The physical exam that Goldfrank performed disclosed no abnormalities. Vikki Chong was a bright and articulate young woman who stood five feet four inches tall and weighed 137 pounds. It wasn't until Goldfrank affixed the monitor leads and began to watch the tracings Vikki's heart generated that he saw marked abnormalities. The prolongation he was seeing of the segment designated the "QT" interval

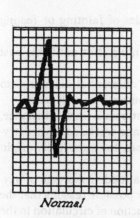

*Normal*

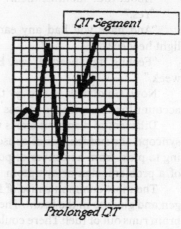

*QT Segment*

*Prolonged QT*

suggested electrolyte abnormalities.

After the physical exam was finished, Goldfrank turned to Vikki: "Do you mind if I have a talk with Wendy, too?"

"No, please, go ahead."

"All right. Make yourself comfortable. I'll be right back."

Outside in the waiting room, Wendy, the roommate, looked worriedly at the doctor as he approached. "It's nothing to be concerned about," he told her. "But it occurred to me you might have some information that would help us figure out what's causing Vikki's blackouts."

"I don't know."

"Do you know anything about past medical problems? Any medications she takes—she didn't mention any to me—or any change in habits in the recent past."

"No, not really." There was a momentary pause, then a brightening of the girl's face. "Except the diet."

"What diet?"

"Oh, she's on this really weird diet. Has been for—I don't know—all summer and so far this school year. About three or four months." She went on to say that Vikki had lost nearly forty pounds since starting the diet and in fact no longer ate or drank with her friends.

When Goldfrank went back to the examination room to ask Vikki Chong about her dieting, she readily admitted using a modified liquid, high-protein diet to lose weight. "I've lost about forty pounds since the end of last semester."

"About four months, then?"

"Yes."

"And have you had any earlier episodes of fainting or feeling light-headed?"

"Feeling light-headed, yes. But not fainting. Not until this past week."

Now, Goldfrank has to go through a "differential diagnosis" to account for the episodes of loss of consciousness in such cases.

Differential diagnosis starts with the end result (in this instance, syncope, the sudden loss of consciousness) and works backward, trying to pick, from a number of possible causes, the most likely origin of a problem, sign or symptom.

The most common cause of fainting is a sudden shortage of oxygen and glucose in the brain. There are four possible reasons that the brain runs out of fuel: There could be obstruction of circulation in the brain. Something could trigger a sudden drop in the output of blood pulsing from the heart. Falling blood pressure in the arteries could be another cause. And another could be an insufficiency of oxygen or glucose carried to the brain, which could be likened to a sudden decline in the soil nutrients that support a plant's well-being.

After tracing out those rather abstract possible causes, then the art of differential diagnosis seeks plausible physical origins. In this instance, Vikki Chong's fainting spells could derive from a clot or a buildup of plaque (fatty or fibrous material stuck to an artery wall), but that is an unusual cause of fainting, Goldfrank knew, particularly for someone of her age. Then, various medications could so reduce blood pressure as to affect the entire "vascular tree." But that did not square with the medical history the young woman had recited for the doctor. She was taking nothing whatsoever, not even aspirin, according to her own report. So far, the tracings on her ECG gave the likeliest clue. As indicated in the illustration here, the prolongation of the QT interval could have given rise to an arrhythmia called *torsade de pointes* ("points of a twisted ribbon"), which caused Vikki to pass out.

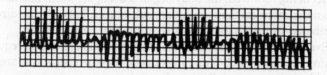

The rest of the analysis of the potential origins of the young woman's fainting episodes ran through the other possible causes of depressed blood pressure and hypoxia or hypoglycemia. In rummaging through the vast store of knowledge and experience he has accumulated in his years of practice, Goldfrank could also recall some of the more colorful asides in the literature of differential diagnosis (and specifically from the tradition of fainting).

In *French's Index of Differential Diagnosis* there is a description of another sort of fainting, *micturition syncope*, a mishap that "is not fully understood. It occurs typically when, after heavy beer-drinking, the subject rises in the middle of the night to pass urine," and it is due, apparently, to the sudden upright posture and the alcohol's powers to dilate the veins. Then, in discussing hysterical attacks, this reference book notes that "Swoons are out of fashion unless one includes the hysterical faints of teenage girls at 'pop' sessions. In the past, revivalist meetings had the same effect; the effects of the preaching of Wesley in the eighteenth century have been paralleled in the twentieth by the performances of Presley."

Finally, after weighing all the possibilities, Goldfrank concluded that it was her liquid protein diet that was causing his young patient to faint. She said that she herself had wondered if that might be a factor.

As he later said: "Our patient was luckier than others who have used various liquid protein diets. A significant number of patients on three hundred to six hundred calories a day of protein-hydrolysate have developed fatal arrhythmias. After extensive weight loss (average ten kilograms or twenty-two pounds), a number of these patients died while they were on the diet or within two weeks of their resumption of regular eating habits." The fatal sequence of events was set in motion by the extreme need of the body for essential amino acids. Because the body was not getting these protein building blocks from the diet being followed, the body began scavenging itself, dismantling its own tissues to get these necessary substances. This process directly contributed to the chaotic heart rhythms that caused death.

Even though the U.S. Food and Drug Administration has required warning labels on such dietary products, these preparations are still available and are especially risky for pregnant women, nursing mothers and those taking medications.

Vikki Chong didn't like the thought of regaining the weight she had so dramatically shed, but she did accept the diagnosis, unwelcome though it was to her. She was admitted to the Emergency Ward, observed to have no cardiac disturbances and was later released to resume her usual routine. She had no further incidents of fainting, but did regain some of the weight she had lost.

As a result of that and similar cases, Goldfrank has made it a particular emphasis in his teaching and writing to urge his fellow Emergency Department physicians to be alert to the great potential for harm arising from fad dieting techniques.

Diets that emphasize such things as thyroid extract, amphetamines and phenylpropanolamine (an amphetamine relative) all bring unbidden complications in their wake. They can cause anxiety, restlessness, irritability, tremulousness, nausea, and a racing of the heart and lungs that can lead to sharply increased blood pressure, seizures and central-nervous-system disorders. Starch blockers, one-food diets and liquid protein diets all arouse Goldfrank's concern and skepticism on similar grounds. In fact, he has set down in his book a roster of ten conspicuous tenets of "Nutritional Nonsense":

It is *not* true that . . .

- Foods grown with organic fertilizers are better than those grown with inorganic fertilizers.
- Natural foods cure while additives poison.
- Daily vitamin and mineral supplements are necessary; "natural" vitamins are better; megadoses of vitamins are better still since they prevent and cure diseases.
- Healthful foods are only found in "health food stores."
- Processed foods lack nutritional quality.
- "It's not the pancake that is fattening; it's the syrup."
- Salt is almost as strong a drug as sugar, "because it is such an intense stimulant."
- "B-15" (pangamic acid) and "B-17" (laetrile) are vitamins.
- Most common diseases are caused by nutritional deficiency.
- Hair analysis allows for heavy-metal screening of bodily needs and creation of a nutrition-deficiency profile.

At the end of the day Sharda McGuire seeks out Goldfrank to tell him she is resigning. He is genuinely shocked, but understands when she says that the pressures, the overwork, and the offer of a better job closer to home on Long Island have all weighed in her decision. She will work through the end of the month, just a few days more than two weeks. In fact, this is the second such announcement: Carey Le Sieur has also given her notice.

Head nurse Pat Kunka has been at Bellevue for nine years. She came east (from Pennsylvania) with her husband when he accepted a position in New York. Now Kunka is taking a moment off, sitting in her office having a cigarette and cup of coffee and telling one of the new nurses some of her favorite stories. "I'll never forget the day we had two men brought in here from the Empire State Building. One had a broken leg, the other a heart attack. We saw them both. The MI (myocardial infarction) was the worse, really, and the broken leg was fixed up and we had the patient referred to Psychiatry within an hour. The other one remained in the Emergency Ward and then upstairs in the coronary care unit for, I think, a week."

"Psychiatry?" the new nurse asks.

"Yes. The man with the heart attack was an accountant, working on the eighty-fifth floor of the Empire State Building. He was bent over his figures, I guess really concentrating, when the window right next to him shattered. You see, the other man had jumped off the Observation Deck, and the high wind blew him right through the window on eighty-five, and he landed right on the accountant's desk! Gave the accountant a heart attack but only broke the jumper's leg!" Both nurses shake their heads and laugh. "Just another day in the Emergency Department at Bellevue," says Kunka.

# 23

# About AIDS

It's a sexually transmitted disease that was unknown ten years ago. It's a new kind of virus that has come, like an alien invader from space, to devastate its human victims. It used to be called GRID—for "gay-related immuno-deficiency." Then it was observed that the disease did not limit itself only to homosexual victims. Heterosexual men as well as women and children victims also began to show up at Bellevue and at other major medical centers. So the new disease was renamed AIDS—"acquired immune deficiency syndrome." By whatever name, it represents one of the most implacable foes ever encountered by the human biological organism.

Antibiotics can kill bacteria. Nothing kills viruses. The best that can be hoped is that the body can fight off viral infection by marshaling its immune system and the antibodies it produces to rebuff the invader. But this new invader is so stealthy that it attacks first of all this very immune system, so as to deny the body its normal defensive powers.

The invader's clandestine style is truly frightening. There are no early warning signs. Some victims, shortly after infection, have a spell of illness that resembles mononucleosis, with swelling of the lymph nodes and generalized aches and pains. But there is nothing more ominous. No chancre, no ulcerating lesion, as with syphilis. No genital distress, no burning sensation, as with gonorrhea.

Then health returns, and as many as five years may go by. During each of those sixty months—or more—the sexually active adult may have encounters with as many partners as his proclivities and persuasiveness may provide for him. Each, all unknowingly, can be infected by him.

By the time that a man or woman has AIDS, it may be impossible to reconstruct the chain of infection. In many cases there's no way to alert those who may have also been infected to warn them that now they, too, are passing on a deadly virus to new and unwitting victims.

More of these victims have presented themselves for treatment at Bellevue than at any other hospital. A visitor walking into the main portal of the Emergency Department any day or any night, at virtually any hour, might well encounter a scene like this:

Two medics enter guiding a wheeled stretcher that bears a thin man, breathing in shallow draughts, sweating, unable to sit up, perhaps with an oxygen mask strapped over his nose and mouth. There's someone with him, holding his hand as his stretcher waits just inside the ambulance doors—a friend, a lover, or perhaps a strong-minded relative. The patient has been brought to the place of last asylum for the indigent sick. And now he is both—impoverished and mortally ill.

When the chart work is begun, the medics recede, finished with their part of the task. The stretcher is pushed through the double doors and down the corridor past the doctors' station, then the nurses' station, and into Room 5, where a nurse and an aide help the patient out of his street clothes and into a hospital gown.

A doctor appears in a few minutes. There is blood sampling, a brief examination and x-rays. The friend asks what the outlook is. Not good, the doctor has to say, in all candor.

Until 1981, before the discovery of the virus, physicians working in New York and California were themselves first amazed and intrigued by the sight of previously healthy men—in those days exclusively men—who were turning up with rare kinds of pneumonia or the unusual Kaposi's sarcoma skin cancer. Equally perplexing were those appearing with thrush (a fungal infection of the mouth typical of infants, not adults) and rapid weight loss.

No matter what the doctors did, their patients died. Then women began to show up with the same disorder. And then it was noted that once established, the illness seems to lead inexorably to an early death.

There are more than fifty AIDS victims at Bellevue on any day, including as many as a half-dozen children, mostly infants, who have acquired the disease from their mothers.

As the disease has matured, many of its victims have contracted it as a result of dirty needles used in intravenous drug administration. Goldfrank has learned from the addicts he has treated that friends of drug addicts who have AIDS are especially vulnerable because the drug-taking ritual often requires the friend to receive an injection of heroin or cocaine mixed with blood as a token of devotion to his or her partner.

The addict "boots" his syringe with liquefied heroin, which has been heated and blended with water. This is in the chamber as the addict stabs the needle into a vein and pulls out the plunger to draw up a quantity of blood, which then mixes with the heroin. As the mingling of blood and drug takes place, the needle remains embedded in the vein.

Then, as the plunger is pushed forward, the addict receives the self-administered dosage of the drug. When the chamber is perhaps still half full, the needle is withdrawn and the remaining part of the mixture is then injected into the partner's vein. This widespread practice is blamed for most of the contagion of AIDS that now occurs in New York, where there is an addict population estimated to be 250,-000.

San Francisco actually has a higher rate of AIDS, but in that city the disease is almost entirely confined to the homosexual population, with significant consequences in the costs of treatment. It is far easier to marshal volunteers to lighten the burden among the gay community than among the IV drug-abuser community. It has been widely observed that the gay community has a much stronger social bond than that prevailing among the often antisocial IV drug users.

At a symposium of the National Association of Public Hospitals at Bellevue, an intensive review of the current knowledge about AIDS made it clear that the word "plague" was not overstatement, and that the outlook for the rest of the century is so bleak as to rule out most optimistic estimates. The disease is now threatening millions of Americans regardless of age, sex, race or place of residence. So long as there is a pattern of active, multipartner sexuality or intravenous drug use, there is a catastrophic likelihood of infection in millions of people. The real total of AIDS may be significantly higher than any estimates now current, because of the reluctance of some doctors to diagnose their private patients.

There are strong indications that AIDS is a tiger out of its cage, and that it is now prowling among the general populace. Often the first awareness that a patient has of his dilemma is the onset of one of the opportunistic infectious diseases with the symptoms that are so commonly seen: persistent fever, cough, weight loss. In Africa, they call the disease "slim," because that is one of its most visible symptoms, the emaciation of its victims.

Other symptoms include diarrhea, general weakness, chills, bruising, vomiting, memory loss, painful rashes, confusion, lack of coordination, incontinence, shortness of breath. But not all AIDS victims arrive at the hospital with such advanced symptoms. Some still have the outer signs of vigor and health.

One of the alarming discoveries about this implacable disease is that it seems to have a particular affinity for the brain, and to affect as many as 70 percent of its victims with some kind of mental problem—ranging from dementia to multiple-sclerosis-like disorders (and perhaps other central-nervous-system syndromes that have not as yet been given names). The virus has been detected attacking brain cells as well as special cells that make up the surface area of blood vessels in the brain, and that are a part of the blood-brain barrier. The virus gets past this barrier, which most medication can't penetrate. That is yet another of its wiles.

Dr. Christopher King, despite his youth, speaks with the assurance of someone who has had to do this more than once. "Just strip down to your shorts and have a seat here, if you will," he asks the patient in Room 3, a well-built young man dressed in a tweed jacket, flannel trousers and tasseled loafers, who looks a bit raw-boned and apprehensive.

"You've already had blood drawn?"

"Yes."

"Good."

King grew up and was educated in Michigan. He came to Bellevue after finishing medical school at Ann Arbor. "For my interest in infectious diseases, Bellevue has been the ideal place. You actually get to see more varieties of disease, more different kinds of patients, than, I'd guess, at any other hospital in the country." When his residency is completed, King will move on to join the staff of the National Institutes of Health in Bethesda.

King begins his examination. The patient is clearly apprehensive. He's thirty-one, divorced and not used to illness. His name is Noel Barcomb. He's a skilled graphic artist and typographer who makes a good living as art director for a group of trade journals. In his leisure, since his divorce, he has been spending a lot of time in the bathhouses and shooting galleries of Greenwich Village.

Shooting galleries, virtually unique to New York City, are rudely furnished apartments or storefront quarters where a drug addict goes to be among his fellows in a setting of congregational drug taking. There he routinely shares needles with his fellow addicts. For a fee he may even have one of the more practiced members of the group administer a shot into an otherwise unreachable vein in a body that has been punctured so often that it is hard for the addict to raise a vein for himself (as most have shrunk back in protest at the repeated violations by the needles). It is estimated that there are nearly a thousand such shooting galleries in the metropolitan New York area.

Having sent Barcomb for a chest x-ray, King goes down to the film viewing box at the doctors' station. There he has to apply pictorial skills as impressive in their way as those that Barcomb employs each month in laying out his magazines.

Gazing into the blue-white light of the viewing box at the transparency of Barcomb's chest, King analyzes what, to the layman, seems a hopelessly confused jumble of spectral imagery—the air-filled lungs (looking very dark), the blood-suffused, muscular heart (appearing bright), the calcium-matrixed calipers of the ribs (brighter, still) and any number of fine details that are conveying information so subtle as to escape all but the expert eye.

Like everyone who has gone through four years of medical school, King has spent long hours familiarizing himself with chest structures and the way they appear in a conventional x-ray. These shots are taken as the patient faces the film, and the camera is positioned behind him—the PA or postero-anterior chest film. King is skilled at comparing structures appearing in that view, and in its contrary (the AP or antero-posterior, which is confined to those too sick to stand up and is taken as they remain abed). Also he knows the peculiar perspectives of a side view, and has built up a fund of knowledge from this variety of viewpoints that permits quick identification of anomalies, such as King is now seeing in Barcomb's lung fields. There are areas of bright white, like granular clouds, against the dark background of normal lung tissue. These he will enter on Barcomb's chart as "diffuse infiltrates in both lung fields" consistent with a virulent kind of pneumonia that has increasingly come to be typical of AIDS.

Now the doctor will have to walk the few steps to Room 3 and tell his patient what he has seen: probable signs of *Pneumocystis carinii* pneumonia. At first Barcomb had denied being a recent intravenous drug user. He maintained that he had used methadone for several years. But then he switched his story and admitted that yes, even as recently as yesterday, he had injected heroin. But that was the first time in months.

Barcomb has a temperature of 102, and his respiratory rate is tachypneic (speeded up) at thirty-four breaths a minute. His lymph nodes are swollen, and he suffers oral thrush—a milky-white fungal growth on the interior walls of his mouth. Besides, he is having difficulty swallowing. The laboratory figures, which arrive as King is reassessing Barcomb's pulmonary exam, show both white-blood-cell and platelet counts well below average.

The resident leans back in his chair at the small examination-room medication cabinet–writing desk. "You've got what looks like a serious diffuse infection in the lungs, Mr. Barcomb. My best bet is that it's something we call pneumocystis."

"That's pneumonia, isn't it?"

"Yes. It could also be TB."

"I see," Barcomb says. His voice is firm, despite his labored breathing. "I've read that pneumocystis is often associated with AIDS."

"Yes, that's true."

"Then perhaps I have AIDS?"

"It's possible. But we'd like to do some more tests."

"Such as?"

"Such as a bronchoscopy—and possibly a transbronchial biopsy of the lung." King explains to a worried-looking Barcomb what this will entail.

Shortly after he is admitted to Medicine, a pulmonary expert will lay Barcomb on his back and, after sedation, insert into his throat a bronchoscope, a hard-rubber-coated, fiber-optic, tubular implement with a small light bulb at the far end to facilitate visualization of the airways.

When King sees him again several days later, Barcomb says evenly, "My resident says I do have AIDS."

A heavy vehicle is moving somewhere close enough to cause a vibration to run through the curtain wall of the building and into this sixteenth-floor room. "Yes, he told me, too."

Noel Barcomb casts down his eyes and makes a wry expression. Then he looks up and meets Dr. King's gaze, sniffling and coughing as he does so. "So what do you think it means? How long do I have?"

There's obvious concern in King's eyes, and his words come with gentleness. "AIDS isn't just one disease. It's a condition that gives opportunity to an entire constellation of diseases."

"Yeah, I've heard that."

"And there's a lot of research going on. New discoveries every day."

"So there's reason for hope—some hope?"

"Sure. You could well have a few years. And by that time there could be a breakthrough in treatment.... Even the new AZT may pan out. We just can't tell, so much is going on. You have a fine team of doctors here. And already the management of your pneumonia is improving—giving your immune system a chance to regroup."

There's talk at morning rounds of the two-page advertisement for condoms in *New York* magazine. "It's something I never expected to see," Neal Lewin said later. "I've seen AIDS, now, for six years, since 1980, and its impact keeps spreading."

Lewin's private practice in Murray Hill, some ten blocks from Bellevue, claims his afternoons. "I've lost a half-dozen of my patients," he says, "and every one of them has left me feeling emotionally devastated. The first was an artist, an unusually gifted young man in his twenties. He had been following a self-destructive pattern, abusing alcohol and drugs and living a sexually promiscuous life. He developed an unusual type of pneumonia and then a central-nervous-system infection. I just felt baffled, because there had never been anything quite like it before.

"I had tried to warn him, a year before, that he was living too hard, was jeopardizing his health, but I didn't make much of an impression. That was the first case of AIDS I saw. And it was still being called GRID back then. He was a very talented man, and a real tragedy, because he was dead within just a few months.

"Then, in rapid succession, three or four others in my private practice died of various opportunistic infections. One day they were ill, and within months they were gone—slipped away. Of course today we can try AZT or interferon, and sometimes get dramatic remissions—for a time, anyway. But back then there was just nothing. I began to feel that I understood how it must have been in the fourteenth century when the plague swept through Europe and Asia.

"Then slowly we got more information, which helped explain

some of the devastation I was seeing in vigorous young men suddenly enfeebled and overwhelmed both in mind and body by AIDS. But, despite all the research findings, we're still up against a horrible disease. And it's caused by a virus. That's the kind of infection that medicine has never had much luck with in the past.

"And now there are three distinct AIDS viruses, and we have to face up to the fact that we have the disease of the decade—or perhaps the century—on our hands." What to do? "From an epidemiological point of view, we should test everyone. But clearly there are ethical problems there. So we go slow because anyone testing positive is immediately stigmatized. Mandatory testing will result in the branding of many people—will make them modern-day lepers.

"Today I encourage any of my patients in a risk group to get tested. Just yesterday, in fact, one of my patients, who is a prostitute, was shocked to hear me tell her that she was vulnerable to AIDS infection. She is in a risk category and could spread AIDS to her clients—her johns. The evidence is beginning to confirm that there *is* heterosexual transmission, from females to males, as well as the other way. And, as many prostitutes in New York are IV drug abusers, they fall into two risk categories—sexual promiscuity and IV drug use. This is not a fantasy scenario. This is what is actually happening. And to my mind it's only a matter of time until there's universal testing. It's become that much of a threat to public health."

Not all share the somber expectations that predominate among the staff of the ED. Another point of view has been advanced by Dr. Gerald Weissmann, who is professor of medicine at NYU-Bellevue. Weissmann thinks there may well be grounds for optimism. In a collection of essays, Weissmann has written that there are historic precedents for the fear and alarm that have greeted the sudden advent of AIDS:

"In its infancy, syphilis presented a much more virulent and immediately contagious clinical picture than in succeeding centuries." So horrible were its various symptoms—pustules, ulcers and actual loss of parts of the nose or throat—that people abandoned their stricken friends in terror and revulsion. "But in a short fifty years, the disease changed its clinical presentation. By 1546 . . . the painful, ulcerative disease had yielded to a chronic, less devastating illness."

Weissmann adds,

The history of infectious diseases presents many similar instances of amelioration over time in the absence of specific therapy. Leprosy, a dreadful and commonplace ill in the medieval period, became—for reasons that remain obscure—remarkably less crippling as its prevalence waned. . . .

The consequences of scarlet fever, which had decimated the children of the Industrial Revolution, were already diminished in severity by the end of the nineteenth century, and the form of typhus known as Brill's disease—which was found among Jewish immigrants of Polish or Russian origin in New York—bore only faint traces of its epidemic precursor in Eastern Europe.*

King is sitting at the doctors' station, responding to a visitor's question. Yes, he still worries about AIDS. He thinks of it every time he gets a sore throat. "I mean, we're supposed to be inured to something like that. To stabbing yourself with a bloody needle from an AIDS patient. But it happens. You get sore throats and you stick a needle in yourself by accident. And there've been, I don't know, three or four people who have suffered finger sticks and then developed pre-AIDS antibodies as a result. But so far none has died.

"When I get home my wife says, 'Oh, you're home, take a shower!' I can't blame her. It's a bit unsettling, sometimes. But, on the other hand, working with the acutely ill—life or death issues—you cut out the superficial, the bullshit. You get close to people who are gravely ill and you're trying every way you know how to help them to get better. It deepens your humanity. And each death you feel personally. The first was devastating. But the ones that follow don't get much easier. You share things with strangers—these patients you are seeing—that you otherwise wouldn't."

From the sadness in the young resident's expression, it's clear that he doesn't give Barcomb much of a chance. Now he heads for a quiet corner to catch up on his chart work.

While the Weissmann view offers hope for the future, in the present the short history of AIDS is well known to everyone in the Bellevue Emergency Department. And, like combat dispatches, news of the plague is disseminated almost as soon as it is known, through a series of "Emergency Medicine Reports," originating in San Francisco; "City Health Information" bulletins from the New York City Department of Health; and the "Mortality-Morbidity Weekly Report" of the U.S. Centers for Disease Control in Atlanta—as well as every leading medical journal from the *New England Journal of Medicine*

*The Woods Hole Cantata: Essays on Science and Society, Dodd, Mead, New York, 1985, pp. 67–68.

to the *JAMA (Journal of the American Medical Association)*. The story they continue to chronicle had its beginning in the seventies, first was identified as a new form of virus in 1981, and has now entered into the general population, despite its heavy concentration, so far, among homosexual men and intravenous drug abusers who share needles.

The prognostications for the future are bleak. One spokesman for the Centers for Disease Control, Dr. Ward Cates, was quoted in *Newsweek* as saying, "Anyone who has the least ability to look into the future can already see the potential for this disease being much worse than anything mankind has seen before."

Other comments on the disease suggest why it has terrified so many: once you get it, you have it for life, and you remain a carrier for life. So far, it is estimated that two million adults have been infected with the virus, and that the virus responsible has become widely distributed now among men, women and children in many nations, in just a few years. Until recently, the virus bore two names: in France, LAV (for lymphadenopathy-associated virus); and in the United States, HTLV-III (for human T-cell lymphotropic virus, type III). The virus now has a new name, ecumenical in its purpose, so as to dampen the unseemly rivalry of the two competing research groups that independently isolated the virus. The new name is HIV—human immunodeficiency virus. And only recently has its method of operation become clear.

Each of your cells, like a small garrison, has its own police force. Defending the cell's perimeter is its top priority. Let any unauthorized bacterium or virus enter, and this security force springs to the alert, intercepts the interloper, takes him out to the edge of town and kills him—by surrounding him and attacking with tactical chemical toxins. Then they throw the "corpse" into the river—your lymphatic system and bloodstream—where it is carried off and disposed of. But the stealthy AIDS virus creeps over the wall of the garrison and captures one, and then another, of your security forces—the so-called "helper T-cells." These cells normally act the part of a police auxiliary that is especially adept at alerting your immune security forces to the incursions of unauthorized bacterial or viral elements. But HIV takes over these helpers and converts them into active collaborators with this alien force. HIV seems to alter its form continually, all the while taking over more and more of these auxiliary helper cells. Eventually your entire security force is ridden with traitorous elements that have for their first order of business increasing their own numbers, and for their inevitable effect, the subversion of your whole garrison.

Now your cells are without adequate defenses against invasion. And even the most easily rebuffed forces in normal times now come pouring over the palisades. In swarm fungi, parasites and viruses. In come things with names as menacing as their purposes: crypto-sporidium (secret spore), cytomegalic (big-cell) viruses, pneumocystis (lung parasites), and normal saprophytes (putrid plants), the "housekeeping" bacteria of the intestines. In addition, there is candida albicans (white on white), a fungus that causes oral thrush and apparent leukoplakia—a fuzzy excrescence at the edge of the tongue; "cotton-wool" spots in one's field of vision; and varied changes in mental status. For the brain seems to be especially vulnerable to the invading HIV virus as well as its many undesirable camp followers. Soon the odds are quite hopeless, and a lingering war of attrition begins, in which the defenders exhaust themselves against a relentless horde of invaders.

Among all the health-care workers at Bellevue, there is close knowledge of the box score, so far, on AIDS among their peers, as presented in a well-known article in *JAMA*, by thirteen co-authors, led by a group from the National Cancer Institute. There, 361 health-care and laboratory workers from various metropolitan areas (including Washington, New York and Boston) were studied. Among the subjects there was a small minority who had manifestations of the HIV virus, presumably from needle sticks. There was evidence of heterosexual transmission—female to male—in the case of a nurse who had no symptoms but nevertheless had the AIDS virus. She suffered two puncture wounds, without injection of blood, but with needles that had been used on AIDS patients. Her long-time sexual partner, who also had no symptoms, nevertheless was found to harbor the virus. The case was thought to confirm female-to-male infection, "a route of transmission strongly implicated in other studies," the *JAMA* article said. Then other studies appeared in the British medical journal *Lancet* and the *New England Journal of Medicine*, reinforcing these findings.

A year after the *JAMA* article, there came another piece that was closely read by Goldfrank and his staff, and discussed among them. It was by Dr. William S. Howland, Chairman of the Department of Critical Care at Memorial Sloan-Kettering Cancer Center in New York. Dr. Howland offered a "primer" on AIDS as well as "AIDS

Guidelines for Health-Care Workers." Among the points in his
"primer": "The incubation period for AIDS is long. In one study of 18
patients with transfusion-associated syndrome (TRAIDS), the median
time between transfusion and diagnosis was 28 months. . . . However
the development of AIDS or AIDS-related complex after transfusion
has been reported to be as short as 7 weeks and as long as 5.5 years."

   The Sloan-Kettering "Guidelines" urge health-care workers to
take precautions similar to those they would take with hepatitis patients—avoiding all body fluids, keeping themselves gloved and
gowned, putting patients on "enteric [intestinal] precautions," because of that possible source of infection. All reusable items should
be carried in clearly labeled bags; all contaminated linen is to be
double-bagged; needles and syringes go into puncture-resistant containers. And "when an AIDS patient is to have surgery . . . the label
'CAUTION—AIDS' should be used . . . and the room should be treated as
it would be with a contaminated case."

   In the presence of such a plague atmosphere, it seems remarkable
to the visitor at Bellevue that there is hardly any evident expression
of apprehension. One attending physician did voice concern that the
high proportion of AIDS patients in the hospital was diminishing the
variety of disease that young doctors could learn about. And the result
was that the pool of intern applicants was thought by some to have
become less impressive than it used to be. But others hotly contradict
such views.

Dr. Kevin Smothers grew up in a professional family on West End
Avenue in Manhattan, went to Brown University and then to the
Medical School of the State University of New York at Stony Brook.
He did his residency at Mt. Zion Hospital in San Francisco, and then
came east to become the first physician at Bellevue assigned specifically to develop a program and assemble a staff to confront the emerging plague of AIDS in a pre-hospital setting.

   Smothers describes what must have been a difficult year: "AIDS
continues to alarm almost everybody. But it's the 'worried well'—
people who are vulnerable but as yet uninfected—whom we see at the
clinics that I run. In the past, these people may have been homosexu-
ally promiscuous, or even shared needles with drug users. Now they
realize that, even though they may have changed their lifestyles, they
are still living with a possible time-bomb inside because of the long

incubation time of AIDS. Often it takes more than three years after exposure for the disease to develop.

"We have two clinics now—one here at the hospital in the Department of Ambulatory Care and the other in Greenwich Village."

As Smothers explains it, the clinics attempt to encourage changes in diet and in sexual behavior without appearing to be judgmental. "Apprehension can do more to disable people than you'd believe," he says. Those who fear they may have AIDS or the so-called ARC (AIDS-related complex) oftentimes just fall apart emotionally. They find they cannot sustain a normal life any longer. They stop going to work. They break off their friendships. "The greatest disease we see is fear of AIDS, which itself may increase vulnerability to the actual disease, in some way."

In the year that Smothers had been developing the clinics, his staff had grown to include two other doctors, nurse practitioners, social workers, a health educator and a clerical worker.

Cost to a patient for a complete work-up ranges from $7 to $42. It includes a complete physical exam, counseling with a social worker (who does a kind of psychological triage) and skin-testing the body's immune system by means of small mosquito-bite-like needle sticks with various antigens. Two days later a patient returns to learn the results of the various tests. If there's no obvious sign of swelling or inflammation at the sites of the antigen injections, that means there are no lymphocytes surrounding the deliberately introduced antigens—a sign of a weakened immune system.

"We have to be especially sensitive to the patient's needs," Smothers says, "because of the tremendous stigma attaching to AIDS. People are losing their insurance, their employability. Their principal economic assets are immediately at risk. So we do everything we can to defend the patients' rights.

"Sure, people die. But what are the things we can do to focus on life? How can we make the best of the situation? How can we help a patient get his psychological defenses as well as his physical defenses up and working? That's what we spend our energies trying to do."

The Health and Hospitals Corporation and the Ambulatory Care Department at Bellevue have approved the new budget requests that Smothers has drawn up for the next fiscal year, reflecting his solid success in establishing a climate of trust with a chary populace of prospective patients. These "worried well" fear the utter ostracism that many of their friends have suffered when it becomes known that they have AIDS or are merely worried about having it. Each month, more walk-in clients have been coming into the clinics than in the previous month. That seems to be clear confirmation that word of mouth is running strongly in favor of the new outreach effort.

Now, nearly a decade after the first signs of the growing plague surfaced in Africa, then were noted in Haiti, and then in the United States, the disease has spread to all fifty states and nearly as many foreign countries. Of the acknowledged cases of AIDS, more than half have resulted in death. And of the nearly two million Americans who have been infected with the virus, as many as 30 percent of them may be expected eventually to develop the disease—by the early 1990s, at the latest.

Furthermore, the rate of its occurrence has not waned in the four centers of infection: New York, San Francisco, Los Angeles and Miami. Nor have the new occurrences spread significantly from the principal risk groups—homosexuals and IV drug abusers—despite the dramatic shift in homosexual practices toward "safer sex." (Rates of rectal and throat gonorrhea in Manhattan have dropped nearly two-thirds since the early eighties. In San Francisco, there's been a 75 percent drop in venereal disease among homosexuals.) Still, the frequency of infection is expected to persist in its rising trend for some time yet, because the incubation period for the HIV virus can be so long.

Among the cruelest of AIDS' exactions is that imposed on the children of AIDS-infected women. Contrary to widespread belief, however, not every child born of such a mother is bound to become infected; in fact, the rate of such perinatal transmission varies from about a quarter to about two-thirds. And there is one report of a child, who became infected via a transfusion, subsequently infecting his mother, who did not wear gloves or always wash her hands after coming in contact with the baby's body fluids.

As if a culminating paradox in a disease that has proved a preternaturally clever adversary, AIDS is said to be difficult to contract because the virus is not robust. Any temperature above 158.6 degrees F. kills it. And it is unable to survive for long away from a warm and fluid environment. So, in the near future, when doctors everywhere will be obliged to confront this disease among their patients, there is solace of a sort in the emerging certainty that those who give care, like the doctors and nurses at Bellevue's Emergency Department, are not at risk for infection to nearly the degree that it was once feared that they were.

✦

A new issue of the *Johns Hopkins Magazine*\* is passed around at Bellevue, with a section highlighted as an especially good summarization of the kind of educational message that ought to be imparted to as many patients as possible. It's headed "Ten Rules to Protect You and Your Family from AIDS":

Stay in a mutually faithful relationship, in which neither partner uses intravenous drugs.

Use condoms to prevent exposure to semen or blood. . . . Even if both partners are already infected, condoms are essential because they prevent reinfection, which may worsen the prognosis.

When entering a new relationship, remember that it is seldom possible to be sure what someone else—and all of his or her previous partners—has done since 1977, when the AIDS epidemic began. You do not have to be a homosexual man, a prostitute, or a drug addict to get AIDS.

Remember that most people who carry the AIDS virus have no apparent symptoms and probably do not know they are infected.

If you have reason to think you might be infected, tell your partner. Make sure your partner is honest with you. Infected people need not lead celibate lives, as many fear; but they and their partners do need to be very careful, observing the precautions spelled out above.

If you think you may be infected . . . do not donate blood to find out. Ask for the AIDS hotline to find out where your community testing site is located.

Tell your children: "AIDS is a new disease, AIDS kills, and there is no cure. . . ." Be sure your children are aware of AIDS and condoms long before you think they might become sexually active. The Surgeon General urges schools and parents to begin educating children about AIDS "as young as possible."

If you shoot drugs, don't share needles. Don't share syringes.

If AIDS comes to your family, remember that you can only get AIDS if you have unprotected sex with this person, or if you share a needle or syringe with the infected person. Hugging and caring for the person do not transmit the infection.

\*December, 1986, "AIDS: Just the Facts . . . from Specialists at Johns Hopkins," by Ann Finkbeiner, Elsie Hancock and Susan Schneider, pp. 15–27.

An infected woman who gets pregnant is risking both herself and the baby. If you get pregnant and think you may be infected, consult your doctor immediately.

$$\sim\!\!\!\wedge\!\!\!\sim$$

It is midnight of a chilly October Friday and Goldfrank is through for the week. As he signs the last of his week's correspondence and clears his desk of the accumulated paper, he gathers up a stack of journals and reprints on various toxicological and medical subjects, puts them in his briefcase and then rises to exchange his white clinic coat for a tweed jacket. He has a 12:20 train to catch at Grand Central, so sets out briskly to make what would normally be an eighteen-minute walk. As he locks the door of his office, he gazes one last time at the waiting area opposite the triage desk. He sees a half-dozen patients waiting. Four of them are young men, looking gaunt. One of them is breathing with obvious difficulty. Two others are coughing—one with a desperate barking intensity. The fourth young man, sitting by himself in a rear seat in the corner, has plum-colored splotches visible on the exposed neck above his crew-neck sweater. Each of them, from the doctor's practiced gaze, doubtless harbors the HIV virus. He sadly turns toward the lobby and makes his way swiftly down the corridor toward First Avenue. The night crew—guards and orderlies, elevator operators, a member of the Housekeeping Department running a mop near the coffee machines—wave and nod as he goes by, making his way toward the light drizzle that is falling on the still-busy streets outside.

# 24 A Peanut and a Mercury Injection

Elizabeth Swanson is about to carry out a threat. She comes back to the triage desk, picks up the phone and begins to dial 914, the Westchester area code. She glances up at the clock. It's nearly 10 P.M. A harried-looking blond man in a white jacket bursts through the double doors from the Adult Emergency Service and leans over the counter toward her. "All right, Swanson, all right. I'm just overworked. We'll admit the patient." Then he turns and goes back through the double doors, in something of a huff.

Swanson returns the phone to its cradle, nods toward the retreating form of the resident and then walks over to the waiting area to a patient. She begins taking the vital signs of a short man named Abdul Yusuf, a thirty-five-year-old of swarthy coloring who complains of having just passed out.

The quiet contretemps with the doctor involved an issue that especially annoys Elizabeth Swanson: the occasional cold indifference that young residents show to patients—particularly if the patient is helpless, homeless, annoying or hostile. This case involved a prisoner who had been brought in by the police from Central Booking, where he was being held on a felonious-assault charge.

As Swanson later recalled, "The prisoner hadn't been taking his medication—Dilantin—for seizure disorders, so maybe he was a little whacko, but when we saw him he clearly had two broken arms and this young resident treats him for a seizure disorder and then is going to release him back to the custody of the police. I mean the man was being really obnoxious, but still, two broken arms?!

"So I asked the doctor, 'Doctor, shouldn't we keep him until we're sure we've organized some way for him to eat and care for himself?' And he tried to brush me off because we were so busy. And I just began to do a slow burn. I've worked in four different emergency departments, and this is by far the best of them all—and for one reason, I think. Because Dr. Goldfrank is a patient advocate himself. He really practices what he preaches. And he backs you up—a nurse, a nurse's aide, it doesn't matter. If you go to him and tell him something isn't right in the way we're treating a patient, he backs you up. Even if you call him in the middle of the night. So I said to the doctor, 'I guess I'll just have to inform Dr. Goldfrank that we have a patient here with two acute forearm fractures who's about to be sent back to Central Booking!'

"For a moment it looked like he didn't think I'd do it. That's when I walked out of there and came to the desk and started to make the call. Most places such a threat wouldn't mean anything. Here it does. Anyway, the prisoner was admitted. Tried to spit at me later, that's how obnoxious he was. But you've got to try to look past that when you're in this line of work."

Ang Vacharasiriyuth, a twenty-four-year-old Thai, is brought in by taxi, his right hand wrapped in a blood-soaked kitchen towel. He is seen immediately by a resident, who gives him 0.5 cubic centimeters of diphtheria-tetanus toxoid and 250 units of Hyper-Tet medication (human hyperimmune gamma globulin against tetanus). Then the resident sutures up the badly lacerated thumb with a few deft stitches of monofilament nylon.

Vacharasiriyuth is a kitchen helper at a food shop on East Forty-fifth Street, who cut the base of his thumb on a slicing machine. After the stitching and bandaging, it is decided to keep him under observation for an hour, checking vital signs to make sure of a satisfactory response. Then, just after midnight, the patient is released and told to come to the suture clinic in seven days to have his wound checked and the stitches removed.

As the paramedics later got the story, a young heir of a prominent family had gone with some friends to a steak house in Chelsea, in the West Twenties, and was reminiscing with them about his recent trout-fishing holiday on the San Juan River in New Mexico. Lamont Saterlee was twenty-five; his younger sister Deirdre was with him.

Monty had something of the kamikaze in his personality. As he well knew, he had an acute sensitivity to peanuts. Over the past ten years, since an incident at a family vacation lodge in Colorado when he was just fifteen, he had had a number of violent reactions to salted peanuts, peanut brittle and peanut oil used in Chinese cooking. These reactions involved intense itching, wheezing, swelling of the nose and acute asthma-like breathing attacks, each just a bit more pronounced than its predecessor. He had had numerous other life-threatening allergic reactions also, and carried a self-injecting syringe of epinephrine with him on his allergist's prescription. And yet he was testing himself again.

He reached across the table to a bowl of mixed nuts, carefully segregating the peanuts from the cashews, pecans and walnuts. His sister tried to grab the bowl from him. And at that point he picked up a peanut and held it poised to his mouth.

"Monty, you better not! You know how sensitive you are!"

He ignored her, while talking with his dining companions, and playfully took a bite of the peanut.

"Monty, this is quite mad, you know!" his sister said, in alarm.

Almost immediately, his body started to rebel. Within moments he started to wheeze and to gulp for breath, and his skin grew discolored with red blotches. His sister abruptly ran for the maître d' and urgently asked him to call 911 and tell them her brother was having an anaphylactic attack.*

"What kind of attack?"

"Just say an allergic attack."

At the table, Monty Saterlee was already at the point of losing consciousness, so swiftly had his throat swollen shut. Now, with great effort, he was able to extract a small sip of breath through the encroaching and engorged walls of his trachea. From his pocket, he withdrew the small automatic epinephrine-injectable syringe that he carried with him, and gave himself 0.5 milligrams of the potent bronchodilator. Within a few more minutes he lost consciousness and tumbled to the carpet. His dinner companions could do no more than make him comfortable and wring their hands.

The call to 911 was routed directly to Maspeth, which almost immediately dispatched the nearest ambulance in view of the priority-one level of emergency such an attack presents. The luck of the

*The word in the original Greek meant "overguarding," from "phylaxis," a guarding, plus the intensifier "ana."

draw went against Monty Saterlee this evening, however. The nearest ambulance was staffed with emergency medical technicians rather than paramedics, so the degree of intervention they could effect was restricted by their more limited training.

When they got to the restaurant, Monty's skin was already cyanotic (blue) and he had only the threadiest of pulses. The medics attached a Venti-Mask with 100 percent oxygen and quickly removed Saterlee to their ambulance. Then they raced across town to Bellevue.

By the time he got there, some seven minutes later, Saterlee was in status epilepticus—a state of perpetual seizures, with his muscles flexing and relaxing at a regular rate. If permitted to go on, these continuous convulsions simply wear out the musculo-skeletal system much as a motor racing at unchecked speed can burn itself out. Death comes by hyperthermia—a heat-stroke-like state induced by these seizures and resulting in irreversible muscle destruction.

It was midnight at Bellevue when the loudspeaker announced an incoming respiratory arrest and status epilepticus. Saterlee and his distraught sister went right into Trauma Slot 1, where the immediate emphasis was on restoring breathing, which had ceased entirely. It was with the greatest effort that an endotracheal tube was inserted between the badly swollen—edematous—vocal cords. But at that time they were getting no blood pressure, no circulation.

A bolus dose of epinephrine was given and then an IV drip of that drug was begun, and that got Saterlee's heart started again. But there was possible—even probable—evidence of brain damage owing to the length of time with an insufficient supply of oxygen—"post-hypoxic encephalopathy." His pupils were fixed and dilated and it looked as if brain death were imminent.

After the epinephrine took effect, the lungs were clear, so the doctors followed with IV corticosteroids (200 milligrams of Solu-Medrol) to control the swelling.

At that point, nurse Sharda McGuire took Deirdre Saterlee outside to the waiting room, and accompanied her to the coffee machine.

Saterlee's body continued to jerk, in a dancelike "chorea"—bizarre movements that swept through the muscles, apparently at random, making the young man's motions seem horrifyingly clownish.

These chorea-form movements so alarmed the resident on duty that he called up Goldfrank, even though by now it was nearly 1 A.M. Goldfrank heard the recitation of events and the signs now visible and urged an immediate IV administration of Valium to calm down the involuntary muscle movements. He recommended administering whatever doses were necessary—even to the point of giving general anesthesia or paralyzing the patient with curare-like drugs—to stop this self-destructive motor activity.

"Because the patient had been suffering those movements so long," Goldfrank later explained, "he had a temperature. If the seizures and chorea-like movements continued, he would do himself irreversible cellular damage. You can't get enough oxygen into the cells in those circumstances. In particular, muscle cells are destroyed—rhabdomyolysis—and the destruction product, myoglobin (which is to the muscles what hemoglobin is to the blood), ends up going out into the bloodstream." These remnants of muscle tissue finally lodge, like oversize logs in a small stream, in the highly involuted nephrons, the filters for the blood. As the chunks of myoglobin cannot flow through these small-bore filters, there is acute tubular necrosis, death of these critical filters and ultimately kidney failure.

There was also the possibility that Saterlee might have had additional peanut fragments in his intestine, any one of which might possibly set off a further overreaction of the body. So he was given Sorbitol (a cathartic) and activated charcoal—the first to promote rapid evacuation of the bowels, the second to adsorb whatever peanut antigens or other toxins might be in his stomach.

Several other calls were made to Goldfrank during the night, as it became necessary to administer barbiturates to calm down the continuing myoclonic (twitching) reactions. But the patient had to be sedated so heavily that it became unclear whether any brain function remained. The question of whether he was indeed brain-dead as a result of his seven (or more) minutes without oxygen would have to be addressed after the treatment of the myoclonic seizures had been completed. Once that problem was solved, they would have to wait for the barbiturates to be metabolized. Only then could an electro-encephalographic reading be made to establish how much brain damage might have been sustained.

Through all of this, those involved with the case were moving rapidly forward in their understanding of anaphylaxis, in a way that no amount of book work or lecture time could have equaled. Later, many would read in the new edition of *Toxicologic Emergencies* the chapter on anaphylaxis, which describes in close detail how it is that the body can turn on itself with such savagery as had been seen that night.

As the book makes plain, this kind of attack can be triggered by a wide variety of foods, drugs, insect-venom proteins, sweeteners (such as polysaccharides), dyes and other things (such as human seminal fluid, cold and perhaps even exercise). Or, as Flomenbaum and Goldfrank have written, the three major categories are proteins, polysaccharides and haptens. This last is an antibody-forming substance that is a partial antigen itself, but needs to team up with a protein to generate an antibody response.

At the turn of the century in France, anaphylaxis was described in dogs that had had sea-anemone toxin injected under their skins in laboratory experiments. Then the dogs were reexposed to the toxin. Normally, the first injection, like a vaccination, should have set up defenses against severe further reactions. But the dogs were not protected. "Rather, they died cataclysmically," the book notes. It adds, "Human anaphylactic reactions were recognized in some patients undergoing 'prophylactic' (protective) injections with horse serum and similar materials."

There are many "shock organs" on which the body's own allergic defense mechanism rains its anaphylactic missiles. The mucous membranes of the eyes, nose, tongue, throat and vocal cords all can become enormously engorged with fluid. Such swelling can swiftly obstruct the upper airway so completely that, unless a timely tracheostomy or cricothyrotomy is performed, the victim is immediately pushed to the edge of doom.

Meanwhile, the skin can become as sensitive as if stung by myriad nettles (urticaria) and flushed with a rosy blush (erythema). Then there may be nausea, abdominal cramps and diarrhea, all reflecting general disturbance of the smooth-muscle system in the wake of the violent allergic reaction.

Anaphylaxis is one disorder close to Goldfrank's daily experience, as he himself is asthmatic, and well knows the beleaguered feeling of the victim of a sudden attack of bronchial swelling, during which every breath is labored and the lungs seem to ache continually. In cold weather, if he is in his yard chopping firewood on a weekend, he will sometimes be struck with bronchial spasm, of the sort that Lamont Saterlee suffered. But, fortunately, sometimes a glass of cold water, hot tea or—in more aggravated attacks—an inhalation of a bronchodilating mist will clear up the attack. Then there is the sudden dramatic clearing of the constricted airways, making easy what a moment before seemed so difficult.

By the time Goldfrank got to Bellevue the next morning, the Saterlee family lawyer was there seeking more information. Goldfrank had to tell him that the young man's condition was grave, that there was probably no hope, but that until the barbiturates had been metabolized they would not know for sure.

Later there were visits to Goldfrank from the patient's grandparents and sisters. For each there were further reassurances that the young man had enjoyed the best care, in the circumstances, but, sad to say, little could stand in the way of such a virulently determined allergic attack. They said they knew. Monty was always such a daredevil.

That was essentially how the case ended. Lamont Saterlee emerged from the secondary barbiturate coma, which had been induced in an attempt to save him. He remained in the primary, hypoxia-induced coma, which had been underlying the drug-induced one all along. The electro-encephalographic readings showed no brain function, flat lines. By definition, that is brain death, and the respirator support was terminated, the standard procedure at such times.

There's a new member of the staff at the morning meeting when the Saterlee case is presented. Maria Amodio is a darkly handsome young woman in her twenties who was born in Sicily and came to New York at age seven with her parents. She is now a graduate pharmacist within a few months of her Pharm.D. and is working on her research at St. John's University in cooperation with the Poison Control Center. In that capacity she is participating in morning rounds. Goldfrank likes to involve pharmacists in the clinical events of the Emergency Department, to deepen their understanding of the real-life impact of the substances whose chemistry and pharmacological lore are already well known to them. It's his view that it's never possible to have too much specific information about patients and the way their bodies handle various toxins.

Later, Amodio will explain how incomprehensible she still finds much of the American attitude toward life. "How can you just throw your life away like that? I just don't understand," she says, still brooding on the Saterlee case.

She also finds herself nonplused by the whole intravenous drug culture, with which she has become more closely acquainted as a result of a graduate study she is conducting on pharmacology and toxicology. She is gathering a number of patient histories and participating in examinations of aspirin users, many of whom have other habits far less benign.

"I know people suffer terribly. In high school I watched my own mother die from cancer, and the pain of it. . . . That's why I decided to study pharmacy, so I would know better about managing pain. I find it hard to comprehend how people can just destroy their own lives when they have choices. I used to try to allow for drug abuse, for a chance that people could get out of it somehow. One woman I talked to said she really wanted to quit using. She was divorced, became a prostitute to support her drug habit. Her family threw her out. Now she's worried about AIDS, and I tried to help her, to tell her how dangerous sharing a needle is. And she said she was going to quit. Then, after six weeks, I saw her again. And she was back on. You sort of lose your tolerance. Or there was the young woman I saw yesterday, who has just had her fourth abortion. Like pulling a tooth to her, and then she goes back to eating sweets again!"

Amodio plans to specialize in pain management in cancer treatment and hopes to continue her training at Sloan-Kettering.

One of her supervisors during her orientation to Bellevue is Dr. Diane Sauter, a petite, auburn-haired physician, a toxicology fellow at Bellevue Hospital and the Poison Control Center who has just had a bizarre series of cases involving mercury poisoning. Sauter grew up in the small town of Speculator, New York, did her undergraduate work at the State University of New York at Albany and her medical schooling at Stony Brook.

She is reviewing with Goldfrank the photos and x-rays that accompany one of those cases and later she presents the case at morning rounds. It's a rehearsal for a presentation of the case that she has been invited to make before the American Association of Poison Control Centers Annual National Meeting in Santa Fe.

The case involves a nineteen-year-old Manhattan resident named

Dorothy Dugdale, who determined to kill herself nearly six months ago. She went out and bought a syringe and eight mercury thermometers. She broke open the thermometers, poured the mercury into the syringe, then injected herself with the liquid metal. The result was a blotch about the size of a half dollar on her upper arm, which became discolored and then infected.

"I heard of it because it was heavy-metal poisoning," Sauter later said. That is one of her areas of interest and expertise. "She was admitted to Psychiatry and then operated on by Plastic Surgery. The greatest part of the mercury was removed. Fortunately, her self-injection missed all her veins and arteries, but looking at the x-rays, I couldn't help noticing these three large clusters of mercury surrounded by many little pinpoints, tiny beads of mercury." The patient didn't return for her prearranged follow-up.

Sauter kept thinking about the mercury that might remain in the wound. "So six months later I got out her chart and called her up and asked her to come in for some tests.

"She came in and we found some amazing figure for mercury—17,000 micrograms per liter of urine, where part of the injection ended up as a result of absorption and metabolism. Normal is 100 to 300. Plastic Surgery went in and removed the remaining scattered beads, which were still clustered around their original point of injection." She shows an x-ray and highly magnified pathology slides in photo form that clearly portray the phagocytic cellular response of the body, forming abscesses around the small particles of mercury, as if to consume the surrounded poison. Good as these natural defenses were, they were not enough to cordon off the poison entirely, for it would seek refuge in other organs—brain, kidney, liver.

General Surgery had not wanted the case because of its origins in a psychiatric-related incident. Trauma Surgery declined on grounds that it was surgery on an otherwise uninjured extremity. Hand Surgery said "No," as the injection was above the elbow, and Hand Surgery is below the elbow only. "So we tried Plastic Surgery, and they agreed and now it looks like they've got all the residue out and the patient has a better psychological outlook, too. She plans to go to school to become a photographer. I didn't want to mention it, as she seems to be well now, but photographers are one group at risk for mercury exposure."

So are ceramic workers, dentists, jewelers, taxidermists, embalmers and farmers. In fact, the young patient who injected herself picked a relatively benign form of mercury. Mercury vapor is far more dangerous. As Goldfrank has written, "The central nervous system bears the brunt of this type of exposure." He explains that the mercury easily crosses the blood-brain barrier and becomes bound to various protein molecules in the body, which in turn inhibit the activity of various enzymes (which act as catalysts) in the body. One of the most obvious symptoms of mercury poisoning is an "intention tremor," which, *Toxicologic Emergencies* explains, "occurs most frequently when there is purposeful movement of a limb."

In addition to this kind of tremor, the victim of mercury poisoning can also suffer anxiety, depression, irritability, emotional instability or degenerative behavior such as memory loss and intellectual dullness. As the book points out, "Some South American boxers have injected elemental mercury intravenously or intramuscularly to increase strength." All they got from their experiment was much the same thing Dr. Sauter's patient got: inflammation, abscess formation, blood starvation in the area of injection and subsequent gangrene. (Dorothy Dugdale was treated before the last condition developed.)

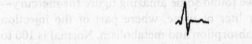

They're bringing in Eighth Street Eddie again. This time he was found in Penn Station complaining of inability to walk. The chart language: "63 W ♂. Brought in by ambulance. Slightly AOB. Legs appear swollen bilaterally. Past medical history: cardiac."

Somewhere, there's an entry being made on the cost sheets that will include this call among many others. The dollar cost of the trip across town with Eighth Street Eddie: $125.

# 25

## Moulage Day

One of the first things Goldfrank did when he was asked to head up Bellevue's Emergency Services in 1979 was to establish the Emergency Care Institute. This adjunct of his department had for one of its missions the training of every physician and nurse on the Bellevue Hospital staff in cardio-pulmonary resuscitation (CPR) techniques and each ED staff member in advanced cardiac life support.

Walter LeStrange, a marathon runner and a nurse by training, is director of the institute and oversees the instruction of more than six hundred people a year in advanced and basic cardiac life-support techniques.

Among those taking instruction each year are seventy or so who are in training to become instructors themselves. Thus the ripple effect of the teaching of cardiac life support that goes on here can reach thousands of heart-attack victims and save many lives.

About four-fifths of the institute's efforts are directed to such training, which is carried on in association with the New York Heart Association. Most days for LeStrange and educational coordinator Susan Callaghan are devoted to reviewing course materials, recruiting doctors, nurses and medics to act as instructors, preparing and coordinating various courses and assessing what additional kinds of training might be needed.

The other fifth of the time the institute is engaged in advanced trauma life support (ATLS) instruction. Then, about five times a year, there is a special day—"moulage day"—in which doctors from all over the Eastern United States come to the third floor of the old Administration Building at Bellevue for intensive instruction in handling injuries of the head, thorax, spine, abdomen and extremities and other kinds of trauma. Many of these doctors are from smaller hospitals that do not have ready access to large academic centers where research results on life-saving techniques might be swiftly disseminated.

At the end of the course, these student doctors take a test in hopes of winning certification in advanced trauma life support from the Committee on Trauma of the American College of Surgeons, under whose sponsorship the Emergency Care Institute's instruction is given. It's all deadly serious, but there's still a lot of laughter, especially in the culminating part of the training session—the moulage segments.

*Moulage* is a French word meaning "a molding." As used at Bellevue, it means the simulation of injuries for instructional purposes. Here, moulage involves the realistic makeup of three volunteers, usually fourth-year NYU medical students, who have earned their right to participate by electing a clerkship in emergency medicine and participating in course lectures previous to the moulage exercise. The students are prepared by LeStrange, who on these few days a year practices his considerable avocational skills as a makeup artist.

Each of the victims is dressed in surgical greens and given the semblance of horrifying injuries. The student doctors have the task of identifying the problems and correctly describing how to save the severely injured patient's life in order to pass the course and gain accreditation as ATLS qualified physicians. The laughter comes from the realistic methods used to create some of the common distractions of an emergency room in the full clamor of daily activity.

But first there is classroom work. It begins at 7:30 on one morning and ends at 5 P.M. the next day. Trauma surgeon Gene Coppa greets the gathered doctors, describing the vast amount of material that will be covered, explaining a bit about how the course arose. Its origin has been traced to Nebraska, where the feeling was that such formal instruction would be especially useful to those who did not enjoy proximity to major centers of medical education.

This morning is the second day of the course. LeStrange, who has an air of easy expertise, sets up a slide projector and prepares a pair of medical antishock trousers (MAST) for a demonstration that will be given later. These MAST trousers are wrapped around the legs and abdomen of gravely wounded victims and inflated, to maintain blood pressure by increasing peripheral resistance.

The first lecturer is Dr. Dana Gage, whose subject is extremity trauma. To the nonmedical visitor, hearing her presentation seems very much like watching a freight train full of facts passing at seventy miles per hour. Much of what she says seems to be an extension of what her listeners already know. In addition, there is an unending succession of practical revelations—like the fact that arteries tend to go into spasm in a traumatic amputation; that the mechanism of injury is of key importance in managing a traumatic event, as is the medical history of the patient; that the ABCs remain of first importance. Unless the airway is unobstructed, the breathing is intact and the circulation is satisfactory, nothing else can take priority. Then, once the ABCs are attended to, you align and immobilize the injury and invariably give a tetanus shot, to head off infection.

Also of critical importance is getting an accurate narrative of what happened in each injury—what actually took place at the scene and what medical history the patient can give to provide a context for evaluation. And then you complete a full examination in timely fashion, to assure that every possibility has been considered. It isn't always straightforward, Dr. Gage says. She cites the case of a woman in the ED some weeks ago who had literally been run over:

"The paramedics brought her in and said she had been 'brushed by a car.' 'Yes,' the woman said from the stretcher, 'I was brushed, all right. Brushed me down and ran right over me!' And when we got her disrobed we actually saw tread marks on her chest and abdomen!" The woman, nevertheless, made a satisfactory recovery.

The story is used as further illustration of the importance of preserving the "chain of information" from the actual scene of the accident right into the Emergency Department, or the doctor may draw the wrong inference from the occasionally incomplete report passed on by an ambulance crew.

The correct procedure for preserving a severed leg or arm is reviewed with the class—first sterile gauze moistened with saline solution, then a bag to enclose the limb and finally a second bag encircling the first with sterile ice water, to keep it chilled. Skills at preservation are improving all the time, Dr. Gage tells the class, in part because of a recently distributed video tape from Bellevue's replantation and microsurgery team. The video shows police, firemen and paramedics the correct methods for preserving a severed extremity until it can be brought with the patient to the hospital.

"Sometimes," Dana Gage says, "the part gets here before the person." She describes an instance when the police rushed in with a small ice bucket enclosing a plastic bag in which was a finger, sliced from a pedestrian by a falling sheet of plate glass. "Fine, we said to the police, but where is the person?" "Oh, he's coming," the police explained. He had to be strapped to a backboard as there was fear of spinal injury. That took a little longer than it took the police to dash across the street to a coffeeshop for some water and a bucket of ice. They themselves had the gauze needed to wrap the finger.

Was the finger successfully replanted? Yes, it was.

As Dr. Gage moves on to a discussion of head trauma, the class is reminded that the key cranial nerve to test for early signs of severe cerebral injury is the third cranial nerve, the oculomotor. So, they are told, if there is any limitation in the person's ability to turn the eyes inward, or any sign of pupillary dilation, that indicates probably serious internal brain injury.

Cervical spinal injury is involved in 5 to 10 percent of all head injuries, Gage says, indicating on a diagram that the seven uppermost segments of the spinal column, the cervical-spine area, are of critical importance in x-ray interpretation. If these seven fragile-looking blocks of bone are not in normal alignment and correctly oriented, paralysis can follow. Furthermore, all seven topmost vertebrae *must* be visible.

There have been cases when an inattentive examining physician did not take the trouble to view all seven segments, because it would entail a follow-up x-ray. Or he hesitated to inflict on the patient the discomfort involved in tugging down on the arms as the patient lay on his stretcher, so that the seven segments all would become visible on an x-ray (a technique needed to assure that all seven show up in adequate clarity in a muscular or overweight man, especially). If the arms are broken, it seems a cruel thing to yank such an injured extremity, even if the justification is the patient's own long-term well-being. But, if such steps are not taken, the patient might be released from the hospital with an unrecognized dislocation of the "C-spine." He could very well go on home, bend over to tie his shoelace and become a quadriplegic, Walter LeStrange would later observe.

There is a pause as a new set of slides is inserted into the projector carousel, affording the class the first lull in the torrent of information that has been presented. Some look around to study their surroundings and their fellow students. The walls are covered with beige-colored tile, reflecting the chamber's original use as an exercise and rehabilitation room. At intervals on the walls, long panels of light blue fabric have been superimposed on the tile, as if to soften what might otherwise be harsh acoustics. At the rear, near the entry door, there are racks of current emergency medical periodicals and research papers. The students themselves are attired in jeans and informal shirts, the standard footwear being running shoes or boating moccasins.

The new series of pictures begins to appear on the screen. The rapid commentary that accompanies them delivers an enormous amount of diagnostic information to the participants. When they see the shocking photo of a patient with a knife plunged into the bone just below the eye, and the following x-ray of the same weapon, they make the right inference: based on the x-ray, the patient probably survives.

Yes, he does, Gage says. But the principal point of this demonstration is that "You should never remove a foreign body from a wound. Put packing around it. Even though paramedics well know it, let them be reminded afresh on this. Pack around it and bring it in that way. Otherwise, you have almost a guarantee of heavy—perhaps uncontrollable—bleeding, especially in a penetration wound as severe as this one."

The x-ray picture remains on the screen for a moment longer. The knife blade has been buried entirely in the cheekbone of the victim, reaching almost back to the C-spine. And there, jutting out of the victim's face, is the silvered handle of the knife, looking with its slight curvature somewhat like the rim of a new moon. "This, we think, was a result of a dispute over drugs," Dr. Gage says.

The express train roars on. . . . Gunshot wounds to the head entail edema—swelling—of a sort that calls on the doctor's highest skills, as there is a hard bony box enclosing a gelatinous organ that takes unkindly to any intrusion, and swells readily, pushing itself against the unyielding walls. The trauma surgeon or neurosurgeon may have to open a flap in the skull to relieve the pressure—and then pray for quick reversal of the swelling, so that the brain can be returned to its normal protective garret of bone.

But complications abound at such times. Dana Gage has two illustrations to clinch the point—one from a book, with a happy ending; another from life, with an unhappy ending. The book she has read tells of a surgeon at an uptown hospital who had a twenty-one-year-old girl, an auto-accident victim, on the operating table, suffering severe edema of the brain. He had taken all the steps a doctor can to reduce the swelling. As time ticked away and he despaired, there was still a visible hump of brain bulging out of the window of bone he had rongeured by means of a few burr holes, out of her skull—a modern trephination. He finally simply gouged out enough brain tissue to reduce the mass enough to permit him to replace the flap of bone and close the opening, ignoring the shocked disbelief of colleagues and onlooking attendants. The patient recovered with only minor neurological consequences. "You'll hear some crazy stories about operating rooms," Gage says.

The story with the unhappy ending: There was a lovers' quarrel in Chinatown that involved a young man and his beloved, who was having lunch with her boss, in all apparent innocence. Her jealous lover came into the restaurant where they were eating, shot both of them in the head and then shot himself, also in the head.

All three arrived simultaneously by ambulance at Bellevue. "We tried to get their brain swelling under control—after the ABC protocol, of course. One case, the boss, was dead on arrival." The other two were barely alive. And, within a few moments, she too died from massive edema leading to herniation (protrusion of the organ). This pushes the brain against the tentorium, a structure that Thomas Henry Huxley called a "shelf of parchment" within the brain, strangling the brain stem and cutting off her breathing. The third one—the perpetrator, as the police say—lived for a few more hours, with massive amounts of steroids and diuretics controlling the swelling somewhat. But he also was too badly wounded to survive, and he succumbed.

Dr. Gage's lecture is through, and she lingers in the back of the room to answer questions in front of a collapsible table on which LeStrange has ranged a coffeepot, Danish pastries and a collection of oranges and apples, which the student doctors now enjoy during this short break.

Next there is a demonstration of the correct way to remove a victim from an auto wreck. Volunteers from the group play the part both of the victim and of the two rescuing paramedics. The victim sits slumped in a straight-back chair, to emulate the posture of a driver who has slammed, at high speed, into his own steering wheel as his car smashes into some obstruction.

LeStrange has a sure appreciation of the unresolved tensions between the new breed of pre-hospital care givers, the paramedics, and the doctors themselves. Now the young doctors, of about the same age as their paramedic opposite numbers, will play the role of paramedics. "We need macho names for these people," LeStrange says, and so, with more than a bit of enjoyment, it is decided that the pair will answer to Bud and Dick for the role playing that they are about to stage.

They arrive at the accident scene and work to achieve an in-line immobilization of the driver, which involves using a Kendric extrication device. The "Kend" has a series of Velcro straps and a bolster behind the head, and a forehead strap in front, to assure that the critical C-spine area is not disturbed until an x-ray can assure that no injury has been suffered there.

They gingerly move the victim out of his car—the chair—and gently place him on a long board, with additional straps. Then, on being told that the victim also has a fracture of his left leg, from slamming into the emergency-brake handle, "Bud" and "Dick" get a chrome-framed Hare-splint device, which immobilizes the leg. This splint also has a harness-and-ratchet arrangement, which puts traction pressure on the leg, to assure that there will be no pinching of nerves or inadvertent inner injury from bone fragments digging into soft tissues.

When the patient is successfully lifted into the "ambulance"—a stretcher parked against the wall nearby—LeStrange then thanks "Bud" and "Dick," and moves on to demonstrate another innovation. Taking the still-immobilized student-doctor volunteer on the stretcher, lifting his arm, he affixes an inflatable, clear-plastic cuff around it. With a few puffs, LeStrange blows the plastic full of air. Then he explains that this is an air splint.

"A few years ago," he says, "when I was nursing on a midnight shift downstairs, I didn't know what this was. A patient came in and I was getting the vital signs and I took my scissors, and snipped it right off his arm. A little later the paramedic comes in and says, 'Hey, anyone seen my air splint?' 'Oh,' I said, 'is that what it was?'" LeStrange looks sheepish. There is a round of laughter.

Taking notes along with the others in the room is Lawrence Glassman, who is one of the chief residents in general surgery. A muscular young man whose beeper has summoned him from the room twice already this morning, Glassman is about to start a two-month rotation as a trauma chief surgeon. Much of the material presented is not new to him, but he finds it stimulating and energizing to hear of the new methods and materials available. Still, student though he may be, he maintains an assurance in his bearing that suggests that his calling and that of the matador are not all that dissimilar psychologically.

Burns are the next subject on this crowded agenda, and the lec-

turer is a slender young doctor who decided, after starting as a surgeon, that he really didn't enjoy operating that much. Instead he
declared for a specialty in emergency medicine. Dr. Carlos Flores has
one of those pleasant expressions that suggest he is on the verge of
amusement, yet he keeps a demeanor of impressive dignity. He lectures in jeans, running shoes and an open buttondown shirt.

He rapidly recites the various inhalation burns that require careful discrimination in the Emergency Department—facial burns,
which are confirmed by eyebrow and nasal hair singeing, and carbon
deposits on the inside of the mouth, which are more alarming still.
Most alarming of all is carbonaceous sputum, which is a confirmation
that smoke and soot have entered the airway and left a residue of
burned debris that can compromise orderly maintenance of an airway—which remains the first priority.

A slide on the wall screen shows the zones of the body of an infant
and an adult, each segmented with a numeric value, under the rubric
"The Rule of Nines." This is a method to make a quick assessment of
the amount of skin area that has been burned, using the open hand
as a rough index. The palm and fingers of a hand represent 1 percent
of total body area.

Who gets transported to a specialized burn unit? Not all burn
victims. The rule that prevails is, using the diagrams to establish how
much of the body skin area (BSA) is burned, anyone with more than
25 percent burned BSA will be sent on to such a specialized medical
service. Flores points out that this is the only specialized EMS-desig-
nated service for which Bellevue is not a center. Severely burned
patients are transported elsewhere—typically to New York Hospital–
Cornell Medical Center.

Electrical burns present a specialized set of problems, and their
own distinctive lore. For example, Flores points out, arteries and
veins, as they are long structures filled with electrolyte-rich fluids,
offer the least resistance to electrical current, and tend to suffer the
most severe injuries in electrical trauma. Muscles offer more resistance to electrical current, while skin offers the most of all.

Suddenly the section on burns is over and there is a momentary
break while Carlos Flores is superseded by Dr. Gene Coppa. In addition to being the medical director of the entire course presentation,
he is lecturing today on pediatric trauma and trauma in pregnancy.

Coppa is a runner—five miles a day—and something of a high-
hurdles man, as well. He lives across the FDR Drive from Bellevue at
the high-rise Waterside, one of the most dramatic new apartment

complexes erected in New York in the recent past. Often, when a summons to the Emergency Department or operating room comes in the small hours, Coppa dons his running shoes and makes a choice. As he later explained, "I can go down to Twenty-third Street or up to Thirty-fourth Street to find a pedestrian bridge across the FDR. But, if it's the middle of the night, I come right across the parkway—jumping over the median barrier. It's a lot faster," he adds with a smile.

Now Coppa's subject is establishing an airway in a small child. One problem is that if you insert a tube you have to be aware of the difference in anatomical dimensions. An endotracheal (airway) tube can, by misapplication, do severe injury to a child. And, just as the bodily dimensions are distinct for little patients, so, too, are blood pressure and pulse rates. The younger the child, the faster the pulse but the lower the blood pressure, is the general rule.

"Performing pediatric surgery is a good way for a surgeon to lose weight," Coppa says, because the room is kept at 90 degrees Fahrenheit to guard against hypothermia in the young patient. Small body mass contributes to rapid loss of heat, sometimes to a life-threatening extent.

Slides and accompanying comments continue at the same relentless pace as the lecture ranges over the telltale signs of child abuse and the special problems of trauma in pregnant patients. A detail heard in passing: the blood volume in an expectant mother grows markedly over the course of her pregnancy and the "tidal volume"—the amount of air that enters the lungs with each breath—is significantly higher than in the nonpregnant woman of equivalent size.

Finally, the class is reminded that, in examining a pregnant woman, the mere presence of a sizable fetus can be enough to depress blood pressure if the examining physician does not take care to assure that the vena cava (the major return conduit of blood from the lower part of the body to the heart) does not become partially blocked by the weight of the unborn resting on it. The best way to avoid this complication: have the mother lie on her side.

Now it's time for lunch. And after that there will be a written test on all that has been imparted in these two days, followed by a session in actually assessing a patient with a simulated traumatic wound. Then in another room, with a moulaged medical student on the stretcher and an examining doctor to test the practical skills of the course participant, the student doctor will be formally tested on newly acquired skills.

Gene Coppa bids everyone reassemble in the rooms specified on the individual schedules that each student doctor has. Then Coppa and LeStrange and Susan Callaghan go off to the hospital coffeeshop for a quick lunch, at which they discuss their preparations for the afternoon's presentation. They talk, also, of the cardiac-arrest committee, on which LeStrange serves. This is a group that evaluates each death to see how well the staff responded to the heart-attack crisis, with an eye to developing new course materials to improve performance among medical and nursing staffs.

While the ATLS students take their written tests, LeStrange prepares three medical students dressed in surgical green for their role playing. David Karras is a volunteer who will be made up as a victim of a stab wound in the third intercostal region of the upper chest—roughly on the latitude of the nipple and about halfway to the armpit. As LeStrange rips the surgical green shirt at this point, he picks up a realistically gory patch of plastic from his work table. He inserts this patch into the opening he's just torn and tries to tape it to Karras's heavily forested chest. Dr. Carlos Flores has come into the room to watch. "His chest is too hairy. This wound won't stick," LeStrange says. "We'll just suture it on," Flores suggests, laughing. But instead a heavier course of tape does the work.

Then simulated blood is liberally applied around the tear in the surgical greens. Following that, LeStrange gets some white and blue greasepaint and creates a cyanotic cast to Karras's facial skin that is deathly in its pallor. Karras is placed on a stretcher and awaits transport to a neighboring room, where he will lie for the next two hours as relays of student doctors come in to study him and his injuries.

Next, medical student Ken Stein is made up to reflect the injuries suffered in an auto accident. His car—in the scenario that they are following—smashed into a utility pole, and from his eyebrows up he is to be enclosed in a turbanlike wrapping of gauze. At the focal point of his head injuries, on the bandage over his right eye, another realistic rivulet of blood and gobbet of extruded tissue modeled out of plastic paste-ons give him the appearance of a gravely wounded motorist. The addition of a cervical collar for his C-spine and blue grease paint to impart the look of doom complete his makeup. "I live in Greenwich Village," Stein says. "For the first time I think I'll really fit in." Then he takes his position on a second stretcher.

The third accident involves Jordan Busch, who has suffered, in the ATLS script, an injury as a result of running his motorcycle under the rear of a tractor-trailer truck: a thighbone fracture so bad that it pokes through the skin. To create the effect of such an injury, a large plastic ellipse with the sanguinary details modeled in realistic soft rubber is attached by tape to Busch's leg, through a large tear in the pants of his surgical greens. Then the entire leg is enclosed in a chrome-plated metal splint. As Busch lies there, he says to an onlooker, "Well, anyway, it's better than my last motorcycle accident!" When asked to explain, Busch says that when he was a student at Bowdoin College, as a Henry Luce scholar, he had a chance to travel in northern Thailand one summer. "A friend and I were north of Chiang Mai on two motorcycles on a winding dirt road, and I skidded and fell and skinned my leg pretty badly." Busch had to return to Chiang Mai, a one-hour journey, to get first-aid help with antibiotic ointment. And then he went on a thousand miles to the south, to Kuala Lumpur in Malaysia, to get better treatment. The upshot was that he hasn't ridden a motorcycle since. "Probably never will again!"

At the first of the "skill stations," Gene Coppa greets two ATLS student doctors, Tim Whiteside, from Long Island, and Roy Seitz, from Cincinnati, who are led into the first moulage room. Lying inert on the stretcher in the middle of the room is stab-wound victim David Karras. The pair of young doctors approach Karras and begin the immediate assessment of his vital signs. As they stand there, Adult Emergency Service nurse Carlotta Renwick, who has been in the hall awaiting this moment, enters the room pushing a mock-up of a portable x-ray machine before her. She marches right between Drs. Whiteside and Seitz. "Excuse me, is this the intercostal stab wound? I'm supposed to take some x-ray pictures."

"Ah, not now," says Seitz, a tall and deliberate man. "Later, please."

"No, it has to be now," Renwick says, "because you see I'm supposed to be going on my break and then they want me in Pediatrics Emergency . . ."

Seitz is firm. "Not now, thank you." There is a smile on his lips, as he has successfully rebuffed the first staged incursion.

Others await outside. Next, nurse Teresa Linkmeyer enters officiously, moves right up next to Whiteside and nudges him aside so that she can reach down and swing upward the side railing of the stretcher. "These are *never* supposed to be left down like this!" she says with annoyance. "And who locked these stretcher wheels? They're not all correctly locked."

Whiteside has a tight smile on his face, also. "Please, nurse. You're in the way." Linkmeyer makes a hasty retreat and joins Carlotta Renwick against a blackboard on the far wall to watch the rest of the training session. "How do you test for cardiac tamponade?" Coppa asks, prompting.

"Oh, yes, of course," Whiteside says, and he and Seitz explain what other immediate steps they will take to complete their assessment. Just then there are two fresh distractions. Nurse Sherri Rappoport comes in pushing a broom, which she shoves on a beeline directly between the doctors and their patient. Almost simultaneously, while the doctors are contending with the intrusive broom, Elise Parker, also a staff nurse in the AES, comes bursting in, her blond tresses bouncing as she runs, sobbing out her contrition to her wounded "husband" lying there on the stretcher. "Oh, honey, I'm so sorry! I didn't think you'd just stand there! Will you *ever* forgive me?"

At this last, riotous interruption, both Whiteside and Seitz grin broadly, as does the object of all their attention, David Karras, his head rocking with amusement, despite the restraining cervical collar.

"All right!" Coppa says, calling events back to their more serious purpose, and the examination continues. "So you have engorged neck veins, diminished heart sounds, pulsus paradoxus, and the ECG shows electrical alternans, switching the axis of the QRS—all suggesting tamponade—and what do you do now?" he asks. There is a silence as the student doctors gather their thoughts for the correct response. Only the distant low-frequency rumble of the traffic on First Avenue intrudes as the concentration in the room seems to focus on finding the economical phrasing of the correct answer.

A shriek at the door. A small form darts forward. It's Teresa Rizzo, holding her own gruesome hand. A plastic fork has been run completely through it. The hand is hideous with trailing blood and filaments of tissue that seem to have been wrenched from her palm by the thrusting tines of the fork. "Doctor, doctor! Help me. My hand! Help!"

For an instant, both Seitz and Whiteside appear genuinely horrified. But their skepticism serves them well. First, how could a plastic fork itself survive such a force? And second, real as it all appears, why does the young woman seem so calm? Then there is more laughter, especially from Rizzo herself. Stabbing victim Karras hitches himself up on one elbow and studies the grisly tableau with obvious admiration. Then calm is restored and the student doctors satisfactorily answer the cardiac tamponade questions. A moment after that, the session is declared over by Gene Coppa.

Afterward, LeStrange demonstrates to a visitor how a bit of embalmer's wax, a plastic fork broken in two and a judicious application of artificial blood and plastic gore can create a convincing moulage of an agreeably disgusting sort.

The Advanced Trauma Life Support course goes on to its successful conclusion, graduating twelve newly qualified ATLS physicians, who will be saving lives—better able, now, to cope with the tumult and confusion of emergency rooms in numerous hospitals in many states. They have learned much about the distractions, maladroitness, officiousness, and occasional lunacy that often are part of the daily life of an emergency doctor. In fact, the syllabus materials that the American College of Surgeons distributes in its instructors' manual includes additional recipes for staged chaos—even beyond those that the Emergency Care Institute at Bellevue presented for Drs. Seitz and Whiteside. For examples:

- An emergency medical technician approaches, anxious to get back aboard his ambulance: "Doctor, we have another run and have to leave now. It's a car accident and this is the only cervical collar we have in our ambulance, so we'll have to take it with us."
- "The patient being cared for happens to be the Governor's son. The news media, acting on a 'hot tip,' arrive . . . they finagle their way into the examination room. Once there: 'Joe, turn that spotlight on. Jay, move the camera in and get it focused. Everybody set? Doctor, just speak right into the microphone. Could you tell us exactly what happened here?' "
- "The x-ray technician arrives in the ED with his new trainee in tow. This is the *first* trauma victim the trainee has seen, however, and in the process of holding the lateral C-spine plate, he becomes ill and vomits. Naturally, he extends his sincere apologies to the physician for vomiting all over his shoes while heading for the door and grabbing his stomach." ("To simulate vomitus," the instructors' manual advises, "a tablespoon each of instant oatmeal and water can be held in the mouth until the proper moment for vomiting. The vomitus can be tinted green, yellow or red with appropriate

food coloring. Crushed graham crackers in a watery state can also
be used to represent vomitus.")

- A fastidious nurse "may distract the physician by continually point-
ing out an injury that is bleeding all over her new shoes and/or
uniform by saying: 'Isn't there *something* we can do to control this
bleeding? Just look at my new uniform and blood is *so* hard to get
out of these new synthetic fabrics!' "

- Another intrusion: "A distraught police officer arrives in the ED,
gun drawn (empty), waving it around and stating rather emotionally and hysterically: 'Everybody back! I don't want to hurt any of
you! I'm going to blow this guy away, he just killed my partner!' "

Back in the real world of the Bellevue Emergency Care Institute,
the Advanced Trauma Life Support course is over, and several of the
participants are congratulating Teresa Rizzo in the corridor, as she
takes her leave to return to Goldfrank's office on the ground floor. She
accepts the plaudits for her dramatic efforts with a bright smile and
sparkling eyes.

# 26

# Knocking Back a Few

Imagine timing an armchair binge, drink by drink. Starting time: 8 P.M., no minutes and no seconds.

*8:00:00. You take a drink of a ten-to-one dry martini. As it hits the mucous membrane of your mouth, there's an immediate reaction. White blood cells begin to mobilize to repair the insulted tongue, mouth and throat, which are immediately inflamed by the aromatic liquor washing over these tissues.*

*8:00:03. The swallow enters your esophagus and plunges into the stomach. Some small proportion of the alcohol—about 20 percent— will enter the bloodstream directly by migrating in small quantities right through the mucosal folds of the stomach wall.* But alcohol—also called ethyl alcohol, ethanol or ETOH—has the power to auger a hole right through this wall if too much is ingested on too empty a stomach for too many days. It is a prime culprit in causing gastritis.

*8:00:07. The rest of this swallow cascades right through the end of the stomach and into the first part of the small bowel where the ethanol is immediately absorbed into the bloodstream.*

*8:00:20. The first molecules of this toxin reach your carotid artery and are propelled right into the major lobes of your brain. Within just a few more moments, an immediate effect is manifest—a light-headedness, a distortion of your perceptions.*

 ⌐⌐√⌐⌐

"In the Emergency Department we think of alcohol as America's most prevalent drug. And we see it every day, in numerous ways. Just think of its impact." Lewis Goldfrank is giving a slide-illustrated talk to an NYU School of Medicine postgraduate course in emergency medicine on a toxin that has a greater impact than any other on the lives of people coming through his hospital.

"There are ten million alcoholics in America. Their suicide rate is fifty-five times normal. Their life expectancy is ten to twelve years below average. Ninety percent of alcoholics are also cigarette smokers. Fifty percent of medical admissions to urban municipal hospitals are alcohol-related."

Even as he talks, a forty-five-year-old white male is brought in by ambulance, with AOB (alcohol on breath) and complaining of having had a seizure and needing medicine. His past medical history: ETOH abuse. He is a man with the remnants of a handsome visage. One part that's missing is the left eye—lost, he says, in a barroom brawl some years ago. The present diagnosis: alcoholic blackout. Initial treatment: IV thiamine and 50 percent dextrose.

As Goldfrank has explained in informal talks to student groups on the subject of alcohol as a toxin, most foods are absorbed in the stomach and the intestines and metabolized in the liver in spurts and lags, depending on the need for energy. Not alcohol.

Alcohol remains in the blood, circulating and recirculating, making its transit of all the vital organs, until the lungs expel a small proportion of it and the liver metabolizes the rest—which it does at the deliberate pace of just half an ounce an hour.

Drop by drop, then, the liver breaks the grip of this substance on both blood and body fluids. And, during this process, the body's enzymes and organs create successive metabolites, as the breakdown products are called. Each of these interim products is poisonous to some degree. But, fortunately, so long as the liver is healthy, each breakdown product is only a transient visitor, quickly converted to a less poisonous metabolite.

The first enzyme in the liver that confronts ethanol is alcohol dehydrogenase, which converts alcohol to acetaldehyde (a poison of some virulence). Then the liver comes forth with acetaldehyde dehydrogenase, the second of the body's enzymes to contend with alcohol.

This transaction converts the poisonous acetaldehyde swiftly to acetic acid (vinegar). Then the liver converts this annoying, but not toxic, metabolite into the benign final products—carbon dioxide and water. Meanwhile, the conversions taking place, and the interim metabolites that enter the drinker's system, can cause nausea, vomiting, cardiac irregularities and—in some unfortunate racial groups—flushing and extreme susceptibility to the effects of alcohol. Apparently because of genetic predisposition, certain Eskimos, Chinese, American Indians and others have an insufficiency of acetaldehyde dehydrogenase, making them more vulnerable than most. The anti-alcoholism drug Antabuse mimics the effects of this insufficiency by blocking the effects of acetaldehyde dehydrogenase.

On its transit of the body's provinces, alcohol has the power to derange many of the settled relationships it encounters. Normally, about two-thirds of the water in the human body stays within the membranes of the cells, and about a third remains in the blood and in the intracellular fluids. Alcohol has the capacity to alter this balance. It is like a rogue moon that can entirely upset the climate on land (the cells) and the tidal behavior of the oceans (the blood and body fluids). In doing so, it causes the vital electrolytes such as potassium, calcium and magnesium compounds to be dumped and excreted into the urine, thus impoverishing the body's cells, weakening their ability to function normally, and causing a ferocious thirst to assail the drinker. The usual response to this thirst is to drink more alcohol. And that, in turn, acts to dilate, even more, the arteries and veins in the body, thus affecting blood pressure and overall body chemistry.

*8:00:30. You take a second swallow of your very dry martini. This time, when the fluid gets to your stomach, a drawstring-like muscle, the pyloric sphincter, reacts sharply and goes into an irritated spasm, partially blocking off the small intestine (the duodenum).* This sphincteral valve normally has the task of keeping food in the stomach until gastric juices and enzymes can reduce the food that's eaten to a softened state. Then the pyloric sphincter loosens its tight grip, and permits the viscous, partially digested food to move into the small intestine. Usually it takes from three to four and a half hours for the contents of the stomach to be emptied into the small intestine.

Goldfrank is coming out of his office, having checked the scheduling for the next month with Dana Gage, who is just going off duty. She says she wonders if she has not done this job long enough, as all it does is make enemies for the person who has responsibility for the assignment of work schedules for the forty attending physicians. Couldn't she be relieved of this task? Very well, we'll talk about it, Goldfrank says. Maybe another senior attending physician will take it on. He'll see. Meanwhile, there is a back-up in the ED. Waiting times are longer than they ought to be, so Goldfrank jumps in himself, hoping that his example of brisk and decisive activity will serve to energize all of the staff.

Stephen Simpson is a Vietnam veteran with psychological problems. He is so tall that his feet dangle off of the stretcher. An unruly hank of hair half obscures his features. From time to time he brushes it back up onto his forehead, but it soon falls in his face again. He has been waiting for nearly three hours. He has been triaged into Category 2: "urgent." Goldfrank picks up his chart and approaches Room 1, where Simpson has been wheeled and given a preliminary examination. "Mr. Simpson? I'm Dr. Goldfrank."

"Hello, doc."

"What's bothering you?"

"It's bad, doc. Sour stomach. Tossing my guts. Pain in the throat. And the legs. I mean I'm on the wagon. What is this?"

"We'll try to find out, Mr. Simpson." Goldfrank asks a series of questions and makes notes: "Pt. not able to eat solids for prev. 2 days. c/o SOB [complains of shortness of breath]. PMH: denies diabetes or heart disease . . . Physical exam. reveals cachectic [emaciated] male in moderately acute distress. Radial pulse 120 lying, 140 sitting. BP 130/90 lying and 110/70 sitting. Temperature 99 degrees, rate of respiration 20."

Then begins a quest to establish just what impact the alcoholic habit of many years' standing is having on patient Simpson. Right away the change in blood pressures (lying and sitting) suggests a volume loss. Vomiting has led to a substantial loss of fluid that can't be compensated for when he sits up. So immediately an IV is established to replace some of this shortage with 0.45 percent saline solution and 50 percent dextrose. As always with alcoholics, thiamine is also given. While Goldfrank gathers information and waits for test results, the patient will necessarily spend a good deal of additional time waiting in the observation area.

*8:02. A third swallow. This one puts your pyloric valve into a nearly complete spasm, and no more alcohol will enter the bloodstream through the wall of the small intestine for some time. Your rate of inebriation will slow down.* All the alcohol that is to be assimilated now will make its way, slowly, through to the walls of the stomach. There it will gradually migrate into the bloodstream. Drinking milk or eating a slice of cheese, or even swallowing a raw egg before drinking, will noticeably slow the rate of absorption.

Every day Goldfrank and his colleagues note that alcohol can so inflame the mucous membrane of the stomach that it can cause severe internal bleeding. It can so stupefy the brain as to leave a lasting negative impression on the central nervous system. And besides the alcohol itself there are the by-products of fermentation—the so-called congeners—which add their own measure of poisonous impact on the abuser of alcohol. Among such congeners showing up in whiskeys and brandies is the familiar toxic metabolite of alcohol, acetaldehyde.

Two paramedics wheel in a black male patient, twenty-five years old. There's a bloody bandage on the man's right foot, and a wave of alcoholic odor precedes him. He dropped a knife onto the instep of his right foot during a break-dancing exhibition in midtown. He had been drinking before the incident, the paramedics report, and drank after it, until the ambulance arrived. Dr. Toni Field examines the patient and decides that she'll have to suture him up.

"Wha's gonna happen?" the man asks. "My foot . . ."

"I'm going to stitch you up."

Blood samples from Stephen Simpson have gone to the laboratory on the fourth floor and now the numbers have come back. Goldfrank studies them closely. Simpson's pH is below the ideal figure, so he's suffering from acidosis. Because he also has decreased carbon dioxide in his system, which would usually push a person away from acidosis and toward alkalosis, Goldfrank concludes that metabolism, not an anomaly of breathing, is causing Stephen Simpson's acidosis.

There's a chart that appears in *Toxicologic Emergencies* that gives at a glance the relationship between pH and the amount of carbon dioxide in the blood. It acts as a striking illustration of how delicate is the balance on which each of us depends for ongoing life. The diagram shows crosshairs like those of a finder telescope, on which, at the center, is a small rectangle that represents the zone of normality. The horizontal axis represents carbon dioxide in the blood, the vertical axis reflects pH (alkaline toward the bottom, acidic toward the top).

- If you breathe too slowly, and you accumulate too much carbon dioxide in your blood, your system will drift out of the box of normality into a zone marked "respiratory acidosis."
- If you breathe too fast, you'll drift in the opposite direction—into a zone marked "respiratory alkalosis."
- If something in your cells and fluids interferes with normal metabolism—something like absence of insulin or lack of kidney function, for instance—then you could drift into the zone of "metabolic acidosis."
- If you vomit excessively, and lose hydrochloric acid (gastric juices) from the stomach, this loss will push you into "metabolic alkalosis."
- All of these departures from the central box are life-threatening to a greater or lesser degree.

Because of these findings of too little carbon dioxide, and a derangement in the fluids in and surrounding the cells, Goldfrank lists "respiratory alkalosis (hyperventilation) and metabolic acidosis," on Simpson's chart. Based on this finding, there is strong reason to draw additional blood to measure how much ethyl alcohol, methyl alcohol (wood alcohol), ethylene glycol (antifreeze) and other congeners are in the serum of Stephen Simpson. For now, there is strong evidence that the chemical relationship between his cells and the intracellular space is sharply out of balance. The new bloods are drawn and sent on to the laboratory to be tested for electrolytes, ketones, glucose, blood urea nitrogen and creatinine. These levels will shed further light on what Goldfrank is almost sure will prove to be alcoholic ketoacidosis. (Ketones are acetone-like breakdown products of fatty acids and ethanol, yet another variety of alcohol metabolite that can cause derangement in a person's system.)

*8:13.   You pour yourself a second martini, also at ten-to-one, and begin to sip it in a more determined fashion.*

The young woman shuffling down the hall had been outside, despite the chill, sitting on one of the newly varnished benches near the small ornamental fountain that is the centerpiece of the Bellevue "park"—a small swath of lawn and plantings behind a wrought-iron fence that borders First Avenue. Her name is Tanisha Gabriel and she's been sharing a pint bottle of fortified wine with an old man on one of the benches. He brought the wine, she the 100-millimeter cigarettes. Now she's moving toward the Walk-in Clinic, sauntering past the triage desk.

She is attired in a blue warm-up suit and a deer hunter's cap, worn sideways. For a necklace she wears a bright green telephone extension cord and a string of purple wooden beads. She spots Goldfrank as he comes down the corridor and gives him an amiable wave. "Hello, doctor."

"You're back?"

"I'm back."

"How are you?"

"I wanna be detoxed."

"You want to be detoxed."

"I want to be detoxed, yes."

He talks with her a moment longer, then accompanies her down the hall to where Mary Dwyer is taking the temperature, pulse and blood pressure of a tall Hispanic man. Here the preliminaries to detoxification will start for Tanisha Gabriel, who, as she waits for Dwyer to finish with the man, thanks Goldfrank and calls after him, as he moves back toward his office, "I was born in this hospital, you know?" And then, to the bright poster on the wall, she addresses a few additional comments in an air of self-amusement, "Born right in this hospital, yessirree."

Not far from where these modern scenes are played out, another drama in alcoholism came to its end early in 1864, when in this same institution the same disabling disease, complicated by tuberculosis (a combination that is still frequently seen at Bellevue), ended the life of composer Stephen Foster. A plaque notes the place of his passing, in a building long since torn down. His brief biography in *Grove's Musical Dictionary* notes, as might be said of some of those still suffering from his disease, "Foster's life was not passed in poverty, as has sometimes been said. . . . But the last years of his life were made miserable and squalid by his intemperance."

*8:17.   Your eyes are not as acute on the page that you are reading as they were, and there is a kind of floating sensation as you down the last of martini No. 2. No. 3 is swiftly prepared and a solid draught of it is taken as you continue reading. There is a sudden, involuntary protest from your stomach, as this bolus dose of a strong poison hits the stomach wall. There is a moment of nausea, but you hold your breath and the feeling passes.*

Shortly after noon there was a motor-vehicle accident in midtown involving a car and a motorcycle driven by twenty-four-year-old Fukuda Nakaishi of Montebello, California, who is brought in by ambulance and seen in Room 3 by Dr. Carlos Flores. The young man fell on his left side and complains of pain in his left elbow and left ankle. There is a strong odor of alcohol on his breath, and in the pocket of his leather jacket there are several small airline-style bottles of Southern Comfort.

As Goldfrank returns to check in on Simpson, he runs through in his own mind the steps that have been taken, and which he urges his staff to follow in all such cases. First the ABCs are automatically checked. And when circulation is checked, "occult hemorrhage" is also looked for. This is to make sure no unobserved gastrointestinal bleeding is taking place. Then the patient is placed in a "left semilateral decubitus" position—lying on his left side with his head forward and mouth down so that he will not breathe in his own vomit if he throws up—a frequent cause of aspiration pneumonia among alcoholics.

After the drawing of blood, the ECG leads are attached to Simpson's chest and the state of his heart is closely studied. Alcohol's power to disturb the relationships of electrolytes in the body can cause malfunctions of polarity in the small electrical-muscular pump that is the heart. Dangerous rhythms may begin. Sharp changes in the electrical and mechanical functioning of the pump may take place.

Thiamine is invariably administered to alcoholics at Bellevue because of the common lack of this B vitamin in the systems of heavy drinkers. Without thiamine, a kind of mental scrambling takes place, in which the victim cannot even recall the simplest thing. "Confabulation" is frequently seen in such people: they will supply missing information in hopes of sustaining the illusion of normality. In one common test, Goldfrank offers the patient his choice from an imaginary collection of colored yarn that, with a gesture, he puts before an alcoholic. There are no threads there at all, yet, when asked to pick out the purple one, the vitamin-depleted patient will point to a spot on Goldfrank's outstretched palm and declare that to be purple yarn. That's taken as confirmation that mental function has been lost, at least temporarily.

In one recent case, an old man closely resembling the conventionalized picture of an Old Testament prophet—balding and bearded, with a fierce-looking countenance—could not recall his own name when he was brought in. He was given thiamine intravenously and the next morning was able to demonstrate a sturdy enough memory to give a comprehensive medical history of six decades of alcohol abuse.

A complete physical exam was performed on the naked Stephen Simpson. "Alcoholics," Goldfrank has written, "present atypical manifestations of typical diseases. They should be examined thoroughly in the nude." This assessment includes looking for trauma, checking for meningitis (inflammation of the lining of the brain) or any hemorrhaging inside the cranium. And a check of neurological functions by way of the twelve cranial nerves and especially motor function of all the extremities also is made. In examining alcoholics over the years, Goldfrank has observed a repeated pattern: "When you see an alcoholic, there is better than an even chance that he will have an old rib fracture that shows up on the x-rays. It seems to go with alcoholism, just the way motor-vehicle accidents do."

8:26. *A lethargy marks your movements as you pour yourself a fourth martini. By now, your stomach seems to be in a state of perpetual apprehension, as it is uncomfortable at the mere fragrance of the juniper-laced gin and vermouth. There is something else—the doctors in the ED would call it diplopia. To you it means there are two glasses and two bottles before you. In fact, there seem to be two of everything.*

A badly battered woman is brought in by ambulance. She is a twenty-seven-year-old Murray Hill housewife, who was found struggling with her husband in the vestibule of their apartment house—she trying anxiously to get away, he so obviously drunk that he was taken into custody by the police for aggravated assault and public drunkenness. She has Le Fort No. 1 facial fractures and an apparent collapsed lung, suggesting a rib fracture.

Patient Simpson has done well with the intensive treatment he has received—glucose, thiamine, fluids and electrolyte replacement, over five hours and close monitoring of his progress. As Goldfrank was later to write: "All major electrolyte abnormalities were corrected within eight hours. The patient was admitted for alcoholic ketoacidosis and he also suffered from depressed levels of blood sugar and nutrients."

*8:37. A fifth martini is at hand, but there is now a struggle to keep yourself awake. If only you were substantially bigger physically, you probably would be much more alert right now, as liquor affects physically large people more slowly than their more average-size fellows.* Were there more clarity of mind, the equation would be easily grasped: for every kilogram of weight, 1 gram of alcohol aboard is expressed as 100 milligrams of alcohol per 100 cubic centimeters of blood. So a 155-pound (about 70-kilo) man who had drunk 150 cubic centimeters (about 70 grams) of 100-proof gin (about two martinis) would be at about 100 mg% alcohol (about 70 grams per 70 kilos). At that level, driving is dangerous and muscular coordination is poor. At the level of 400 to 700 mg%, coma, respiratory failure and death ensue. As this is No. 5 drink coming up, the level is somewhere close to 300 mg%.

*8:38. You take a swallow. Then another. And you go into a dark cavern of unconsciousness and pass out right in your own chair.* Muscles relax. The glass tumbles to the floor, spilling its contents. A long-neglected book thumps down beside it. And then the drinker follows, falling out of the chair to the carpeting below. A rough snoring announces the agenda for the next few hours.

Had there been heavy drinking for several weeks, the drinker's body would have activated a second enzyme system—the so-called MEOS, standing for microsomal ethanol oxidizing system—which can double the body's ability to metabolize alcohol, and which accounts for the amazing abilities of old alcoholics to souse themselves and still remain conscious.

*8:58. From time to time you may endure a bout of marked slowing of respiration or even of apnea—a long period when there is no breathing taking place whatsoever. Then, after perhaps twenty or thirty seconds in this state, the carbon dioxide in your blood and fluids builds up to the point where an involuntary reaction—a lifesaving reflex—is triggered, and you convulsively gasp for a breath. In this manner some hours pass—heavy snoring, long interludes of no breathing whatsoever, and sudden explosive gasps for air, as your heavily drugged body fights on its own to restore the balance that your armchair binge has momentarily upset.*

Some, in these circumstances, find that the sensitive tissues of the gastrointestinal system stage a rebellion of their own. Insulted by the quantity of alcohol dumped onto them, they reject the accumulation of this toxin. They do so by reverse peristalsis—progressive muscular contractions occurring in the tubular organs of the body (stomach and segments of the small bowel) that expel the slurry of alcohol and half-digested food in the stomach. Geyser-like, this viscid mixture is propelled right up the esophagus and out of the mouth—if the drinker is lucky. If the coma-like state is a bit too deep, it is possible that this potentially lethal mixture will be coming out just as the body is urgently commanding the drinker to suck in a new breath of air. It can happen that vomitus is aspirated and floods the bronchial tree, down to the alveoli, with a liquid barrier to breathing. If that happens in solitude, the drinker very well may die.

There are other ways that alcohol kills. One of them is depression of the breathing function. Drink too much and the "primitive" brain, which governs such automatic functions as respiration, is put to sleep and death comes by suffocation. In one braggadocio exhibition in a Chicago bar in 1954, as Berton Roueché has written,* "A commuter informed a group of friends . . . that he could drink 17 dry Martinis in something less than one hour. He did, but as he emptied the final glass, he toppled off his stool. He was dead when he hit the floor."

\* \* \*

*The Neutral Spirit*, Little, Brown, Boston, 1960, p. 55.

This imagined hour in the life of a drinking bout suggests how quickly alcohol can take over your system, how dangerous its side effects can be, and why Lewis Goldfrank says that the biggest toxicological problem facing American society today is the abusive use of alcohol.

Patrick O'Toole is forty-one, a veteran of Vietnam and a steamfitter who has been down on his luck. A friend brings him into Bellevue, leaves him in the lobby, and then departs. On his way back from the coffeeshop, Leon Fields finds O'Toole talking disconnectedly with a hospital policeman, and brings him in to the triage desk.

O'Toole tells nurse Jeanne Delaney he is afraid that he is going into the DTs. She quickly gets his vital signs and triages him into the emergent category. Dr. Robert Hessler immediately takes O'Toole into Room 2 and gets the rest of his story. O'Toole is breathing in a labored way, suggesting pneumonia, and says that he has been drinking heavily for nearly twenty years. Never has been dry for more than six months at a stretch, he says. Recently he has been on a nonstop binge, drinking mostly vodka at the rate of about a quart a day.

"When was your last drink?" Hessler asks.

"When I ran out of money," O'Toole says. "Last night."

"How long ago?"

"Twelve, maybe thirteen hours."

He says he is feeling wobbly, sick to his stomach, racked with chills. Hessler knows that he is in a race with time. He knows it is too early for DTs, but he must work rapidly. Delirium tremens can kill. And its onset is oftentimes disguised by other problems—pneumonia, brain injury of some kind, a broken bone. A reading of 150 mg% is legally drunk in most states. Yet, as Goldfrank has said, "We've seen people who can think and function relatively well with an ethanol level of 1000 mg%. Once a person drops to the level about 100 mg% below his normal level, he will begin to manifest withdrawal. You don't have to have a total lack of alcohol; you simply have to have less than usual. That's why the alcoholic has to drink continually, at the same rate as on previous days."

Now O'Toole begins to develop twitching motions of his whole body and he has a tonic-clonic seizure, accompanied by licking and sucking movements of his lips. His pupils become widely dilated and his gaze fixed to his left. Dr. Hessler, helped by nurse Sharda McGuire and aide Leon Fields, immediately begins clearing O'Toole's upper airway and administering 100 percent oxygen by means of a small bag-valve mask. They draw bloods and start an IV line in his left arm with 5 percent glucose and half-strength saline, and then give 50 percent dextrose, thiamine and Valium.

In the torment of his condition, the victim is at grave risk of doing himself serious injury. However, timely medical intervention can reverse the effects of poor nutrition, deranged body fluids and a feverish overheating. Also, as long as the patient has ethanol in his blood, he has impaired glucose metabolism. And glycogen, a starchlike substance that the liver stores and then converts into glucose as the body needs it, is usually unavailable because of inadequate nutrition or, if present, because of an ethanol blockade of metabolism. So in essence he has to be gavaged—force-fed.

The basic strategy followed by Hessler is to put the patient into a peaceful sleep, from which he can nevertheless be roused. As Goldfrank has written, "You want to prevent hyperthermia, agitation, aspiration, physical injury and rhabdomyolysis—destruction of the muscle tissue.

"The reason people die is that their condition is uncontrolled; they're extremely agitated and lack adequate volume repletion [replacement of water and saline fluids]. Hyperthermia—alcohol-withdrawal-induced heat stroke—is a sign of the DTs. The patient needs fluids, glucose, electrolytes and vitamins. And he certainly needs vitamin $B_1$ [thiamine] in particular. We also very commonly see patients with folic acid deficiency, because they don't get enough green, leafy vegetables. And several times a year we see patients with vitamin C deficiency so pronounced they have scurvy."

As Hessler continues to work over O'Toole, to head off further manifestations of alcohol withdrawal, a thirteen-year-old boy is brought into the Pediatric Emergency Service. He is nearly comatose from imbibing what appears to have been rum, according to the paramedics, who were called by the police. They had found the youth on a park bench with an empty pint bottle of rum next to him.

George McVan, a twenty-seven-year-old black male, comes in with a bloody towel over his right eye. He exudes a strong alcoholic odor. "Pt. states he was tripped and hit his head on sidewalk. Denies loss of consciousness. Laceration noted over R eyebrow. Past medical history: ETOH abuse."

Pat Kunka is reminiscing during her break in the Emergency Ward. "Most drunk stories aren't so amusing. At least not to me. But one night in come four policemen with this Hispanic male, about twenty-five, covered with blood and smelling like a brewery. They were rushing him right into the trauma slot about three in the morning. And he was yelling something, but he was so drunk who could understand it? He'd been shot, the police told us, and we got him down on the table and he kept trying to sit up, and we pushed him down and were—three of us—getting his clothes off, cutting them away, and he kept yelling, *'No yo!'* We kept scissoring away, looking for the gunshot wounds. And we kept looking and he kept yelling until we had him naked and then he says, *'Esta mi hermano! mi hermano!'* But we were still looking for bullet holes, and there were none. Then one of the aides said, *'Hermano,'* that means 'brother.' It's not him. He was saying it was his brother who was shot!"

The police went back to look for the brother. Social Work got the man new clothes.

The pressures that remain on Bellevue Hospital's Emergency Department are very largely related to the poison of choice among the populace at large: ethyl alcohol. "If it weren't for alcohol, we could cut our budget by 25 percent," Goldfrank says. "And if it weren't for cocaine, heroin and tobacco, the other leading toxins, we could reduce our costs another 25 percent."

Another ambulance is arriving, and another unconscious citizen of the streets makes his appearance and his olfactory statement at one and the same moment. It's Eighth Street Eddie again. The fragrance of alcoholic congeners hangs heavily in the air over his wheeled stretcher. His is the last arrival of the day, at 11:56 P.M. There have been 340 patients altogether in the ED today. Of that total, 121 have involved alcohol in one way or another.

# 27 ∿∿

# Pokeweed, Headaches and Tribal Medicine

Nine A.M. of an unseasonably chilly Saturday in October. Goldfrank is wearing his sabots, dressed in a corduroy shirt and jeans, standing at the range in the kitchen preparing crêpes for the family. Eleven-year-old Rebecca is helping by setting the table and opening jars of homemade jellies and distributing them around the table. As she does so she tells her father about her school science project, in which the dissection of a chicken has given her insight into something she never knew before.

As the fragrant fumes of his cooking waft upstairs, to announce to the rest of the family that breakfast is about to be served, Rebecca tries out her new knowledge on her father.

"And the intestine is enclosed in the mesentery . . . ?"

"Good. And what's that?"

"A supporting membrane that attaches an organ to the body?"

"Excellent! . . . You got blackberry jam there too?"

"Yes. And the chicken intestine, mine anyway, was five feet six inches long! Isn't that amazing?"

"Amazing!"

"But the teacher said that if you unfolded lung tissue and exposed all its possible surface area, a pair of human lungs would cover a whole tennis court!"

"I've heard that too." He turns the last batch of crêpes. "All right, Beck, you want to go up and tell 'em we're ready to eat?"

"It's the time of year I like best of all," Susan Goldfrank says. "When the steam is hissing and knocking in the radiators and Lewis has three frying pans going, cooking crêpes, and six jars of our jelly are open and waiting and the family gathers . . ." She's had two projects on her mind this week. First, she started piano lessons again, going back to an instrument that she played well as a girl. Then, her restoration work on the house continues to be both satisfying and vexing: removing paint and finishes from the front stairway banister gives beautiful results, but it has begun to seem like an endless task. She wonders if perhaps the progress is so limited that she should abandon her meticulous approach—with surgical tools and sandpaper—in favor of a more frontal assault (paint remover and steel wool).

Fifteen-year-old daughter Jennifer is struggling with reciprocal equations in her calculus homework, and is stuck. She's especially glad for the call to breakfast.

Meanwhile, son Andrew is working over a final draft of a paper he is about to submit for an Introduction to Philosophy course he is taking at NYU on the question "What Is a Mentor?" He has chosen his father as the mentor he will describe. With some diffidence he has shown the paper to his mother, the family arbiter in matters of grammar and literary style. She likes what she reads, telling Andrew he's made a good start.

But now it's time for breakfast. They all troop down the steep back staircase, drawn by the enticing aroma of fresh-cooked crêpes, to which they all devote themselves with gusto.

Afterward, Goldfrank and Jennifer go up the back stairs to his study, where they will review her calculus. She is a winsome girl, who laughs easily, but her spirits are momentarily dampened by auxiliary variables and ordered pairs. So father and daughter spend a half hour going over her work. She is taking an enriched math course in high school, which will have her finishing at the end of her sophomore year work that normally is taken up in the senior year.

Susan can only admire her daughter's mathematical attainments, with a glad heart that Jennifer does not have to confront the mysteries of calculus as Susan had to, back in the sixties: at the University of Brussels in an amphitheater-style classroom with nearly seven hundred other students. "The instructor used to cover the blackboard with these enormously complicated derived functions and you'd struggle to take notes, because he was such a nonstop lecturer, and when he finished all these derivatives and ordinates and slopes, he would turn to the class with a great smile of satisfaction with himself and say,

'N'est-ce pas?' Sometimes I wanted to stand up and say, 'Pas de tout!' but of course never did. I got through that course—but barely!"

Father and daughter have reached a stopping place, so Lewis will drop Jennifer off for field hockey practice while he goes on with Rebecca and does the family's marketing. If it's shortly after payday, Susan says, he goes to the local gourmet-food market, where he will get the staples for the week to come plus an array of especially good cheeses, pickles and the ingredients for dolma—stuffed grape leaves. The grape leaves come off the Goldfrank arbor, which Lewis and Andrew spent weeks rebuilding. Among the stuffing ingredients are green onions, garden dill and pine nuts. Later in the month, his shopping list grows less exuberant.

As the sun warms up the air and drives off the morning mist and chill, Goldfrank takes an axe out to continue with a task he has set himself—removing a large stump of a maple recently felled in a storm. Nearby, Susan is at work stripping creepers and overgrowth from a stand of forsythia.

A neighbor stops by and Goldfrank breaks off his work to greet him. As they stand close to where Susan is working, Goldfrank notices that she has removed from among the overgrown forsythia a number of stalks that bear dark plum-colored fruits about the size of blueberries. They grow from magenta-tinted stems. Goldfrank asks if his visitor recognizes the plant. No, he doesn't. Looks like little grapes. Fox grapes?

"Pokeweed," Goldfrank says. "Did you get the root, Susan?"

"No. It broke off."

"Always does. Andy, lend me the mattock?" he calls across to his son, who is at work on the roots of the maple stump Goldfrank has just been chopping. Andrew brings over the mattock and offers to excavate the root himself. Goldfrank thanks him, and with a few powerful strokes Andrew lays bare the root below the broken-off stem that Susan has just yanked from among the forsythias.

Goldfrank drops to his knees and brushes the earth away from the fist-sized pokeweed root, which he then digs out of the earth. He recites some of the lore of this common weed.

"The berries make a great dye. And, if they're boiled, people say they make a fair jelly or wine. Some just cook them in a pie. But I wouldn't recommend trying it in any form. The root, here, is often mistaken for horseradish or parsnip by health-food faddists." Everyone has moved to the porch for a break.

"A few years ago, we had a twenty-nine-year-old man come to the hospital complaining of constant vomiting, cramps and diarrhea. He had been out in New Jersey—Morris County—with his family for their annual 'rites of spring' gathering. It was a great feast. He told us all the aunts and uncles and cousins each brought their own culinary specialties for the get-together. In fact they had a tradition of not eating the day before, so as to enjoy the food to the fullest.

"He ate some of almost everything served: numerous hors d'oeuvres, freshly prepared; home-grown eggplant; deviled eggs; gefilte fish with fresh, homemade horseradish; liver pâté; and assorted cheeses. He ate several large portions of main courses, which included exotic Chinese, Italian and Middle Eastern preparations. Dessert included several slices of Boston cream pie and rhubarb pie, and he finished up with coffee and dessert truffles!"

"Other people became ill too?"

"Yes. We contacted everyone we could, and located several others who had been at the feast who also got sick. Same symptoms, but theirs cleared up in about two to four hours."

"So what did it?" the neighbor asks.

"Told us he takes no medication, doesn't smoke, only drinks wine socially."

"Had to have eaten something, didn't he?"

"Yes. Pokeweed. This stuff! One of his uncles made the 'horseradish' sauce. The victim complained about the bitter, burning taste of the horseradish and that's what led us to the uncle. Yet another amateur gourmet botanist had dug in error!"

The neighbor is duly impressed, and even more so when he asks for details on the search for the right answer. It was revealed that a painstaking "differential diagnosis" for the young man's acute diarrhea originally presented Goldfrank with more than forty possible causes of the symptoms. The possibilities fit into eight major categories: infectious, structural, metabolic, functional, drug-related, food, metals or others. In the "others" category were digitalis, alcohol and insecticides, among a half-dozen possibilities.

By tracing through the potential causes and seeking out other family members besides those sickened themselves, the search finally narrowed to the uncle, who might well have killed off his nephew. As

Goldfrank later wrote, in summing up the case, "Fortunately, this patient was spared the more severe neurologic manifestations such as blurred vision, hypoventilation, diaphoresis, weakness, convulsions, arrhythmias and death."

After the neighbor's short visit, Goldfrank goes back to his yard work and Susan to hers.

The weekend passes in similar fashion: strenuous physical activity punctuated by several phone calls from the Poison Control Center, and one from the ED on a question about a confusing cardiac problem. On Sunday there were visits from Susan's mother, Ava Harrington, just back from a trip to Israel and full of tidings about what she saw there, and friends of the children.

Late that Sunday Goldfrank read himself to sleep with E. L. Doctorow's *Lives of the Poets*, especially struck by this portrait of homelessness:

We borrowed an ordinary precinct car and went looking for one. Contact on Fourteenth Street and Avenue A, time of contact ten-thirteen P.M. Subject going east on southside Fourteenth Street. White, female, indeterminate age. Wearing WWII-issue khaki greatcoat over several dresses, gray fedora over blue watch cap, several shawls, some kind of furred shoes overlaid with galoshes. Stockings rolled to ankles over stockings. Pushing two-wheeled grocery cart stuffed with bags, sacks, rags, soft goods, broken umbrellas. Purposeful movements. Subject went directly from public trash receptacles to private trash deposits in doorways, seemed interested in anything made of cloth. Subject sat down to rest, back to fence, East Fifteenth Street. This is the site of Consolidated Edison generating plant. Subject slept several hours on sidewalk in twenty-degree weather. At four A.M. awakened by white male derelict urinating on her.*

Neal Lewin is presiding at rounds this Monday morning as Goldfrank has a lecture on hypo- and hyperthermia to deliver at a postgraduate NYU seminar. These are two areas of particular Goldfrank expertise, so while he gives his lecture, Lewin invites a new resident on rotation in the ED to present his case from the night just past. As he begins, the doctors' station fills up with medical students, residents, attending physicians, nurse practitioners, staff from the Poison Con-

*Lives of the Poets, Random House, New York, 1984; Avon Books edition, 1986, p. 84.

trol Center and an emergency physician from the University of New Mexico Medical School and another from the U.S. Army at Fort Hood, Texas.

Lewin is looking natty in a hound's-tooth jacket. He perches himself atop the shelf that holds the telemetry equipment that now has been largely superseded in its function by the Maspeth central dispatching center for ambulances. The resident describes a case that so closely parallels a case that Lewin has earlier encountered and written about that hearing the details is almost like reading his own chapter in *Toxicologic Emergencies* on ergotamines, a medication derived from a grain fungus that helps fight headaches.

Barbara Wilson was an anxious young woman with "the worst headache of my life," when her husband brought her in. She was thirty. Suffered abdominal cramps, leg pain and numbness of the fingers and toes. She related a history of migraine headaches since age sixteen, which always—until recently—she could successfully head off with ergot medications.

Ergot is the dried, hardened black or reddish-brown mass of threads that clings to rye grain after a fungus invades it. In the middle ages, epidemics of dysesthesia (burning of the extremities), delirium and convulsions called "St. Anthony's Fire" were traced to bread made from grains contaminated with this fungus. Later it was discovered that an alkaloid (with the complicated formula $C_{33}H_{35}O_5N_5$), which was isolated from this fungus and called ergotamine, was especially useful in treating migraine headaches. It acted to constrict blood vessels that had dilated too much. Such dilation sets off a migraine.

Several days ago, just before the onset of her menstrual period, the patient developed a severe headache and began to use ergotamine suppositories and to inhale ergotamine aerosol mist. She also took pills that her doctor had prescribed for headache. Clearly she had overdone it. She was now sweating a bit and was pale, with a mottling of her fingertips and toes—suggestive of sharply constricted arteries. Her headache raged throughout her skull—was not just localized, as is often the case, at one point or in a limited zone. As the resident would note, "Her headache is holocranial, throbbing and accompanied by severe neck and back pain."

The physical examination was unremarkable except for tenderness about the cervical spine and guarded neck movement because of muscle spasms in the neck. "I don't want to move my head at all, doctor. The pain is too bad," she told the resident. Further evidence of the extreme constriction of her arteries was provided by the discovery that she had no pulses below the femoral arteries in either leg—beyond, approximately, the mid-thigh.

"So what's the differential diagnosis?" Lewin asks.

The resident is slow to answer, as well he might be. For there are at least thirteen different kinds of headaches that Lewin himself has listed in a table that dominates his ergotamine chapter in *Toxicologic Emergencies.* And there are more than forty distinct causes of headache.

He lets the resident off the hook by reciting some of the salient facts that his researches into headaches have revealed:

Women are about three times as likely as men to suffer migraine headaches, except in the "cluster" category, where men outnumber women ten to one. This is the sudden, nighttime stabbing pain on one side or at the back of the head, triggered typically by too much alcohol, nitroglycerine, or tyramine-containing foods, such as aged cheeses or Chianti wine. Tyramine is a substance found in ergot, cheeses and mistletoe that is "vasoactive"—able to dilate blood vessels.

The many other types of headache that women are most prone to are often linked in some as-yet-unexplained way with the onset of their monthly period. There are also a great variety of other triggers: oral contraceptives, vitamin A, fluorescent lighting, exercise, sexual activity, ice cream, high altitude, too much sleep, wood smoke and allergies, among others.

Lewin has another table in his chapter that lists three major kinds of mechanisms that set off headaches—vasodilation, muscle-contraction, and traction-inflammation (including such things as tumor, swelling, hemorrhages, meningitis, encephalitis, and eye, ear, nose and throat diseases).

"So what about this patient Barbara Wilson?" Lewin asks.

"Her headache has characteristics of both a migraine and a muscle-contraction. I'd call it a 'mixed headache syndrome.'"

"Good. Subarachnoid hemorrhage? Could that be a cause here?"

"No."

"Okay. Why?"

"There were no red blood cells in the CSF (cerebrospinal fluid) when we did a tap."

"Good. How about bacterial meningitis."

"I'd say 'no.' There were only two lymphocytes—absolutely normal. And no increase in the number of cells. The CSF gram stain was negative as was the acid-fast stain and a normal CSF/serum glucose ratio: 100 mg% in serum and 65 mg% in CSF. We're seeing mild hypertension, but not enough to set off a headache."

"Good. When would you expect to see a hypertensive headache?"

"Much higher pressures. Perhaps not until the diastolic pressure exceeds, say, 130. And here it's only 95."

"Exactly right. Very good." Lewin asks the resident to sum up what else they have concluded about Barbara Wilson's throbbing head.

There is nothing to suggest any kind of metabolic or endocrine disorder or anything of a pathological sort inside her cranium. Finally, there is agreement that her headache was chemically induced—paradoxically by the very substances that offer relief: ergotamine and another anti-headache drug, propranolol, which her private physician had prescribed, and which she took in pill form four times a day, a so-called beta blocker. (Beta blockers interfere with the function of the sympathetic nervous system, the body's automatic pilot, which governs the heart, blood pressure and dilation of the pupils of the eyes, among other things. They block the "adrenergic receptors," and inhibit the influence of adrenaline, therefore dilating blood vessels, and thus setting off headaches in some people.)

The result of this complicated pharmacology was a vascular system so constricted that there was no discernible pulse in her lower legs and her fingertips were mottled and numb.

The ED had already begun a regimen of cure, which in this case was to have its wanted result within twenty-four hours. As Lewin has written: "The treatment for ergot toxicity depends on the severity of the clinical findings. In mild cases, characterized by tingling of the extremities but few other symptoms, supportive measures (such as rest, hydration—with oral or IV fluids—and minor analgesics like aspirin) are adequate. In more serious cases, severe peripheral vasoconstriction may produce pregangrenous changes, for which the use of a potent vasodilator, such as nitroprusside, has been suggested." Another possibility: taking the victim for a "dive" in a hyperbaric chamber, to correct oxygen starvation in arms and legs. Also, Lewin notes, yet another medication can be tried: a stiff drink. "Ethanol *is* a vasodilator," he reminds the group.

Lewin is standing now against the telemetry bank with his arms crossed over his chest in a typical professorial stance. He straightens his horn-rimmed glasses and continues: "What other cases did you see during the night?" he asks the resident.

"A couple of prostitutes on crack." The first case was a young black woman of about twenty who solicited a police plainclothes officer, and when they arrested her and began taking her in to the lockup, she nodded off. They rushed her to Bellevue, fearing that she might have taken a lethal dose of something in the back seat of their car. When she arrived at Bellevue she was given intravenous glucose, thiamine and naloxone. There was no reaction to the naloxone.

The resident's recitation continues: "When we began our examination, a broken glass crack pipe fell out of her underpants. She had no track marks or other signs of drug injection. Her numbers came back from the lab; all were within normal ranges. A urine-tox screen sample got lost. We saw Transport take it, but still it got lost."

Dr. Kathleen Delaney, who has joined in at the edge of the crowd, offers the view that the woman could have been nodding off from angel dust or from alcohol or from just plain weariness. "Some prostitutes stay up for thirty-six or forty-eight hours at a stretch," she notes. "And for the naïve user PCP can look like cocaine." It is agreed that those are possible clues to the diagnosis. Meantime, the resident reports, the woman went to Psychiatry at 3 A.M.

"You said you saw a couple of prostitutes?" Lewin persists. "What was the other one?"

"Okay. A twenty-five-year-old Hispanic female got in a car with a customer and smoked crack with him and he punched and robbed her. She was brought in by ambulance with a cervical collar but her C-spine x-rays were normal."

"All seven vertebrae were seen?"

"All seven."

"Good. Go on."

"Okay and we gave her 50 percent dextrose and thiamine because I thought she might be an alcoholic. . . ."

"Good. And then?"

"She responded to naloxone. And she also had macular lesions [blotchy discolorations] all over her skin."

"All over?" Lewin asks. "On her back? Places where she couldn't reach?"

"Actually, no."

"Okay. She responds to naloxone. Meaning what?" Lewin asks anyone in the group. Several answer, "Heroin." "Okay, or another opioid. And she has these blotchy scratches over what? Her arms and legs?"

"Yes, arms and legs," the resident responds. "But that's all we have at this point. We don't have her numbers back yet."

"Okay. What do you think of the scratches? Cocaine bugs?" There is a general nodding of heads. The tactile hallucinations that the white alkaloid can cause have been emphasized time and again in these sessions, by Goldfrank and by his deputies, so it remains on everyone's mind. Now Lewin invites Delaney to add her observations.

"I'd just add, Neal, that you have to be careful about naloxone if there is a chance of C-spine injury. Because if you administer naloxone before you have your C-spine x-rays and aren't sure there's no injury, no subluxation or fracture, and you precipitate a withdrawal reaction, with vomiting, you get airway compromise because of aspiration. I like to wait (if I have time) to see the C-spine film."

"That's a very good point," Lewin says. "I hope everyone is clear on that." Again, a general nodding of heads around the group, which now has grown to nearly thirty, with the recent arrival of Drs. Richard Weisman and Mary Ann Howland from the Poison Control Center.

No sooner do they arrive than Dr. Carlos Flores is called from the meeting to attend an incoming Hispanic woman with an acute asthmatic attack. Dr. Robert Nadig also leaves, almost simultaneously. An old man moaning piteously is wheeled down the corridor, and Nadig falls in beside him and begins to question him.

Dr. Howland has a letter from the Health Commissioner of the city that she reads. It warns of a new penicillin-resistant strain of gonorrhea that is showing up in increasing numbers in many of the ninety hospitals in New York City. This "PPNG" strain (penicillinase-producing Neisseria gonorrhea is the formal name) marks a dramatic therapeutic change from previous strains that yielded to one shot of penicillin in the buttocks. As Howland explains, that used to be enough to knock out both gonorrhea and any incubating syphilis bacteria (spirochetes) that a person might also have been exposed to.

As she explains, it has been the practice—in the interests of epidemiologic control—to bring in any prostitute discovered to be harboring the new strain, and give her medication so as to prevent an epidemic proliferation of the PPNG. And it used to be the practice to give sailors off visiting ships shots so that gonorrhea could be controlled. Still, nationwide, there were a million cases a year. But the surviving *gonococcus* bacterium—a bean-shaped organism—remodeled itself until now it resists penicillin. The letter concludes with suggestions on the best course of antibiotics—spectinomycin or cephalosporins—to confront the gonococcus. These medications may control the new, hardier gonorrhea, but they don't necessarily knock out syphilis simultaneously, as penicillin used to. So other medications have to be brought in against syphilis, and there's no longer any prospect of killing both bacteria with one shot, as formerly.

"Thanks, Mary Ann," Lewin says. The wall clock says 8:55, and the meeting is about to disperse. But there is time for one more case. Delaney has one.

"An eighteen-year-old black male with a gunshot wound to the left temple."

"This is last night, Kathy?"

"Yes," she says, with an ironic smile and a sad shake of her head. "A pulse of 140 and no blood pressure. He was brought in in MAST trousers and the medics said there was only a terrible head wound. But as soon as I started examining him I knew something wasn't right. With a head wound of that sort I'd expect him to be hypertensive. Perhaps 200 over 100. Or at least have a good blood pressure.

"He was struggling to breathe and he had a fixed deviation of his eyes to the left." She pauses in evident weariness.

"Any exit wound?" Lewin asks.

"No exit wound. We gave him two liters of fluid and suspected some C-spine injury from the impact of the bullet on the head. Also we undid the abdominal section of the MAST trousers, and there was an abdominal gunshot wound, also."

"Where?"

She gestures low on her left stomach—left lower quadrant. "They both were small-caliber bullets and there were no powder burns that I or the surgeons saw. He went right to the OR and they discovered a severed aorta and a severed vena cava. The patient could not be resuscitated."

"Died?" a student asks.

"Yes. In the operating room."

It's nine and time to break. Just then Goldfrank returns, glances in briefly at the concluding session, greets a few participants as they scatter to their respective assignments. He goes with Lewin down the hall to Lewin's office to confer on Robert Hessler's agreement to take over the scheduling task for the department. They also discuss several new candidates for new attending-physician positions in the department for the year to come.

Dr. Hedva Shamir is examining a twenty-seven-year-old black woman, who has been brought in by her four brothers. They may all be naturalized American citizens, but Shamir has a strong hunch that they are still bound by the ethos of their native Kenya, where—she would later learn—as members of the Dorobo tribe they once lived amidst the Masai and the Nandi, close to the earth and bound by ancient tradition.

"Our sister is a virgin," the oldest brother is saying as Shamir looks at the young woman lying on the stretcher in Room 3. The patient appears to be four or five months pregnant and is complaining of severe abdominal pain. Her brothers look on uneasily, and that is all Shamir needs to see. Her intuition mobilizes swiftly and now she knows that the woman may be in peril not only from her disease—whatever it may turn out to be. What Shamir sees here before her closely resembles, in its psychological purport, what she used to see in the Sinai among the Bedouin tribesmen. When she was a fourth-year NYU medical student, years ago, she spent a summer working with the Department of Health of the Government of Israel ministering to the Bedouins.

And, just like those desert folk, this young woman, before she will allow a pelvic exam, insists that she needs her mother's permission. One of her brothers takes a note that Shamir has hastily drawn up and sets out to get the mother's signature on it. That entails a trip far uptown and back. Shamir continues her examination. As she does so, she is aware of the hubbub in the adjoining rooms, as the Emergency Department is, because of its construction, well suited to eavesdropping.

In Room 5, Henrietta Dorset, a fifty-nine-year-old white woman from Winston-Salem, has just been brought in. She was on an escalator at Grand Central Station when it suddenly lurched forward, throwing her down and causing painful, grit-encrusted abrasions on her left knee and on both of her feet. Also, she is complaining of pain in both of her hands.

In Room 2 is a twenty-six-year-old Hispanic woman who has been brought from the Police Department's Central Booking complaining of vaginal spotting. She thinks she has a condom stuck inside of her.

Near the nursing station, outside of Room 2, two paramedics just returning to the field are talking about the New York Mets and the tumult at Shea Stadium the night before. One of them describes the antics of some of the exuberant fans. He concludes, "It's true! May I drop dead on the spot if it isn't." Another voice injects itself: "You wanna drop dead you better go somewhere else. We don't authorize that around here!"

In the outer office of the director's compact suite, Teresa Rizzo has just received notification from the Secret Service that a new presidential aide and a new Secret Service agent will be coming in to the Emergency Department to acquaint themselves with the layout in anticipation of an imminent presidential visit to the United Nations. They want someone to show them around at three.

While the Dorobo woman awaits her brother's return, her discomfort becomes very much worse, and it is Shamir's growing determination to ask a colleague to endorse her plan to sedate the woman, even though it will be against the patient's will. But, as she later said, in her mind was unreeling the images of her Sinai experiences:

"This woman and her brothers both were convinced that a pelvic examination would destroy the woman's claim to virginity, just the way they were convinced in the Bedouin settlements we used to go to. There was me, a nurse and twelve soldiers. Our base was El-arish, and we had a big MERVan-style truck. But when we went out to treat the Bedouins the rules were: One, you could never remove a veil. Two, you could never undress a woman, not even unbutton her garment to get a stethoscope on her chest, without having the husband in the immediate area and having his permission. Three, you could never perform a pelvic exam.

"We saw things like cancers the size of watermelons, eye diseases that I had only read about before—trachoma, when the eyes grow clouded and sightless. My one triumph in diagnosis was when I noticed that the pee of one Bedouin drew flies. That was diabetes. And we started him on insulin.

"But any time a gynecological problem in an unmarried woman arose, you had to take the woman and her mother or brother to El-arish for treatment. The chaperones were to guarantee the patient's continued virginity. And, if a woman surrendered her virginity, it was a loss of what was considered property by her father and brothers. A terrible thing. She could be killed.

"So I had this absolute conviction that we had a similar problem with this tribeswoman. If her brothers thought she was no longer a virgin, it could be a threat to her life."

By now the mother has arrived. She looks uneasy under the bright fluorescence. All of the family is opposed to a pelvic examination. Shamir assures them she will use a baby speculum. "True virginity won't be affected," she insists. They remain unpersuaded. But at least they will permit sedation and agree to Shamir's proposal to call in an OB-GYN resident for consultation.

"I went down the hall to meet him, so he would know that if he discovered pregnancy he had to do something to keep it from the family. 'Just admit her but don't tell the family,' I said. 'It could be a threat to her life.'"

So, after repeated assurances and reassurances, the mother gives her permission for the pelvic examination, and the young woman is admitted to the Gynecology Service. It isn't until the next day that Shamir hears the end of the story: "It turned out to be a massive, twisted ovarian cyst. Just as her brothers said all along, despite her appearance of being four months pregnant, she was a virgin!"

It's 2:30 and Deon Jenkins, twenty-four-year-old messenger for the Omega Nebula Messenger Service, has just been brought in. His chart: "24 yr. B ♂. Riding a bicycle which was hit from behind by motorcycle. Lost balance and fell to ground. No LOC. c/o abrasions R ankle, R elbow, and R hip and c/o R sided neck pain; cervical collar on . . ."

A U.S. Government car arrives at the ambulance entrance and two muscular young men in brown suits flash their IDs as they enter. Their shoes are well shined and the creases in their trousers are conspicuously fresh. They say a word at the triage desk and follow the nodded directions down the corridor, second door on the left.

"Are you Teresa?"

"Oh, right, you're with the government?"

They show their IDs. "Okay," Teresa says, "just a minute." She dials Kay O'Boyle's number, then escorts the pair around the corner to O'Boyle's office. O'Boyle will show them around as she usually does during one of the Secret Service's frequent update visits. As O'Boyle said later, "They're very thorough. Every new member of the presidential security staff familiarizes himself with the layout here. And we show them what we have ready."

Bellevue is the hospital of choice when the President is in town. In addition to special arrangements here in the Emergency Department made during presidential visits to Manhattan, the coronary care unit, the Medical intensive care unit, the Neurosurgical intensive care unit and the General Surgery intensive care unit all are on stand-by alert when he comes to New York.

"At first," Rizzo said later, "it was a big thrill, the first time they came in. But you get used to it. In my five years we've gone through, I don't know, maybe a dozen times the President has been in town. And we always stand by. And it's exciting to be, you know, given the recognition. But so far, thank God, the President himself has not come in."

The tour complete, the pair stop in to thank Goldfrank, who is over at University Hospital. So they ask Rizzo to convey their thanks for them.

# 28 No Place to Go

Taurus has come to sit high over the East River. And the Pleiades, the seven daughters of Atlas, glitter near the zenith. But, like the C-spine vertebrae of a fat man, only six out of seven of them are readily visible.

These constellations look down on a city in which scores of fifty-five-gallon oil drums have been pressed into service as outdoor stoves. Scraps of lumber and newspaper are burnt in them. Around the glowing drums huddle the homeless who have the physical vigor to stake a claim close to the heat and hold their places against all comers. There are frequent fights for the choice spots.

"We see patients who have inhaled the smoke and carbon monoxide from these fires, who have fallen asleep and had their clothes burned, who get too close—or are pushed too close—and have second- and third-degree burns. And of course at this time of year we see people who are badly chilled, sometimes frozen to death," Goldfrank says. He has written extensively on the effects of the sudden cold on people of all ages.

One such case involved thirty-year-old Linda Wellcome, who was brought to Bellevue by ambulance after being found unconscious in Stuyvesant Park. It had been raining, the outside temperature was forty degrees and the wind was blowing at fifteen miles per hour. The woman was wearing light clothing soaked with muddy water. A clinic card found in her pocket indicated that she had been under psychiatric care for several months and gave her address as "undomiciled."

When she got to the hospital, she was breathing only eight times a minute. Her breath did not smell of alcohol and there were no signs of injury, but she did have the typical abnormalities of hypothermia —an irregular and slow heartbeat. The big worry in resuscitating such patients is that, if you don't go carefully, you can provoke ventricular fibrillation, to which the heart of the hypothermic patient is especially vulnerable.

As Goldfrank has written, possible causes of hypothermia besides exposure to the elements include endocrine disorders (such as diabetes), chronic diseases of various sorts (starvation and cirrhosis of the liver among them), harmful changes in the central nervous system (such as a stroke) and poisons (including alcohol, opioids, carbon monoxide and tricyclic antidepressants). But in this instance the problem was clearly linked to exposure, cold rain, thin clothing and the constant winds.

Of the various methods of reheating such a dangerously chilled patient, the Bellevue Emergency Department chooses to rely solely on the common hospital blanket. They employ no electric mattresses, no heating of the blood or body fluids by dialysis or of air by warmed inhalation techniques. Further, Goldfrank has made it a rule to give no medications that elevate temperature or alter heart rates, either by mouth or by injection, until the patient has returned to near-normal temperatures. This is because of the unpredictability of the body's metabolism of all substances, including medications, when very chilled.

Several patients seen at Bellevue with body temperature below seventy degrees have survived such passive rewarming without going into fibrillation. One of the key tactics used is to limit the amount of patient activity or stimulation, which could increase the demand for oxygen by the heart and thus set off fibrillation because of the imbalance between available oxygen and the body's demand for it. Hemoglobin becomes more miserly with oxygen as temperature drops, more unwilling to surrender its cargo of the life-giving gas to the eager cells. Hence any muscular activity could precipitate fibrillation.

Linda Wellcome went into Room 8, where the charts would note: "On examination the patient was cold and wet but well-nourished and anicteric [without signs of jaundice, which would suggest liver failure]. Her systolic blood pressure [outgoing pulse of the heart] was 75 mm Hg [millimeters of mercury] by palpation [by pressing against the radial artery in the wrist with a thumb; a routine blood-pressure cuff determination had not given any reading at all]." The pulse was irregularly irregular—suggesting atrial fibrillation. This means that the collecting chambers of the heart (the atria) were squirming instead of pulsing synchronously with the pumping chambers (the ventricles) at fifty-two beats a minute.

Her temperature was not recordable with a clinical thermometer. A thermocouple was introduced to get her rectal temperature. (This is a flexible length of rubber tubing covering a wire coil attached to an electronic meter, which obviates risk of possible trauma from a glass thermometer broken by an agitated or seizing patient.) Wellcome's temperature proved to be 30 degrees C. (86 degrees F.).

After Goldfrank's team assured her airway, breathing and circulation, the patient's wet clothes were replaced with a hospital gown, and she was covered fully with hospital blankets. As Wellcome was being dried, staff members took blood specimens and then began an IV line with Ringer's lactate, a saline solution given as maintenance fluid.

In addition to the fluids, the only pharmacologic agents given Wellcome were 100 percent oxygen and, by IV, 100 milliliters (about a cup) of 50 percent dextrose plus 2 milligrams of naloxone and 100 milligrams of thiamine. There was no reaction to the naloxone, so opioids could probably be ruled out as a factor contributing to her nearly frozen condition.

As the temperatures in her body slowly rose, the results of the various tests that she had been subjected to came back from the laboratory and showed no underlying explanation other than the apparent one—environmental—for Linda Wellcome's comatose and cold condition. That was a relief, because underlying diseases can make recovery from extremely low temperature far more unlikely.

In screening her blood in the laboratory, technicians found no ethanol, anti-psychotic agents, opioids or chloral hydrate (a common sleeping pill). But Wellcome's phenobarbital level, in comparison with a normal therapeutic level, suggested an overdose. As Goldfrank has written, "Hypothermia is a likely finding in the barbiturate-poisoned patient, particularly when the patient has been exposed to inclement weather. These patients often become poikilothermic." That's a Greek word meaning "variably thermic," or like cold-blooded creatures, in that the heat of their blood varies with the outside temperature, instead of remaining under the control of inner heat-conservation mechanisms.

After they got Wellcome's temperature up above 34 degrees C. (93.2 degrees F.), there was another discovery besides the phenobarbital. That was the finding of a "serum amylase of 350 IU." Amylase is an enzyme that normally is found in the saliva and pancreatic juices and acts to break down a slice of bread or a forkful of French fries into usable sugar in the digestive process. Normal levels of this enzyme are no more than 200 IUs (international units). So Linda Wellcome had some 150 units above the norm, suggesting to Goldfrank that she could have kidney disease, or diabetes, or other possible disorders: salivary gland malfunction, fallopian-tube, small-bowel disease or pancreatitis.

As he watched her body temperature climb, Goldfrank could remind his junior staff members of some of the complexities of dealing with extremely low temperatures. At temperatures below 25.6 degrees C. (78 degrees F.), death may ensue, if the condition is prolonged, through ventricular fibrillation. While the brain may not need as much oxygen as normal in such conditions, it still needs more than a fibrillating heart is likely to be able to pump during its state of squirming disorganization.

After a prolonged time at this temperature, the end of life is virtually inevitable. "Under controlled circumstances," Goldfrank writes in *Toxicologic Emergencies*, "patients have survived with temperatures as low as 16 C (60.8 degrees F). . . . The old adage that a patient cannot be considered dead until he is warm is the crux of management."

In "significant hypothermia," where the patient is between 25.6 and 32.2 degrees C. (78 to 90 degrees F.), the category into which Linda Wellcome fit, "shivering stops and is replaced by muscular rigidity. Delirium, stupor and coma may be present." The patient may be able to carry on a conversation, yet temperatures below 30 degrees C. (86 degrees F.), the level that Wellcome had reached, are particularly dangerous. "At this stage, the basal metabolic rate is less than 50 percent of normal." Or, in plainer terms, the total exchange of energy that is going on in liver and muscles, which govern the conversion of food into usable energy for the body—the actual production of its preferred fuel, glucose—is so slow as to deny the needed level of support for any physical activity. Any sort of exertion generates lactic acid as a result of anaerobic metabolism; the machinery of the liver and muscles keeps cranking away, but without oxygen the result is as potentially ruinous as running an engine without oil.

Mild hypothermia is from 32.2 degrees C. to 35 degrees C. (90 to 95 degrees F.) and may show up in muscular awkwardness (ataxia), clumsiness, slow response time and garbled speech (dysarthia). Shiv-

ering starts at this stage and the veins and arteries of the body constrict to conserve heat.

Cooling depresses both kinds of cells in the heart: the muscular
and the electrical. The beat becomes slower and slower. As the chilled
patient is warmed, the possibility of straining the heart rises, especially if the warming takes place too rapidly. Low temperature, low
blood pressure, limited oxygen supply (because of hemoglobin's par-
simony), hypercapnia (elevated carbon dioxide levels) all make the
heart more irritable. To keep this from happening, it appears that the
body, in its wisdom, slows down the heart.

Other organs and systems are also affected by extreme cold—
lungs, central nervous system, kidneys, blood and gastrointestinal
tract. Among the most dramatic changes are those affecting the blood
itself, which increases its viscosity owing to body proteins activated
by the cold—cryofibrinogen and cryoglobulin.

Meanwhile, increased oxygen-poor metabolism in the liver and
muscles, owing in part to the depressed breathing rate, leads to metabolic acidosis.

As Linda Wellcome was revived and admitted to the Emergency Ward, her case inspired a spirited discussion among the staff
on the true incidence of hypothermia. For years it had been impossible to assess, for the simple reason that the thermometers in
use—until the fairly recent past—showed temperatures only down
to 94 degrees F. A decade ago, when he was at North Central Bronx
Hospital, Goldfrank and others started campaigning for the city to
adopt thermocouples, which would give a true temperature reading
down to 0 degrees C. (32 degrees F.) and a top temperature of 50
degrees C. (122 degrees F.). Hyperthermia was also being missed
because the maximum temperature recordable in the old days was
105.6. Today, such thermocouple devices are standard in New York
City, with the result that there are far more reliable figures on hypothermia.

Now that the cold weather is at hand, it is clear that this disorder
will afflict many of the alcoholics and street people who will be
brought into the ED in the months to come.

One of the New York City Human Resources Administration
men's shelters, on the Lower East Side of Manhattan, is typical of the
institutions that have sprung up to cope with a problem that seems to
grow more visible in New York as new building construction and new
hotels with ever more luxurious furnishings proliferate. The fiscally
revived city, robust once again in its budgetary health, still has to live

with piles of excrement on the stairs of Grand Central. It still must endure odious reminders of squalor in the persons of ruined men and women dozing on its heating grills and in its unguarded doorways, lying in its parks and in its abandoned buildings. The glowing city has dispersed its Bowery bums and the state has virtually closed its mental hospitals. That's the good news. The bad news is that there is no longer much asylum. Those who had been kept out of sight in their own parts of town, their own institutions, now are everywhere, in plain view. In the frame with every vivid picture generated by the great metropolis is the unwanted reminder of those who have not found a secure place for themselves in the food chain: There in the same snapshot with the radiant new high-rise luxury condo is the vomitus-flecked chin stubble of the repellent minority: those who drink too much, eat too little, sleep in the streets and don't have the price of admission to the toilets that serve the rest of society.

Goldfrank is well acquainted with the historic precedents for the ordeal of homelessness that New York now endures. He finds it notable that there are strong parallels between George Orwell's own autobiographical account of his days at the bottom of the social pyramid in the 1930s, *Down and Out in Paris and London,* * and contemporary accounts that Goldfrank has read. When he reads Orwell, he is reminded time and again of the stories he has heard from patients he sees at Bellevue and the things he has seen with his own eyes that define the daily life of the homeless in the 1980s.

Down on his luck, Orwell was led to a "spike," a workhouse, at about a quarter to six one evening:

> Already a long queue of ragged men had formed up, waiting for the gates to open. They were of all kinds and ages, the youngest a fresh-faced boy of 16, the oldest a doubled-up, toothless mummy of 75. Some were hardened tramps, recognizable by their sticks and billies and dust-darkened faces; some were factory hands out of work, some agricultural laborers, one a clerk in collar and tie, two certainly imbeciles. Seen in the mass, lounging there, they were a disgusting sight; nothing villainous or dangerous, but a graceless, mangy crew, nearly all ragged and palpably underfed.†

*Harcourt, Brace, New York, 1937.
†Orwell, *ibid.*, pp. 142–43.

As is the case in the 1980s, in the 1930s many of the homeless resorted to begging to support themselves. Wrote Orwell:

> If one looks closely one sees that there is no *essential* difference between a beggar's livelihood and that of numberless respectable people.
>
> Why are beggars despised?—for they are despised, universally. I believe it is for the simple reason that they fail to earn a decent living. In practice nobody cares whether work is useful or useless, productive or parasitic; the sole thing demanded is that it shall be profitable. . . . A beggar, looked at realistically, is simply a business man, getting his living, like other business men, in the way that comes to hand. He has not, more than most modern people, sold his honor; he has merely made the mistake of choosing a trade at which it is impossible to grow rich.*

In his concluding passage, Orwell wrote of "at least 15,000 people in London . . . living in 'common lodging-houses,' in which great rooms full of men sleep a yard apart from each other in conditions that make rest difficult," as "nearly all the lodgers have chronic coughs, and a large number have bladder diseases which make them get up at all hours of the night. The result is a perpetual racket, making sleep impossible."†

Today in New York City, there are at least 36,000 homeless people, by common estimates. The number of single-room dwellings in New York City decreased from 170,000 in 1971 to 14,000 in 1986 because of tax abatements for condominium conversion of such structures. These figures come from Chester Hartman of the Institute for Policy Studies in Washington, who was quoted in a *Science* article on the origins of homelessness. Between 1978 and 1984, the article says, there was a reduction of 715,000 units renting for $300 per month or less in the city.†† To Goldfrank, such developments have given New York something of the quality of Port-au-Prince in Haiti, "where there are palatial quarters for the few, like the Duvaliers, in the old regime in Haiti, and the rest of the people live in the rubble. Here in New York there are 30, or 50, or 100,000 homeless—however

---

*Orwell, *ibid.*, pp. 173–74.
†Orwell, *ibid.*, pp. 210–11.
††"Homelessness: Experts Differ on Root Causes," by Constance Holden, *Science*, May 2, 1986, p. 569.

you want to look at it. There is no question that the problem is much greater than it was ten years ago."

Of the three hundred or more patients that the Bellevue Emergency Department sees every day, at least 25 percent are homeless, and typically the complaints involve alcohol and drug abuse, cellulitis, hypothermia (or, in summer, its opposite, hyperthermia—heat stroke). "The victims of hypothermia are those who don't have enough competitive force to get out of the worst of the elements. Typically, they don't have a grate; or the grate they did have is now closed off by ribboned barbed wire—the sort that has razor-blade-like barbules embedded into it. Now they're over in the corner of an abandoned building, and reach a state at which they can't move. Project HELP calls the ambulance. (This is a group founded by the Mayor to help those too impaired to stay on the streets, those resisting all other attempts to help them.)

"We do what we can, but often have to discharge patients who, without a regular home setting, can't get bed rest, can't put their feet up, don't take their medication. So what do we do? It's a constant problem."

Releasing a homeless patient with the name and address of a shelter written on a piece of paper causes uneasiness. But, on the other hand, without timely discharge the hospital would soon be converted itself from an institution for the care of acute illness to a custodial institution—an unthinkable metamorphosis. And an additional problem is posed by the failure of the support system that was supposed to have arisen after the "deinstitutionalization" of mental patients in the 1960s and '70s. The mental patients were given medicines and sent out into the world, where they frequently do not take their medicines and just as frequently run afoul of the urban environment to which most of them find themselves gravitating.

Toby Wilts has just left the Bellevue ED once again, and is trying to find some way around the regulations at the Bellevue Shelter, housed in the old Psychiatry Building, which is considered one of the best of such shelters. But Wilts doesn't like the rules there. No gambling. No drugs. No stealing. No loud behavior. Toby doesn't steal, but the rest are among his daily activities, and restrictions on his freedom are intolerable. So he will remain in the streets. Or perhaps try the

men's shelter on Third Street, as the nights are now so cold he is in
real danger of freezing.

Another writer, in the tradition of Orwell, found himself down
and out in Manhattan. Writing in *Christianity & Crisis*, Tom Kelly
reported that he wound up, homeless, at the Third Street Shelter in
New York.

It was about three months before I would be totally out from under the
aegis of the Human Resources Administration; ten days or so of the full daily
routine at the shelter, the rest as a "ticket man," the city paying my rent at
a Bowery hotel/flophouse.

The concentration of a few thousand homeless men in one poor neighborhood in lower Manhattan sure plays hell with street survival mechanisms.
. . . So you've got three or four thousand broke and homeless guys pouring
out of this building wedged between the Bowery, the Lower East Side, the
shooting galleries of Alphabet City (Avenues A, B, C and D) and the . . . East
Village, and it's 7:30 in the morning and it's 20 degrees out. A few tips: It's
tough to panhandle from people wearing gloves (mittens are worse); a brief—
one phrase—tale helps, but don't claim you need the money for transportation unless you're near a bus or subway stop; stay away from other panhan-
dlers. . . .

There are very few places in a city that are both warm and free. Street
and shelter people tend to favor donut shops—the heavy sugar dose assuages
various addictions—but that means money, which means panhandling or (the
other major free enterprise in the shelter system) peddling cigarettes, one at
a time.

Libraries are your best bet: The elderly already know this, though, so
there are sometimes small wars over seats between the ladies and gents eking
out the Social Security and the shelter crowd. The powers-that-be know it too.
They've seen to it that most neighborhood libraries are open four days a week,
afternoons only, so you've got to have your wits about you and be quick on
your feet. . . .

Nights were another story. After the evening feed, the . . . system changed
gears to bed the monster down for the night. Shortly after 6 o'clock, big yellow
schoolbuses began to double-park on Third Street. Inside, by the front desk,
someone would call out "Brooklyn," or "Queens," or "Keener"—the destina-
tions of the buses and the options available for bed-space. . . . Brooklyn,
meaning a converted junior high school building in East New York, a neighborhood that has the distinction of being just as "bombed-out," but not nearly
as famous for it, as the South Bronx. It would not have been my first choice.

There were about 20 beds to a schoolroom, but a number of them were claimed on a semipermanent basis by locals (Brooklyn-homeless, as opposed to Manhattan-homeless), virtually all young and black or Hispanic, who strove with considerable success to maintain a pecking order in "their" rooms. . . .

Queens, or more properly the National Guard armory in Flushing (near where the Mets play ball), became my bedroom suburb of choice. The best thing about the place was its size: It was made to accommodate a large number of men (not for sleeping, I grant you), which meant that the showers were vast and the toilet stalls numerous and functioning—neither of these things being true at Third Street. . . . Of course the buses turned up again in Queens at 6:00 in the morning to take you back to Third Street; bourgeois Queens wanted no homeless on their streets after sun-up. . . .

When my appointment with the shelter social worker came due, I paraded my impeccably middle-class credentials. I was rewarded for my "normal" and compliant behavior, and my more-apparent-than-real employability, by being authorized a month's stay in a top-of-the-line Bowery flophouse and a meal ticket for the same period. My quest for shelter would no longer begin anew each day."*

An old man has just been brought into Bellevue. He goes into Room 3, where he lies in a disheveled huddle. Despite appearances, patient Edward Riley is just thirty-four years old—chronologically. Physiologically he is perhaps seventy. His left leg is severely ulcerated. The paramedics, lingering in fascinated horror, told Liz Reynolds, who now is taking Riley's vital signs, that he had been lying on a sidewalk on Avenue C for some time. Passersby assumed he was sleeping off a bender.

As an attending physician examines the ulcer, he can see right down to the ankle bone. He cannot easily regain his composure, as he calls for a surgical consultation. When the General Surgery house officer comes in, he agrees with the attending physician that the only course for Edward Riley is amputation of the leg. There's too much necrosis, too much evidence of irreversible injury, too much gangrene already advancing, to save it. Riley is whisked off to an eleventh-floor operating room. "I'm glad I didn't have to do it," the attending physician later said. "I once did such a procedure, and won't ever forget it.

"It was my first year as a surgical intern. I remember every detail:

---

*"No Place Like Home," *Christianity & Crisis*, April 21, 1986, vol. 46, no. 6, pp. 130–32, *passim*.

how you mark off the leg with a blue felt marker, where you are going to amputate. You allow for a flap of skin to cover after you're finished. Then cut in with the scalpel, tying off the arteries and veins, and removing the muscle around the bone. Then the saw. It's really more of a metal wire with fine teeth. Surprisingly efficient. I don't know, maybe forty or fifty strokes back and forth, and it's over. Then you trim back the muscle around the stump so you can make a loose enough closure so that it will heal smoothly—and drain while it's healing. Too watertight and it can result in necrosis."

At any rate, Riley's relative youth is indicative of a change among the homeless. Today, the average age of the adult homeless has dropped to the early thirties, according to Dr. Philip W. Brickner, Director of the Department of Community Medicine at St. Vincent's Hospital and Medical Center. Together with Bellevue, St. Vincent's brackets the lower half of Manhattan and also takes a leading role in caring for the homeless. "In 1969," Brickner has written, "it was nearly impossible to find a homeless woman. The epithet 'bag lady' didn't even exist."

In addition to her Emergency Department and Emergency Care Institute duties, Dana Gage also acts as medical director of the Belle-vue Shelter Clinic, where she works with Flo Botte and others. The shelter cares for as many as one thousand men a night, and suggests in its roster of common disorders what the toll of homelessness is. Wherever there are homeless there also is tuberculosis—still highly contagious, still a mortal illness. This disease is virtually certain (if untreated) to destroy lung tissue and to ravage almost any part of the body—bladder, chest, spine, liver. . . . In recent years, Gage says, there's been a dramatic increase in tuberculosis among younger homeless patients, among whom are alcoholics, IV drug abusers, the psychiatrically impaired and AIDS victims. All are commonly victimized by this plague that has been almost eliminated from the rest of society.

Furthermore, the new victims of TB are getting the disease for the first time, not suffering a recrudescence of an earlier bout. One problem that vexes the doctors who try to contain TB is that the medications that can control it are not compatible with the disorganized life of the homeless. Such medicine must be taken for long periods—a year, typically—which requires stability that the homeless do not have. And, aside from that, many patients—even the middle-class and nondrinking populace—don't comply very well with their doctors' medication recommendations. As Goldfrank has said, only 35 percent of patients actually follow the instructions on their pill bottles. "The rest take a tablet or a few capsules for a few days, and then if they feel better, skip the rest. Or others take the whole batch at once, and wind up with drug toxicity. For another thing, the two drugs most effective against TB are isoniazid or rifampin. They react badly with alcohol and can cause hepatotoxicity—liver poisoning." Dr. John McAdams of St. Vincent's, who works with the homeless, has written, "I've told patients they can't drink while they're on a particular medication, and they've believed me. But what they've done is continued drinking and stopped taking the medication."

Diabetes is another frequent disease among the homeless, and its potential costs to society are incalculable. The correct treatment of diabetes presupposes a settled environment in which a proper diet and medication, such as insulin injections, can be administered, and alcohol absolutely avoided. "What we do for patients who aren't too fragile," Dr. Gage says, "is to try to maintain them on pills instead of injections. But any ulcerations, such as commonly afflict those who have to spend most of their days standing and most of their nights sitting, can lead to the loss of a leg for the diabetic." That's what happened to Toby Wilts some years ago, and that is what now is costing the city at least $30,000 a year in his frequent visits to Bellevue.

Malnutrition, alcoholism and mental disorders all are rampant among the homeless, wearing down the equanimity of the doctors and nurses who try to minister to their needs.

At Bellevue Shelter, Dana Gage and Flo Botte and the rest of the staff make every effort to maintain a personal bond with their patients in the clinic, thus hoping to avoid many hospitalizations. "It also does a lot to dispel the distrust that many have for doctors and nurses," Gage says. "They know that this is a place for them, this clinic. They aren't going to have to wait eight hours to be seen. In fact, we've found that they cease to be 'the homeless,' and become true personalities, just like you and me." But the Bellevue Shelter is not typical of other city shelters, from all accounts, and seems to owe much to the leadership of its director, Norman Trosten, for the atmosphere of relative calm and warmth that pervades the place. Trosten spends his Thanksgiving and Christmas holidays with the men in the shelter before going home to celebrate with his own family in Queens. In fact he knows the names of 90 percent of the men in the shelter, and makes active efforts to learn as much about the families and personal interests of the men in the shelter as he can.

Even more wrenching to the emotions than the sight of ill and self-destructive men is the problem that until the recent past did not intrude on public awareness at all—homelessness of families. New York is trying to shelter some 4,000 homeless families, including 9,590 children. The conditions under which families are housed make even hardened citizens shudder in dismay. Among the consequences of the public housing of such single mothers and their children— which is the typical configuration of a homeless family—are these:

The concentration, in the midtown area of Manhattan, of numbers of these families in old hotels, now known as "welfare hotels," seems to have acted as a magnet for hangers-on, who are often alcohol or drug abusers. Street crime, including the mugging of middle-class residents of those neighborhoods where the families have been relocated, has become commonplace. Letters to the editor of the *New York Times* convey something of the desperation that these unfortunates inspire in their middle-class neighbors. Wrote one woman:

> I am in no way against helping homeless people, but there must be a better way . . . than to squander millions on the fees (at least $500 a week per room for a family) that these hotels charge New York City. . . . If one travels along Sixth Avenue late in the evening an appalling number of children from the welfare hotels can be seen begging or demanding money from drivers and their passengers by washing their car windows.

The Pediatrics Emergency Services at Bellevue can document the number of misadventures to which these children of the homeless fall prey, the diseases that mark their lives, the traumatic injuries that come to them because they do not have a personal habitat that is secure and warm. Yet the doctors also hear what often is said, especially by admirers of Reaganomics: that somehow such misfortune represents a shortcoming on the part of the homeless, that poverty is in significant measure the fault of the poor. Therefore it is not really the responsibility of the rest of us.

"When you can learn to care for these men and women, to see them as persons and share their problems, you appreciate the most basic aspects of humanity—what's left when all the trimmings of society are removed," Goldfrank has written. "We're fighting to preserve the integrity of the individual and uphold each person's potential strength. When you understand this and deliver the best care you are capable of to these men and women, you grasp the human experience in a way that I think equips you to work with all human beings."

He doesn't deny that working among the homeless sometimes taxes your equanimity to the uttermost, that you have to subdue your own negative feelings, oftentimes. But there's human potential in everyone, a history well worth hearing, something to learn to better understand ourselves. Hope is never completely dashed so long as heart and lungs and mind continue to work, he feels. "To most of society, the homeless are hidden people living in their hidden part of the city," he says. "To us here at Bellevue, they're highly visible. And audible. We can do our share to make life a bit better for them, and I'm glad to have the opportunity."

Clearly Goldfrank and his team have infused their work with the spirit that Sir William Osler gave voice to: "While doctors continue to practice medicine with their hearts as well as with their heads, so long will there be a heavy balance in their favor in the Bank of Heaven—not a balance against which we can draw for bread and butter, or taxes, or house-rent, but without which we should feel poor indeed."

# 29

# A Day and a Night to Remember

7:30 A.M. The patient in Room 6 was brought in by ambulance with a wrist abscess so advanced that there is real danger of losing the hand. Kelly Dowling is her name, and she's thirty-five. Dr. Goldfrank has just arrived for the day. Dr. Robert Hessler is briefing him on the patient as he looks at the form lying there with heavy makeup and streaked blond hair. She wears wrap-around sunglasses, even though it's a depressing November dawn without a rumor of sunshine.

"Good morning, Miss Dowling. I'm Lewis Goldfrank."

"Hi."

"How long have you had this?" he asks, gesturing to the abscess, a large, fluctuant mass of pus, having a soft, semi-liquid center. There is an inaudible reply. The mnemonic for abscess diagnosis, Goldfrank recalls, involves the Latin words *rubor, dolor, calor, tumor*. Gauging the qualities of the abscess, whether red, painful, hot or swollen—or all four—helps clarify how serious a condition the victim suffers.

"How long?" he repeats.

"I don't know. But I have to get home to take care of my guinea pigs. I have two."

"Miss Dowling, your infection is serious enough to cause the loss of your hand. Could even cost you your life."

There seems to be anguish behind the sunglasses. "But they need me, Willi and Carla. They'll starve."

"We'll see if our social workers can't get a neighbor to help out. Try to cooperate, now, for a moment."

Goldfrank and Hessler confer. The question that Hessler needs clarified is what type of bed to admit the patient to. Clearly admission is necessary, doubtless Plastic Surgery to deal with the abscess, which has to be lanced. What isn't clear is how to treat a transsexual in mid-passage, so to speak. Which ward does the patient get admitted to—male or female?

Hessler has told him that there has been an orchiectomy, removing the testicles, but the penis remains. There has also been a vaginal opening fashioned in stages. Now it appears that further work will be needed to make the transition to female complete.

Secondary sexual features—sizable breasts—attest to a course of hormonal medication that has been largely effective.

"Her medications have been Provera, estrogens and methadone," Hessler murmurs.

"So what do you think, Bob?"

"I'm leaning toward admitting her as a female patient."

"I think that's the right call."

"Thanks, Lew."

Later, Social Work was able to locate a neighbor who was acceptable to Kelly Dowling to take care of her guinea pigs during her hospitalization, which, after the initial lancing incision and drainage of the open abscess, and several days of elevation and antibiotics, did clear up her badly abscessed and ulcerated arm.

9:02 A.M. As Goldfrank goes out by the triage desk he hears officially that Carey Le Sieur has given her notice. First Le Sieur and then Sharda McGuire, two of the best nurses in the department.

As he later explained, "It's one of the worst problems we face. Just not enough nurses. But another problem, from my standpoint, is that nursing has its own chain of command. They are great in our department—the soul and conscience of it. But for years we have been getting by with about 60 percent of the staff levels that we need. So we have to go out to hire agency nurses frequently, to get part of our coverage. They often lack the commitment. Don't know the system well, really."

The conditions that inspired McGuire and Le Sieur to quit originated in some respects fifteen years before, when both nurses were still in school. Then the city, to economize during days of fiscal crisis, closed its nursing schools, including the Hunter College–Bellevue Nursing School, adjoining the hospital. That cut off the supply of student nurses who had spent part of their training cycle at Bellevue and had come to know it, and, often, to form long-term attachments to it. Bellevue, being a city hospital, cannot pay as much in dollar income to its professional staff as the city's private institutions. But what it gives in compensation, and continues to disperse at a scale probably above any other institution in town, is psychic income: the peculiar remuneration of knowing that you serve a tradition unequaled in its longevity and richness in the United States, and undiminished in its absolute commitment to help anyone who asks for help.

Those nurses who remain on duty feel the stress of having to break in a large group of new nurses every year as well as those hired from agencies to fill in from time to time. And they each are often expected to do the work that in other hospitals is allocated to two nurses of similar training.

9:23 A.M. A stretcher moves in from the ambulance entrance with a large thirty-four-year-old black female whose neck is enclosed in a white cervical collar and whose torso is lashed to a backboard as a precaution against possible spinal injury. Felicia Verchild is moaning and holding her head and her upper chest. Goldfrank begins to move toward the stretcher but his progress is interrupted by a small woman in a bright blond wig.

"Hello, doctor, do you recognize me? When is somebody going to see me?"

"Hello, Miss Neville. Has no one seen you?"

"It's been three hours."

"And no one's seen you?" He gazes over at Elizabeth Swanson at the triage desk, who makes a subtle elevation of her eyebrows.

"Only the nurses. No one around here cares about me."

"So someone has seen you?"

"They took my blood pressure and pulse." At that moment the loudspeaker overhead speaks: "Incoming pediatric trauma: eighteen-month-old. Severed arm. Fifteen minutes." Goldfrank catches Kunka's eye, and she nods, to let him know that the Pediatric trauma-team calls will go out directly. He turns back to Mary Helen Neville.

"All right. We've got some serious cases, and we're backed up."

"You don't call cellulitis serious? High blood pressure? This ulcer here?" She points to a leg, bandaged and unhappy-looking.

"Someone will get to you as soon as possible."

"Nobody cares."

"We care, Miss Neville. We know your name."

She looks up, surprised by his response.

"Just have a seat. We'll get to you as soon as we can," he adds gently.

"All right."

Felicia Verchild is in acute distress in Room 7, where Dr. Toni Field is examining her. The portable x-ray machine is in place and a technician is taking cervical-spine x-rays now. A glance tells Goldfrank that the case is not serious. Later the chart will say:

Patient statement/assessment: MVA driver hit from behind at fast speed. C/o pain L side of body—LOC. Full immobilization maintained.

Doctor's comment: 34 y o B ♀ in MVA this AM rear-ended while moving in traffic on Queensboro Bridge. States hit chest against steering wheel. No LOC. C/o L hip pain. PMH=PSH=Med=All=Θ Smokes 1/2 ppd. ETOH rare.

Diagnosis: MVA/c soft tissue contusion. T & R.

To novices who don't understand the abbreviations, Mary Dwyer, whose office is close to the chart racks, can be called upon to explain. In this case, MVA is motor vehicle accident. LOC is loss of consciousness. C/o is "complains of" and Θ can mean "negative" or "noncontributory." "PMH" and "PSH" are past medical and past surgical history. "Med" is medications and "all," allergies. "T & R" is a welcome abbreviation, as it means treated and released.

9:25 A.M. Another MVA victim is coming in. The chart that her brief visit will generate gives her name as Allegra Cranmer and mentions a fashionable address.

Patient statement/assessment: "I fell off my bike."/Cervical collar placed on pt.

Doctor's comment: 49 w ♀ riding a bike hit by a car. Fell to ground *not* hitting head. Rolled on L side. Θ LOC. No other c/o. Minor contusions L forearm, L zygoma [cheek]. Pt. extremely AOB.

Diagnosis: Bike/MVA, no cervical injury. T & R.

9:41 A.M. The pediatric trauma case is coming, they learn, by helicopter from the Kittatinny Mountains near Panther Valley, New Jersey. It is just now in sight of the Thirty-fourth Street heliport. Paramedics are standing by in their ambulance waiting for the chopper to set down. As soon as it does they rush forward to coordinate with the medics aboard, who see to the swift offloading of the tiny form of eighteen-month-old Calvin Deshawn, who with his distraught mother gets into the ambulance for the quick transit the seven blocks south to Bellevue. Siren screaming, the ambulance races down the service road paralleling the FDR Drive.

Members of the ED trauma staff are waiting as the vehicle pulls up to the entryway. Through the short exchange with the New Jersey paramedics the trauma team hears that little Calvin was sitting next to his mother, Lynette Deshawn, in the front seat of her car on an access ramp to Interstate Route 80, when she was slammed into from behind and her car was sent sideways into the flow of traffic. Her Toyota was struck by several oncoming vehicles. The baby was thrown out the window by the force of the impact. Medics who got to the scene found his right arm nearly cut off just below the elbow. Doctors at a regional hospital took one look and ordered the boy flown by helicopter to Bellevue.

The chief pediatric and surgical residents head the team that makes the assessment, and establishes that there is no internal bleeding, no C-spine injury, no airway compromise—in short no reason that the little boy cannot move into Microsurgery immediately. They have the baby upstairs within ten minutes. There, Dr. William Shaw and his replantation-microsurgery team of twelve specialists successfully reattach the arm in a seven-hour operation.

**11:12 A.M.** There are three construction workers at the triage desk. They complain of stinging eyes and rashes. They are at work on the finishing of a twenty-two-story curtain-wall structure. For Goldfrank, there's a mystery here that has been nagging him for some time. Others before have been grumbling; painters, carpenters, electricians, carpet and telephone installers who have been preparing the new building for occupancy have been showing up at the ED. They come one or two at a time, complaining of shortness of breath, tearing, stinging in the eyes.

As Goldfrank recalled later, "This was the day when it all came clear, and we had an answer as to what was poisoning them. We had been seeing these people ever since the work on finishing off the upper floors on their building began. The venting of the air-conditioning system went only as far as the nineteenth floor. I thought about it a lot because this building is typical of the newer structures in Manhattan. It's a sealed-window environment. The idea is to keep out the toxins from the outer world. But these tight buildings capture and recycle their own toxins—fumes from varnishes, polishes, breakdown products of synthetic fabrics.

"In a way it's a quarantine-minded design—or a suburban-minded design. Let's wall ourselves off from all the ills of the world, keep all those poisons away from us here inside. Well, it's ironic. Of course you can't build any such environment that you'd want to live in. I can't keep myself free of infection by walling myself off from the rest of the world, because we're all in this together. And no building can wall itself off, either. In other words, Pogo was right. We have met the enemy and the enemy is us.

"The workers were using a number of paints, thinners, adhesives and solvents with fumes that are quite toxic. But the venting only went to the nineteenth floor and all their noxious fumes were being recycled, because their air-conditioning did not connect up with the roof.

"So we had a microcosm right there of what the environment at large is suffering: their fumes were being vented right into their own sealed-window environment. They had a sophisticated air-handling system—a controlled HVAC system, it's called. (The initials stand for heating, ventilation and air conditioning.) But those fumes have a dynamic relationship with all the new materials that fill every skyscraper on this island—polymers, polyvinyl chlorides, acrylics. And, as the fumes settle on those materials, you get metabolites that are very damaging, causing rashes, wheezing, all the symptoms we were seeing from these workers."

In months to come, the Occupational Safety and Health Administration and the National Institute for Occupational Safety and Health both confirmed that the big problem was the incorrect venting. "Once the vent got hooked up to the outflow on the roof there weren't any more problems."

12:21 P.M. "Trauma. ETA 4 minutes." Again the quickening at the doors of the Emergency Ward as the trauma team musters to accept the next arrival. Burly nurse Ivan Quevedo, a man of average height but imposing breadth, moves with ursine grace to clear the x-ray machine out of the way of the incoming patient. The stretcher speeds down the entryway and through the pale red doors of the trauma slots. The paramedics bark out the details: twenty-six-year-old tourist Roger Abt of Evanston, Illinois, was browsing in the gift shop of the

United Nations when a man believed to be a released mental patient walked up to him and plunged a knife into his chest.

As the chief surgical resident, Dr. Lawrence Glassman, assesses the flow from a chest tube that he has deftly inserted, Dr. Kathy Delaney tries to assure that there are no more threatening disorders among the ABCs. One lung had collapsed under the impact of the knife. Now both lung fields are "expanded" (reinflated). Also, negative pressure has been reestablished on the pleural, or outer, side of the lung, which holds air without leaking on reinflation because of the surfactant (a word made up by combining "surface" and "active" with "agent") on its inner surface, which acts to seal the wound in a fashion similar to the inner sealant in a puncture-proof automobile tire.

Glassman fears that the knife has nicked the right atrium of the heart. Bad as that might be, it is still the least dangerous part of the heart to be pierced. Later the doctors explain the structure of the heart:

Imagine the heart divided into four chambers. The top two are the atria, which act principally as reservoirs to hold returning blood. The bottom two are ventricles, the powerful pumps that send blood out into the body's successively smaller conduits: arteries, arterioles and capillaries. Had one of these ventricles, these high-pressure chambers, been pierced, blood would have jetted out of the punctured ventricle into the space between the adjacent lining of the heart and the ventricle itself. As this adjoining space became flooded, it would have acted as a counterpressure on the ventricles, suppressing their function, creating a cardiac tamponade, and threatening a swift death.*

Glassman and a medical student start to push the stretcher, and the small procession makes a rapid trip down the hall to the waiting elevator, which will take Abt to the eleventh floor to establish for sure if his atrium has been torn, and, if so, to repair it.

1:15 P.M. The fire got started from a cigarette spark in an overloaded ashtray dumped into a wastebasket full of narrow strips of perforated paper. It spread rapidly through the sixteenth-floor storage room of the Park Avenue marketing firm, filling the corridors with

---

*"Tamponade" comes from an Old French word meaning "a plug." As Glassman later explained, tamponade can also occur from an atrial injury. As little as 50 cc. of blood— about 1½ ounces—in the pericardial space can produce tamponade, in a somewhat less dramatic fashion than ventricular injury will cause.

smoke and sending badly frightened office workers and executives to the emergency stairwells.

The 911 call sent several city agencies speeding to the scene, and activated the MERVan at Bellevue. An emergency medical technician climbed behind the wheel of the van and made the circuit of Bellevue's inner courtyard to pick up Dr. Robert Hessler and nurse Jeanne Delaney at the ambulance entrance of the ED within about five minutes of the first notification by the watch commander in Maspeth. An alert also went to Dr. Richard Weisman at the Poison Control Center, who organized information specialists prepared to give advice to any callers, particularly other hospitals such as Lenox Hill and Roosevelt, who would be admitting victims, also.

"We've been lucky in New York," Goldfrank has said, "in that there has been nothing like the disaster at the MGM Hotel in Las Vegas, where many guests were trapped in a closed structure with heavy smoke and fumes driving people to the roof. So far, in New York, there have been only minor incidents. But in this age, with so much synthetic building and furnishing material—capable of generating really lethal gases in the event of fire—it is only a matter of time until a major event of this sort materializes. There was Bhopal and there was Institute, West Virginia, and Sevesco, Italy, and Chernobyl—all reminding us that we live in a vulnerable ecosystem and we ourselves are an ecosystem that can stand only so much insult. We depend on 21 percent oxygen and suitable temperatures and humidity to survive. Every single high-rise built since the advent of new construction technology in the sixties, sealed against the outer world, and furnished with new plastics and petroleum-based synthetics, is a toxicologic emergency waiting to happen."

While Goldfrank continued to man his station at the ED, Hessler and the MERVan crew went on to the Park Avenue skyscraper, where the fire department had already sent its first contingents to battle the blaze. Hessler, Delaney and the van would be based to treat patients as they were brought down from the fire—firemen, occupants of the building or passersby who might be felled by the smoke and flames. By this time, the fire had spread generally on the sixteenth floor. Several firemen already had been disabled by smoke inhalation and were being treated by Hessler and his team with 100 percent humidified oxygen and swift transport, via EMS ambulance, to Bellevue.

* * *

1:25 P.M. A very badly beaten and comatose two-year-old has been brought in by ambulance. His heart has stopped. He is taken right into the Pediatrics Resuscitation room, where the chief pediatric resident and nurse Denise McLean try to save the boy, Derrick Haskins. His head is injured and there are signs of considerable distress, also, in the mottled bruises around his lower back, as if he had been heavily whipped with a belt or another hard-edged flail of some sort. There is real doubt whether advanced cardiac life support will succeed. The police are attentively watching as they have the mother's boyfriend under arrest for aggravated assault. In about twenty minutes, the pediatric resident comes out to the officers and tells them that little Derrick has not made it. It appears that he has succumbed probably to subarachnoid hemorrhage and epidural hematomas—bleeding in the brain likely to have been started as a result of a violent shaking before the beating that probably he received at the same time. His heart could not be restarted, in spite of extensive medications.

1:35 P.M. Diane Sauter is examining a young woman named Catherine Nicholas who has been treated by a "holistic clinic" for diarrhea. They gave her an arsenic-based compound, on the theory that she is suffering a parasitic infection contracted on her recent vacation in South America. But, rather than helping her condition, the medication has caused her diarrhea to become fulminant—sudden and severe—and Sauter has picked up signs of significant loss of fluid and also some deterioration of kidney function.

After depending on essentially cultic sources of treatment, Kate Nicholas realized she had made a mistake. Now Sauter thinks they can get rid of the arsenic successfully, and control the fluid imbalance in the meantime, but it will require hospitalization and chelation therapy. That involves an IM (intramuscular) injection of BAL (British Anti-Lewisite) compound that will intercept the arsenic in the bloodstream and body tissues and bind it in such a way that the heavy metal can be harmlessly excreted.

This type of problem annoys Goldfrank, and in fact he has pictured in his book an odd mixture that an herbalist prescribed for a four-year-old boy with meningitis: a combination of unknown weeds, leaves and locust exoskeletons (the outer shells). "It's discouraging to see someone well educated and sophisticated, like the young woman

Sauter was treating, resorting to highly questionable, sometimes dangerous, kinds of medication because it bears the name 'holistic.' Her misplaced faith may have cost her permanent loss of kidney function, and can significantly shorten her life."

1:45 P.M. Reports are coming back to Goldfrank from the fire. It looks as if they will be bringing in two firemen and a number of employees. They will be shuttled down to Bellevue by ambulance from the MERVan, where Hessler reports they are handling some twenty cases in all. The Holding Area is already being readied as the site for disaster care.

As Goldfrank said later, "I was thinking about the possibilities as we waited. The acute chemical pneumonitis, the hypersensitivity reactions, the asphyxia." When some of the new fabrics and plastics burn, their combustion products range from phosgene (the gas that was responsible for 80 percent of the World War I poison-gas deaths) to carbon monoxide.

He has written on the effects of such toxic inhalants in *Toxicologic Emergencies:* "Poisonous gases need not damage the lungs directly to have adverse effects on the body. Displacement of oxygen from the inspired air, as occurs with carbon dioxide poisoning, will cause passive asphyxiation. Exposure to carbon monoxide or cyanide, on the other hand, could lead to active interference with oxygen uptake, and to active asphyxiation." Whatever the circumstances, if you are caught in a sealed-window environment and there is a major fire, you could well be getting less than 10 percent oxygen, and "death may result on even slight exertion," Goldfrank writes. So your very attempts to escape such an atmosphere could doom you.

1:55 P.M. As they await the first arrivals from Park Avenue, yet another trauma case arrives, this one from across the river, Queens. Eighteen-year-old Lance Swoboda has been severely injured in a hit-and-run accident near a gas station on Northern Boulevard in Sunnyside. He was rammed by a pickup truck that knocked him from his Yamaha 750cc. bike with such force that the machine and he went skidding along the pavement and his left leg was almost completely severed. The truck kept on going.

The leg has been wrapped in saline cloths and young Swoboda is being given IV pain medication as a resident examines him. Blood pressure, 100 over 70; rate of respiration, 16; temperature 97.8 F.;

pulse 110. He is cleared for escort to Microsurgery and Replantation within minutes.

1:57 P.M. The first pair of firemen appear, both pale gray from smoke inhalation and both on 100 percent oxygen. At times like these, there's a feeling of desperation that the closest hyperbaric chamber readily available for use is on City Island. But there is no time for pondering the rejection by the city, on economic grounds, of the appeal for a hyperbaric facility here at Bellevue. There has not yet been the disaster that will, in the minds of the keepers of the public fisc, justify such an outlay. Goldfrank hopes today's fire is not that unwished-for event. Now he assigns himself the task of active coordinator of all that goes on in these 60 rooms for the duration of the fire.

2:02 P.M. Information from the disaster site is strong in detail: From Dr. Hessler via radio hookup and from the paramedics who accompany the firemen, it is clear that sulfur dioxide (from wool fabrics), nitrogen dioxide (from plastics), carbon monoxide, formaldehyde and acrolein (from wood and paper) all are present. And all are causing severe difficulty in breathing. There are also traces of hydrogen cyanide and organic cyanide, both resulting from pyrolysis (chemical decomposition by heat) of laminates in the paneled conference room where the fire rages hottest. Even the Styrofoam coffee cups, which stand on desks throughout office suites, become a threat to life. Styrofoam, when it burns, produces acrolein (an aldehyde, resembling formaldehyde—a colorless gas good for pickling and disinfecting, but no good for breathing). A distinct hazard is presented by the pair of nylon pantyhose, in the bottom drawer of the executive secretary's desk near the main lobby of the office suite. Melted and vaporized by the high heat, they produced ammonia and hydrogen cyanide. Acrilan (in carpeting, draperies, upholstery—as well as in fake furs and wigs) spews forth its own quantum of acrylonitriles, which can do permanent central-nervous-system damage: atrophied muscles, spastic bodily motion, lost mental function—all leading to the risks of hypoxia (oxygen starvation) and then post-hypoxic encephalopathy (brain destruction).

2:11 P.M. Drs. Stephen Waxman and Kathy Delaney each take a fireman for intense diagnostic work-up. They see in both the signs of extreme irritation to the mucosal membranes, eyes and upper airways, giving every indication from the heavily labored breathing that the small passages of the lungs are suffering obstruction as severe as that of a lethal attack of asthma. They worry that various gases have set in motion the build-up of fluids in the alveoli. If not promptly reversed, such edema can cause rapid death. The A of the ABCs is still the scene of the major struggle.

The bolus dosage of aminophylline (a bronchodilator) by the IV line that Hessler had established in the MERVan has begun to ease the breathing of fireman Beaufort Callahan. He's a burly twenty-eight-year-old veteran of firefighting in the Fort Apache section of the South Bronx and in midtown Manhattan, as well. He looks up at Waxman, who murmurs a word of encouragement: "Nice work. You're getting things under control. Nice work."

2:35 P.M. Now there are seventeen victims from the fire—firemen, executives, secretaries, delivery boys—arrayed in the Holding Area, among whom Leon Fields moves swiftly taking vital signs and penciling his findings—blood pressure, pulse, respiration and temperature—on the sheets at the head of each stretcher. Calls that have gone out have resulted in the sudden appearance of extra attending physicians, residents and specialists from Internal and Chest Medicine as well as General Surgery, who blend swiftly into the intensive activity in the Holding Area.

2:37 P.M. A report comes in from the EMS watch commander: the fire is now under control, and hospitals throughout midtown Manhattan have received some of the victims. At Bellevue, only fireman Callahan seems to be in serious difficulty. Plans are now being made to send him to City Island for a hyperbaric dive, to hasten his recovery.

2:38 P.M. A tall and very photogenic brunette is asking Mary Dwyer at the Walk-in Clinic how long she will have to wait. She has already been here an hour. "Miss, we've got a major fire in midtown and all our staff is taking care of that right now. You'll just have to be patient." The brunette goes back to take her seat, watching, along with the others in the waiting room, the blaring progress of a game show on the TV set high on a shelf in the corner over the Walk-in nursing station.

3:15 P.M. By a quick succession of visits to the various examination rooms and corridor stretchers, Goldfrank continues to energize the staff to a brisker rate of decision making.

Later, in tranquillity, he can reflect, "You have to convince people they can become expert right now," he says of the young attending physicians who are still somewhat diffident about following their own intuitions. "They're well trained. Well motivated. All they *aren't* is well experienced. So I tell 'em: Look, you've got to become an expert, *now*. We need your skill today. Right now! You can't go at a pace that will result in developing the needed expertise next year."

3:50 P.M. Hessler, Jeanne Delaney and the MERVan crew are back. They don't even take time to debrief, but fall in to take up the backlog of less urgent cases that remain to be handled. The tall brunette with the provocative eyes is seen, after waiting nearly two hours.

Her name is Susan Cleshette and she's twenty-nine years old. She's a singer, born in Kentucky, now on her way to Paris. Her complaint is chronic asthma. Her chart:

29 W ♀ requesting medication renewal for 1 yr and a health certificate prior to 1 yr long trip to Paris. Pt c/o asthma since age 12, on daily medications since age 14 and requiring approx. 1 hospitalization per year for asthma. Denies intubation and prednisone use /c each hospitalization. Pt. was in Paris from 3/1/84 until 6/85 and had 6 hospitalizations for asthma. She states the medications are not as good in Paris and she now requests 1 yr supply of meds in preparation for her trip.

PMH:—⊖ cardiac, TB, emphysema, hepatitis, renal. Maintenance medications: Theo-Dur, terbutaline, Vanceril, Alupent. Single singer born Ky. ⊖ cigs ⊖ IVDA + ETOH: socially . . . previously heterosexually active; recent "cold" /c running nose . . .

Diagnosis: Asthma T & R.

The attending physician explained the names of European equivalents of American asthma medications, then gave her prescriptions for her first month's needs.

4:10 P.M. The ambulance entrance doors burst open and a comatose and emaciated young man is wheeled in to Room 6. The loudspeaker calls Goldfrank to the triage desk, where Elizabeth Swanson hands him the note that came with the patient:

To Whom It May Concern:

Kyle Tippett has AIDS and has exhausted his benefits and is therefore conveyed to the care of the Health and Hospitals Corporation. His current diagnosis is Kaposi's sarcoma and *Pneumocystis carinii* pneumonia. The patient says that his regular physician, with whom I am not acquainted, is out of the country.
—Garamond Thorne, M.D.

"All we can do in a case like that," Goldfrank later said, "is to do the best we can by the patient and make a full report to the Health and Hospitals Corporation about such dumping. It's supposedly illegal, but we still see a lot of it. The patient never regained consciousness and died the next day in the Emergency Ward."

4:50 P.M. Goldfrank makes another circuit, noting that about half of the fire victims have been admitted upstairs, while a few have been discharged. So far, no deaths have occurred among those brought into Bellevue from the fire, although Fireman Callahan is still in respiratory distress at the City Island facility, but things are looking good for him.

5:15 P.M. There's an angry voice coming out of Room 2, so Goldfrank looks in. There Steve Waxman is telling a forty-three-year-old Hispanic that he can sign out if he wants to, but it will be against medical advice. Well, he wants to. Okay. Sign out, then. Goldfrank gets the gist in an instant: Manuel Ortiz was cooking chicken fat for his lunch at his place on Jefferson Street when he spilled the fat on his legs. He went to Gouverneur Hospital and then was brought by ambulance to Bellevue.

Pt. has been on Thorazine for 20 yrs. for Ψ [psychiatric] diagnosis; he is followed at Gouv. He denies any past medical hx [history]. No med allergies. 2nd degree burns R foot, medial aspect L thigh. 1st degree burns medial aspect L thigh.

Pt. seen to be admitted to plastic surgery—all labs done for admission. Plan: diphtheria-tetanus toxoid, Hyper-Tet, Silvadene oint, dry dressing, admitted; Velosef 500 q6h × 7 days ["q6h" is a Latin abbreviation for "every six hours"; not the same as "four times a day," Goldfrank reminds his students].

"So what's the trouble, Mr. Ortiz?"

"I wait already too long."

"We're ready to help you now."

"I wait too long."

"All right," Goldfrank says, and then makes one more attempt to persuade Ortiz that his best interests will be served by staying in the hospital. But Ortiz is adamant.

Later Goldfrank said, "We felt he was competent to leave against advice, that his psychosis was not so pronounced that we would be justified in keeping him against his wishes, as we knew that Social Work would be urging him to the clinic for follow-up." So Ortiz signs himself out and leaves, AMA—against medical advice.

7:10 P.M. Just when it looks as if Goldfrank will get a few minutes to review his correspondence in the quiet of his office, there's another loudspeaker interruption: Shot cop. He's coming in, they soon learn, with his partner, who, while struggling with a suspect, accidentally wounded his own partner.

Dana Gage is on hand to talk to the partner, while the victim himself goes right into the trauma slot, where Robert Hessler makes sure of the ABCs, while chief surgical resident Lawrence Glassman, having supervised the placement of the chest tube, studies the x-rays, which reveal that the bullet is lodged in a lung. While serious, the wound is not mortal and there is no indication for an immediate operation. To justify immediate surgery, these signs would have to be present: loss of more than a liter of blood (a bit more than a quart), persistent bleeding, marked by lower blood pressure or indications of cardiac injury—none of which is present. So, for now, no surgery is planned.

Brian Filosa is a baby-faced policeman, twenty-two years old. He was struggling with a perpetrator who had grabbed his partner's gun in a drug arrest. Filosa is almost in tears as Dana Gage tries to calm him down, trying to get epinephrine injected swiftly enough so that Filosa's aggravated asthma attack will not worsen dangerously.

Already the Police Internal Affairs Department interrogator wants to talk with Officer Filosa at the first possible moment. But Dana Gage keeps him cooling his heels until she is sure that Filosa is all right.

As Dr. Gage later said, "It's hard to resist the momentum that develops at Bellevue whenever a policeman shoots someone or is shot himself. The place turns blue." There are now an extraordinary number of police officers in the corridors and entryways to the ED.

"Like other uniformed services and the members of the 'caring' professions," she said, "an assault or injury to one stirs a dramatic response in those closest to the victim. It seems to be especially difficult to see a brother or sister become vulnerable if you are used to being in control. That state of affairs stirs up the deepest of anxieties."

8:05 P.M. There is a quiet period at last, and, this being a Friday, when Goldfrank normally works until midnight, he can go to his office and get his dinner out of the refrigerator—a large slice of cheese, half a loaf of Susan's homemade bread, some fruit and a can of apple juice. Then he reads current issues of journals and "request for proposal" letters from several foundations and state agencies opening competition for limited funds for medical care and research of various kinds. He is never at a loss to make a proposal, and regards it as an essential part of his job to assure healthy survival for his department in an age that seems to offer less funding each year. At the same time it is an era that demands increasing intensity of care as AIDS, crack, ETOH and other drug abuse extend their pandemic pattern.

9:47 P.M. Two screaming young men are brought in. One threw a glass pitcher at the other, and nearly severed his nose. The two let it be known that they are lovers. Now the wounded one is being sewed up by Plastic Surgery and they're telling the aggressor what may be entailed in the recuperation: The arteries can be reconnected, but not the veins. They're too small. Even so, they will be naturally reconstructed in due course. Meanwhile, it may be necessary to drop back to the eighteenth century for an interim expedient: leeches may well be applied to prevent the newly reattached nose from becoming engorged until the veins can reconstitute themselves and carry off return blood in their own orderly and natural way. "Leeches! Yeeech!" says the perpetrator. He is surprised to see two police officers hovering at the edge of the examination room, looking, apparently, for the roommate's signature on a formal complaint.

11:25 P.M. There is still a considerable backup in the waiting area, so Goldfrank calls Susan and says he will be working on past midnight. And, should he miss the last train, he'll just do what he has done frequently in the past—catch the first train in the morning. That will be the 6:20.

His innovation over the years in the ED has been to schedule work shifts so that they overlap and so that few people arrive or depart at the same time. That keeps a constant level of energy and avoids the "quitting time" phenomenon, when a whole group might slack off in anticipation of an imminent changing of the whole work roster. As the night advances, he continues to prod and urge people to more brisk "binary decision making": Is the patient healthy or ill? If ill, can he treat himself at home or should he be admitted? If home, should there be a follow-up visit to a clinic? What clinic? If too sick to go home, what critical problem makes him too sick? If admitted, to the Emergency Ward or one of the departments upstairs? . . .

As he looked back on it, this was one of his more memorable days—and nights—on duty since coming to Bellevue. It seemed appropriate that it ended with a pair of trauma cases that called upon the varied resources of Bellevue in a characteristic way:

2:10 A.M. A phone call awakens Dr. Gene Coppa in his Waterside apartment across the FDR Drive from Bellevue. The message is terse: multiple gunshot victim, already in the trauma slot and ready to go upstairs to the OR in perhaps eight minutes.

As Coppa is on the duty roster for trauma surgery this night, he has all his preparations made. His jeans and shirt and sweater are on the chair next to his bed. His running shoes and socks are side by side beneath the chair. He leaps up, turns on the light and slips into his clothes quickly, splashing some water on his face to bring himself to full alertness. Then he grabs his beeper, his wallet and his keys— apartment and car keys all on the same ring. He jams the wallet in his back pocket, his beeper in his right pocket, and then goes out the front door, double-locking the door behind him. He slips the keys into his pocket atop his beeper and rings for the elevator.

Outside, there is a sparkling clear sky, and the thrumming of traffic on the FDR Drive is nearly silent, under a bright full moon.

Coppa goes through the same calculation each time he is called to trauma surgery. He can use either footbridge over the parkway— both of which take him far out of a direct course—or he can risk his life by jumping the fence, entering the FDR Drive, running across the northbound lanes, vaulting the center divider and then traversing the southbound lanes. One more fence and he's right at the Emergency Department entrance, and within fifty feet of the trauma slot.

Tonight he takes the frontal approach. Later he recalled, "As I cleared the median barrier, I could feel and hear my keys hit the pavement beneath me. But there was a car coming southbound and it wasn't that far away, so I just kept going. Figured I'd find my keys when I came back."

Coppa got to the ED just as the trauma team got the patient stabilized. He learned that a fellow obstacle vaulter would be on the other end of the scalpel. An obese man, twenty-nine years old, had been observed by two Transit Police officers jumping over a turnstile in the Times Square station. The police gave chase, ordered the man to halt, and when he seemed to be pulling a weapon on them, opened fire, felling the man with five separate bullets.

"It was a mess," Coppa said later. "Took more than four hours—liver, colon, small bowel . . ."

4:07 A.M. Goldfrank is on hand for the arrival of a victim of multiple stab wounds. There's the bustle of the excited arrival, and the paramedics' estimate that there are three stab wounds—chest, groin and hand. The groin wound appears to be the worst. The victim is thrashing beneath his Ambu-bag, trying to sit up, as his clothes are cut away.

"Stay calm," Goldfrank tells him.

But he sits up, looking at the team at work on him—Goldfrank, Carlos Flores, Glassman, Ivan Quevedo, Pat Sorensen. They push him back down, having by now got down to his plum-colored underpants. He keeps holding his hand there.

"This is no time for modesty," Goldfrank tells him. They remove the underpants as two police and two paramedics look on. Clearly the police have some charges pending. From the underpants comes a large wad of bills and a small vial of what appears to be crack.

Swiftly they identify the sites of the three stab wounds. It appears that the groin thrust could well have ripped into the bladder, which could portend grave complications. Insertion of a Foley catheter is the intended procedure. The plan is to push this catheter with a local anesthetic up the urethra of the man's penis into the bladder and to do a rectal exam to see if the bowel has been injured and to plan an arteriogram to establish if the femoral artery has been perforated. If any of these, major repairs will need to be undertaken surgically. If not, they can count on suturing the three wounds and a quick recovery.

The young man gets the nub of the discussion and sits upright once again, shielding himself with his hands as he does so. Quevedo has just finished counting $385 dollars in cash and two credit cards, which he is going to voucher. The police ask for the name on the credit cards.

Across the room, the patient, meanwhile, is hitched up on one elbow, resembling a small version of Rambo as he glares at the doctors. "You going to put that thing in me?"

"Yes. Lie down."

"In my penis?"

"Yes, lie down."

"But my penis is very important to me. I don't want that thing stuck in it!"

Goldfrank firmly pushes him back down and tells him with urgency: "Your choice may be whether or not to walk. If you got a tear in your femoral artery it could mean you could lose your leg. You could also have internal bleeding right now into your bladder. Or urine pouring into your abdomen. We have to find out. Lie down. Be quiet."

"You're going to push that thing all the way to my bladder?"

"Just calm down. You won't even feel it," says the imposing Ivan.

The man lies down. The procedure is over in a few minutes. He is lucky: there is no blood in his bladder. The police step forward and tell him he is under arrest. They'll stay with their prisoner in the Emergency Ward, and then in the fifteenth-floor Surgical Department, where he is admitted for recovery, until he is able to be formally charged. Meanwhile, he goes for angiography, a further check to see if he has suffered an injury to his artery that might have escaped notice.

5:57 A.M. Goldfrank and Coppa stroll out the ambulance entrance together into the luster of a predawn full moon in a cloudless sky. Goldfrank is bound for Grand Central, and will take the footpath up toward University Hospital, then over to First Avenue and on to the terminal. Coppa has a shorter trip. He turns to face the FDR Drive, leaps the near fence, and approaches the median barrier. There is light traffic, but enough to have kicked up grit and gravel, which have half-covered the object of the surgeon's search. As Goldfrank watches, by the blue radiance of the setting moon, Coppa bends over and retrieves his car keys lost earlier that morning. Then he turns to wave them to Goldfrank, and each goes on his way.

# 30 

# Toward Medical Victory

A Mozart sonata, drifting up from the parlor, awakens Goldfrank at about three that Saturday afternoon. Outside, a steady snow is falling, and already there is an accumulation of several inches on the branches of the great Norway spruces that stand like sentinels around the old house.

When he gets downstairs Susan is just finishing her practice at the parlor grand. They go together into the kitchen, where she has coffee and toast waiting for him. She also has a bit of bad news. The roof is leaking. There's a moistened patch of plaster in the hall ceiling over the back door.

They've been through this before, whenever there's wet snow driven by a strong southwesterly wind.

"Is Andy coming out this weekend?"

"He's here already . . . doing some errands for me. And the girls are visiting friends."

When Andrew comes back, he is bubbling and laughing after a brief snowball exchange with a neighbor. He and his father take snow shovels, go up to the second floor and then through a window and out onto the porch roof. There they carefully scoop the snow off the copper-clad roof, pushing it to the front edge, from which it plops down thirty feet to the flower beds below.

As they work, Goldfrank and his son can glimpse the Hudson in the distance through the swirling storm. They can see the great expanse of Haverstraw Bay, bordered on the south by the dramatic hills of the Palisades—High Tor and Hook Mountain. Further to the south, there is the sweep of the Tappan Zee, and just beyond that, visible as a pink aura in the evening sky, lies the great city.

Goldfrank's portable telephone buzzes at his belt. It's an information specialist at the Poison Control Center who has a doctor on her line from a Brooklyn hospital with a case involving hypothermia and barbiturates. As Goldfrank gives his recommendations, Andrew continues to shovel the snow to the edge of the roof, and then, gingerly, pushes it off into the rising wind, which now is shifting around to the northwest. The cold is strengthening, also.

As the call is finished and then their rooftop task, Goldfrank can reflect on the enormous ecosystem of which, here in their wooded preserve, they are a part. It includes twenty million people just a few miles down the river valley and across the moraines to Long Island and the wetlands to New Jersey. These various parts of the system are bound together by such sure and swift means of communication that one part can help and support another. The telephone links maintain in this metropolitan area a kind of homeostasis, just as the body maintains itself in good health by its incessant communication, one cell, one synapse, one enzyme, one organelle with another.

Goldfrank wonders, from time to time, what may lie ahead for him personally. Emergency medicine has always been a young man's specialty, as it requires physical vigor, vast reserves of energy, the desire to explore new terrain in medical care, and the sort of unquenchable hope usually associated with idealistic youth, not chastened middle age. Not that he has entered the latter category. He still strongly feels that "If too many in high positions agree with you, beware!" If you're working with the underdog, striving to be a worthy patient advocate, you won't go wrong, even though you will often run afoul of the establishment.

He is far from being a tamed rebel. In fact, one of his favorite poems is Bertolt Brecht's "Burning Books," which includes the anguished plea of a writer, who, scanning the list of books to be burned by a repressive regime

> . . . wrote a letter to those in power:
> Burn me! he wrote with flying pen, burn me! Haven't my books
> Always reported the truth? And here you are
> Treating me like a liar! I command you
> Burn me!"*

*From *Poems, 1913–1956*, by Bertolt Brecht, Methuen, London, 1976.

Goldfrank thinks that here, working out in the elements in all seasons, chopping wood, mending fences, he achieves the renewal and finds the inspiration for confronting the hard decisions at the hospital the rest of the week. Perhaps, he sometimes thinks, if the problems can't be solved, if he can't continue to find dedicated associates willing to keep up the struggle against chaos, to try to do the impossible, then he might leave. He and Susan might move somewhere where they could concentrate on their gardening and farming interests—to a place where emergency medicine could be done very simply. But those are fleeting thoughts, not more than occasional states of mind. Now there is a sense of successful coping with medical problems that many of his old school chums and professional colleagues find simply impossible to comprehend or perhaps unworthy of their interest.

Father and son finish their job and go downstairs to the dining room, where a diffident Andrew has something to show his father. It's his paper on the meaning of mentorship for his Introduction to Philosophy course at NYU, which he has just got back with a good grade from his professor. As Goldfrank reads it, he cannot help feeling a glow of pleasure, and a bit of genuine discomfort at the same time. For there is, after all, his lifelong disposition to believe that deeds are worthy of belief while creeds are best approached with skepticism. Now here's his own son making him a part of a creed.

Andrew wrote: "My father is capable of turning anger, or frustration, or helplessness into a source of energy to work harder and further. All of this is summed up by the fact that he is an intelligent and articulate individual who cares more about helping others with his experience than advancing his own personal situation."

"Thanks, Andy. Every man should have such a generous portrait painter as you," he says, handing the paper back across the table.

There's another phone call, this one from Assumpta Agocha, who has the Emergency Department at Booth Memorial Hospital in Queens on the line about a case of acute aspirin intoxication. There is a few minutes' discussion about alkalinization to shift the equilibrium from the tissues to the urine; then he rings off and goes to his study to catch up on his journal reading.

It's an airy room near the master bedroom, with bookshelves lining four walls. "Yes," Susan says, "and last summer I was sitting at breakfast, after Lewis had gone off to work and c-c-crrrash! Down came a whole bookcase, it was so overloaded!" She reerected the bookcase and they decided to move some of his sizable medical library out, to reduce the load on the shelves here in this room. So Andrew, on his way to see the Mets one afternoon, took a load of books, plus fifteen years' worth of the *New England Journal of Medicine* in binders. He drove them in his father's old blue station car down to the Emergency Care Institute. Once there, he wheeled the books on a stretcher up to the third-floor Moulage lecture room.

The snow has slacked off, and the wind has dropped, but the thermometer is still plunging, reaching now the low twenties—a sure precursor of a sharp rise in numbers of patients suffering hypothermia.

$$\sim\!\!\!\!\wedge\!\!\!\!\sim\!\!\!\!\wedge\!\!\!\!\sim$$

Goldfrank picks up the *Annals of Internal Medicine,* where there is an article by Dr. Paul Beeson, past editor of the standard work on internal medicine, the *Cecil Textbook of Medicine.* The article gives a cameo history of the last hundred years in the field. It tells how the old distinction between physicians and surgeons has been lost, and "internal medicine" (borrowed from the German word *innere*) has come to designate—as Osler wrote—"the wide field of medical practice which remains after the separation of surgery, midwifery and gynaecology." Yet it is from the union of all these fields that emergency medicine springs.

As Goldfrank reflects on the field, it seems to combine the best characteristics of medicine and surgery—the thoughtfulness of one and the operating skill of the other. And it also calls for the psychological skills and perceptive intuition of a psychiatrist, the abilities of a pediatrician to communicate with children and the skilled understanding of women that an obstetrician-gynecologist must cultivate.

In concluding his historic survey Dr. Beeson touches on the difficulties of nomenclature. Which one of several possibilities would be a good designation for the generalist in medicine? "At present, several terms are used to describe the field: primary care, principal care and general internal medicine." He prefers the last as the best term.

But, even if the problem of nomenclature is solved, there is still the ongoing problem of these generalists—very much including emergency doctors among their number—in medical academia. "In that setting, professional success, as signified by promotion and pay, depends on the production of new information. This is easier for the specialist faculty members who can use some basic science or some complex technology in clinical investigation. In contrast, the faculty members who serve as generalist physicians, although they perform necessary service as role models and providers of patient care in their institutions, have more difficulty in showing scholarly productivity—that is, publications in prestigious journals." These are all circumstances that Goldfrank has had to contend with throughout his career.

In fact, he and his contemporaries have seen new journals established to meet the demand for reports and developments in a field that does not present enough in the way of academic interest for the more research-oriented medical publications that predominate. Traditional specialties are "organized," and are often scripted somewhat in the manner of a corporate-style job description—to avoid the chaos of spontaneity. But spontaneity is a leading element of emergency medicine.

*Hospital Physician* was one journal that sprang up with the new specialty, and it sought out Goldfrank in the early seventies, to present his toxicology in case-history format. As Goldfrank has written, when he started out to practice emergency medicine, he and his colleagues found themselves unprepared for many of the problems that confronted them. Until the recent past, he noted, there was "almost complete neglect of toxicology in our medical schools," yet poisoning represented a fifth of the admissions to the hospitals they then were serving, and alcoholism was found in more than 50 percent of those hospitalized.

"As late as 1969, a classic textbook on pediatrics stated that the leading cause of childhood mortality was poisoning, yet just 12 pages in the text were devoted to toxicology. Only in the most recent editions have the standard texts on internal medicine begun to give comprehensive coverage of this field," he wrote just a decade ago, in 1977.

In the third edition of the text that grew from the articles that he and his colleagues wrote, Goldfrank and his co-authors summed up the evolution of emergency medicine by noting the growth of federal standards and the establishment of high-quality regional poison control centers. "Compared with the 1960s, when the typical poisoned patient was supposed to be a child and the pediatrician the only physician concerned with running the poison control center, the 1980s demonstrated an awakening concern and interest on the part of internists and emergency physicians as active partners with pediatricians in the prevention and management of toxicologic incidents."

Today, adults represent the gravest problem, inspiring the development of new sources of information. "National . . . systems such as *Poisindex* and *Drugdex*, fine texts in clinical pharmacology and toxicology, frequent symposia, journal articles, and thoughtful analyses have dramatically changed the climate of care for the better."

Elsewhere, Goldfrank has noted that "Medicare and Medicaid really revolutionized patients' rights. Before the early 60s, impoverished people couldn't go anywhere." But, once these people were empowered to seek help, and as they did so, they created a level of activity in many hospital emergency departments that acted as a powerful attraction to physicians who wanted to station themselves in a place where they could care for more people. "If you saw 2,000 patients over a period of 365 days," he has said, "who would want to stay there all the time? But if you got 20,000 or 100,000, then the emergency department could be intellectually stimulating." And on just that succession of events emergency medicine established itself.

As Goldfrank says, a busy emergency department provides a community in which "you have minute-to-minute relationships among the doctors, social workers, nurses, physician assistants, nurse practitioners, paramedics and clerks—all trying to find a way to send the patient back into society with some functional status." But, because the city hospital like Bellevue deals with prisoners, immigrants, street people, they see disproportionately the chronically ill, the failures of other doctors, the outsiders who seem to fit nowhere else.

"Often we're told that certain cases won't be reimbursed because the patient isn't sick—doesn't have a medical illness that you can look up in the back of a textbook. But, if you throw them back on the street, they're not going to survive. You have to make a decision about your responsibilities. Are they to your financiers, your intellectual leaders, or to your patients?

"There's a kind of financial triage going on every day. It's become the practice all over the country to transport indigent patients—sometimes exceptionally long distances—to the sole hospital that's willing to take them. If you say that someone who has money can be treated today but a poor person has to wait until next week, you're shirking your responsibilities, not giving people the care they need."

Such attitudes put the emergency department in the position of the conscience of the hospital system, and as such it often engenders conflict with the rest of the hospital staff. "Here's some work you can do" seems to the rest of the hospital to be the message that the ED frequently sends along when it admits yet another patient.

"The practice of emergency medicine is chaotic, disorganized," Goldfrank has written. "You start to take a history of someone and another patient comes in who is sicker so you jump from one patient to another. In the office, you take the person who comes in at 3, then another at 3:30 and another at 4. The chaos of emergency medicine is determined by real time and human events. . . . You've got to be willing to jump from one event to another. To decide what is the branch-point in your algorithm. There's no place to hide in the Emergency Department."

Outside, the snow has stopped, and the thermometer at his window is showing 18 degrees F. Goldfrank turns to another journal article, this one from the *Archives of Internal Medicine* by Dr. Steven Swiryn, a Northwestern University Medical School cardiologist, "The Doctor as Gatekeeper." It touches on several points close to Goldfrank's heart:

"Patients are not evil-doers," Dr. Swiryn writes, in response to the spirit of the "DRG" age, when cost control becomes uppermost in the minds of many, obliging the doctor to put an exact label on each disorder, and giving him strong incentive to treat it within time limits set essentially by accountants and business managers, not doctors. Swiryn continues:

Oh, some smoke cigarettes, or eat marbled beef, or forget their blood pressure pills. And I scold or cajole or predict dire consequence. But to the old man with lung cancer I do not say "look what resources you have stolen from us with your addiction." He is no thief from whom I must keep costly care. I do not emphasize to the 40-year-old executive with his second heart attack that his company's health insurance costs will rise because of the marbled beef he ate. He is not wasting corporate money that I must somehow save. The old woman with a stroke is not at fault because she missed a few pills. She has not misappropriated tax dollars so that I should DRaG her out of her hospital bed. These are not miscreants to be condemned to fires. I do not refer to them as "health care consumers." To me, they are people who are sick.*

Goldfrank turns to another pair of articles, which Waxman has clipped for him, both from the *Wall Street Journal,* noting that the rise

---

*Archives of Internal Medicine*, vol. 146, September, 1986, p. 1789.

of HMOs—health maintenance organizations—which goes hand in hand with the DRG development, gives the doctor the job of getting the patient in and out as quickly as possible. It's fast medicine just the way McDonald's is fast food, with many of the same shortcomings in amenity, variety, courtesy and human warmth. "An increasing number of doctors are labeling it the substitution of the Veterinarian Ethic for the Hippocratic Oath," says Harry Schwartz, writer-in-residence at the College of Physicians and Surgeons of Columbia University.* He explains that "the veterinarian owes his primary obligation not to the animal he is treating, but to the animal's owner who is paying the bill."

Schwartz concludes: "One interesting piece of information has recently come to light about how some very influential and knowledgeable consumers feel about gatekeepers and HMOs. McGraw-Hill's Washington Report on Medicine and Health has made a survey and reports, 'Not one of the top Reagan Administration officials or key members of Congress promoting prepaid ... health insurance belongs to a prepaid health maintenance organization [HMO].' "

The second article, "Medical Discord: Some Doctors Assail Quality of Treatment Provided by HMO's," says, "HMO's do succeed in cutting health-care costs. They claim savings of 25%, an assertion backed up by a five-year Rand Corp. study of a 325,000-member HMO in the Seattle area. But doctors are turning up increasing evidence that the drive to slash costs is degrading HMO medical care, and they fear that its quality may decline further as HMOs continue to penetrate the health market and compete more with one another."

The article quotes internist Arthur Efros, who is an official of a Southfield, Michigan, HMO: " 'The idea of an HMO is to conserve care, restrict care. The less you do for a patient, the more money you make. . . .' He says he once saw an emergency-room patient being denied admission to a hospital because the patient's HMO didn't open until the next morning. Bleeding internally, says Dr. Efros, the man sat for six hours, his stomach pumped periodically to remove blood from it. Under many HMO plans, patients can use only HMO-affiliated hospitals. As a result, unaffiliated hospitals, fearing nonpayment, will sometimes refuse treatment unless they first obtain the HMO's approval."†

---

*_Wall Street Journal,_ July 9, 1986.
†"Medical Discord . . ." _Wall Street Journal,_ September 16, 1986, p. 1.

The sun is setting now over the Hudson, and a long saffron cloud is illuminated in a field of pewter gray over the Palisades. A few more minutes and Goldfrank will go downstairs to lay a fire in the living room, where he and Susan will return, after dining out, to read and listen to old Seeger or Baez records, or perhaps play Scrabble with the children later this evening.

He worries, reading such commentaries as those in the *Wall Street Journal,* about the rise of gentrified medicine, in which private hospitals and their house staffs will, under the new business-management principles of DRG and HMO, see only easily categorized, "paying" diseases, will handle fewer emergencies, fewer poor patients, fewer varieties of disease, consequently. The teaching environment of such institutions will be impoverished at the very time that their bottom lines will appear to reflect ruddy good health. Have well-insured people only and you wind up being an adjunct of the costly diagnostic equipment typically brought to do battle against cancer and heart and kidney disease. Under such circumstances, the doctor becomes more of a machine tender than a real physician. He doesn't talk to patients, because the patients are so busy moving from one test station to the next. It will be medicine by script, by recipe. It will not be medicine of human warmth and educated intuition.

The last thing Goldfrank browses in is a new book by a gynecologist who describes a huge malpractice judgment made against him. The Bellevue ED has been through a number of such suits, many of them frivolous, and none of them very damaging. Still, it is a subject on every physician's mind. In the past ten years, the ED at Bellevue has evolved in a way that has minimized the problem of malpractice suits. Full-time attending physicians have become the norm. These seasoned doctors sign the charts and supervise the care of the interns and residents. Experienced physicians take part in all the major decisions, thus demonstrating a level of concern and compassion that has done much to remove the basic cause of such suits: a feeling on the part of patients that they are being treated with cold indifference.

An intensely chilly Monday morning; 7:30. A jangling in the entryway as Goldfrank comes out of his office to survey the corridors.

There are laden stretchers everywhere, on one of which, in the resuscitation room, since 1:30 A.M. they have been attempting to revive a frequent visitor to these rooms. Leon Fields greets Goldfrank with the news. "It's Toby," Leon says. "Looks like he's gone," he adds in a low voice. They go into Room 6. There, before them, is a malodorous small black man with just one leg, and no pulse, no heartbeat, no respiration. When he was brought in, Wilts had a rectal temperature of 54 degrees F. No one is known to have survived at that level.

Goldfrank looks down on his long-time patient. His temperature has now been brought up to 96 but there is no sign of life. The paramedics found him, after a 911 call. He was frozen solid in a huddle against a half-demolished building on a midtown construction site. No one knew how long he had been there. Probably at least since Saturday is the best speculation. Clearly there is no hope, and Goldfrank shakes his head slightly to acknowledge that the end has come.

The Green Cart is summoned, and Goldfrank is alone with his thoughts for a moment. He knows, as few others do, what happens when a man dies. The brain cells, those miracles of cross-wiring and versatility, are alive for only four minutes after the heart stops beating. Toby's have doubtless been dead for a day or more. Even though the heart could be started up again, possibly, the endless interconnections of the brain have been dismantled, like a giant switchboard whose trunk lines are ripped out of their sockets.

The heart endures longer, and it's probable that Toby's atria continued to quiver and contract long after the ventricles grew still. The Latin name for the right atrium is *ultimum moriens,* the last to die.

Each province and principality of this defeated kingdom that was Toby Wilts disintegrates on its own schedule. Brain and heart may be ruins, yet the stomach continues to digest food for twenty-four hours after the heart has ceased its pulsations. Elsewhere, little enclaves of autonomy go on about their tasks: fingernails and hair continue to grow. Even hemoglobin—and the other components of blood—maintain their enormously complex structure for a few hours after death. The hemoglobin survives still: a molecular lung, a delivery train for oxygen, waiting to be called back to serve. But it, too, is about to lose its structure, as the arteries disgorge into the veins. Gravity, at last, has the final word: the blood pools in the lowest parts of the body. For Toby, that is his back.

Already the nerves and muscles are setting up in *rigor mortis,* and soon thereafter the first obvious signs of decomposition will become manifest. . . .

The cart has arrived. Goldfrank does not await the completion of the task. He has a call in his office from a TV reporter, who has heard of mass freezing in a student group on an outing in the West. Does he have time to do a 10 A.M. interview on hypothermia? Sure. He hangs up and raises his eyes to see the familiar face of Sharda McGuire, who has come in to say hello and thank you. She's now in a leading position on the Emergency Department nursing staff at Long Island College Hospital in Brooklyn.

"And how's it going?"

"It's going great. And every time we get in a tight spot I quote you."

"That's very flattering."

"You stood for the right principles and set a high standard. I'm going to go see some of the others, and I'll say goodbye on the way out."

"Thanks for stopping by."

He will make a circuit of the department, now—Pediatrics, Psychiatry, Holding Area, Emergency Ward, Radiology, Walk-in Clinic and the Adult Emergency Service. He swings out of his office door into the main corridor. There is sound everywhere: talk and bells ringing and the perpetual jangle of police officers' whistles tinkling against handcuffs hooked to their belts, of leg irons scraping the dark-hued concrete floors and paramedics' rings knocking against oxygen canisters. Through the core area, there's palpable energy on all sides, in the faces of the staff members, in their lithe bodies and animated movements. It's almost like a dance company, incessantly moving, from stretcher to examination room. From triage desk to trauma slots. From nurses' station to the radiology developing room. Everywhere there is the same sure-footed celerity, as if each minute of delay is a concession to the Reaper, who is not allowed to enter unchallenged into these sixty rooms, these 22,000 square feet of intense activity.

The atmosphere is similar to that in an opera house or in the stands of a football game—tremendous psychic energy focused on one activity in a way that seems capable for a time of suspending natural laws, transcending physical limitation, defying gravity.

Goldfrank is well satisfied. He need not speak a word to anyone, so expert is the ensemble, so assured their fulfillment of jobs that no one could fully describe to them, but which they well know how to

do. The last scene he sees on his tour of his domain is in Trauma Slot
2. He later wrote about it in *NYU Physician:*

> Far from being discouraging, the challenges we meet here make emergency medicine exciting and fulfilling. Although many of the patients we see have chronic health problems, we can make others well in a matter of minutes. A scene in one of Bellevue's trauma rooms makes this point tellingly:
>
> On the floor lies the disheveled evidence of the remarkable yet typical event that has just occurred. A man's clothing is scattered wildly about: Each of his heavy work boots has been flung to a far corner of the room; his belt was sliced off him and lies on the floor next to some drying drops of blood; thermal underwear, a tattered jacket, and a grimy pair of pants sit in a heap under half-empty bags of saline solution suspended from an IV pole. He is by no means a rich man. In his 60s, he was picked up off the street, ashen, with no pulse or blood pressure, having suffered a cardiac arrest. Now, 20 minutes later, a nurse is calmly gathering used syringes, gauze packages, and lengths of IV tubing—the remnants of hurried activity that has ceased—and the patient is sitting up (against our advice) and talking.
>
> Only by devoting ourselves to solving the unsolvable and treating the untreatable have we been able to succeed in Bellevue's emergency setting. Many of the patients we serve are members of the population with the least hope for healthy lives. Our successes are a great medical victory and a continuing medical challenge for students, nurses, residents and attending physicians. But most important, our success is essential for a city where all can seek care and have hope.

As he makes his way back toward his office, a siren at curbside is heard, and a stretcher with running paramedics propelling it comes bursting through the sliding doors.

"Dr. Goldfrank, to Room 6, please."

# Afterword

Emergency medicine at Bellevue Hospital is much the same and yet very different than it was in 1987 when this book was first published. Approximately 95,000 patients come to the Bellevue Emergency Department each year. Over one and a half million patients have passed through the doors in the last sixteen years.

Extraordinary times have placed enormous demands on the Bellevue Emergency Department. The World Trade Center was attacked twice; the anthrax attacks of 2001 caused the infected and uninfected alike to come to our door. Yet, even as these disasters unfolded, the crush of patients with everyday crises—from stabbings to heart attacks—never slowed. We've had to deal with it all.

The role of emergency medicine in our society is better understood than it was in 1987. The nurses and doctors are far better prepared for the unknown, but the human experiences are much the same. Many of the staff initially described in 1987 are still working at Bellevue, New York University, and the New York City Poison Center. Some have gone on to be leaders at many of the great institutions in our country: New York Hospital; Parkland Hospital, Dallas's premier public hospital; University of Miami, Florida; Harlem Hospital; and many other hospitals.

The physical space we described as constraining our function in this book was dramatically altered to meet twenty-first century needs. We went to four mayors and to the Office of Management and Budget five times. Finally, in 1996, a new emergency department, three times as large as the one described in *Emergency Doctor*, was opened. This new space was designed to offer privacy, limit delays, and, most important, to improve the quality of the healthcare. It is meant to allow for dramatic action for masses of people whenever it is necessary and to protect us against any known or unknown toxin or new microbiologic agent. Many of the architectural changes were made because of the New York City tuberculosis epidemic that was rampant at the time the new emergency department was built. For this reason, there are isolation rooms everywhere in this new emergency department. The quality of the air exchange is very high to prevent dangerous infections, achieving ten to twelve air exchanges per hour with no recirculated air. The value of this standard for our ventilation systems was immediately appreciated and is now recognized as one of the major preventive health standards recommended nationally and internationally to protect emergency staff facing risks such as SARS in 2004.

The new physical space allows for the care of numerous multiply injured patients simultaneously. The links between the emergency ward, the adult and pediatric emergency areas, the urgent care area, and the psychiatric emergency area were dramatically strengthened to meet whatever imaginable or unimaginable problems might arise in an emergency department. The architectural and functional characteristics of the physical space have been tested numerous times because of crashes, epidemics, fears, and events such as the World Trade Center attacks. The highest quality CAT scan was recently donated to Bellevue by the Swiss Re Corporation, the Phillips Corporation, and the Bellevue Hospital Auxiliary. Patients arriving from the street, no matter how critically ill, can now move rapidly through resuscitation and CAT scan examination in a matter of minutes.

As we've learned, New York City is an ideal site for terrorist attacks. In response, we've built one of the most advanced decontamination units in the world, permitting the decontamination of hundreds of people per hour. It stands at the outside entrance to the Emergency Department—ready to protect both patients and staff.

Just as we fought for the development of emergency medicine at Bellevue in the early 1980s, a battle ensued for the development of residency training for young medical students in emergency medicine in the late 1980s. By 1990, we won approval for the development of a residency in emergency medicine, a residency which has grown to sixteen residents for each year of a four-year program. That residency is highly sought after by people committed to caring for the disenfranchised and to working in a socially responsible, rigorous academic environment. Fellowship training programs in pediatric emergency medicine and medical toxicology have been developed in our premier educational environments. This academic organization is a vital asset in the retention and development of faculty. Not only do young doctors need mature faculty for growth and development, but in order for faculty to maintain and advance their own creative skills, they depend upon the presence of a provocative and inquisitive group of inspired residents.

The roles of our staff at the New York City Poison Center have been fully integrated into the structure of the city's emergency response system. Students, residents in training, and faculty from across the country and world come to learn and develop their skills in managing critical toxicologic events and prepare for the unknown at Bellevue. They come seeking to understand the complex issues that make emergency medicine in New York City so exciting.

Our staff's textbook, *Goldfrank's Toxicologic Emergencies*, describes strategies for care of the poisoned patient. We expanded it from a small text in medical toxicology to a text of approximately 2,000 pages. It is written almost exclusively by authors and editors who developed their intellectual foundation at the New York City Poison Center and Bellevue Hospital Center.

The recognition of the importance of Bellevue as a safety-net hospital has been described by many; the newest immigrants, the poorest patients, and New York City's most desperately ill and injured seek care in our hospital. The continuous onslaught of needy patients, often with the most complex medical problems, has created an ideal site for education and service. This enhanced understanding of the role of emergency medicine and its significant contributions to patient care, education, and research led to New York University Medical School's support of an academic department of emergency medicine in the fall of 2003—a department finally recognized as equal to all other departments in the medical school. This accomplishment is vital for the development of the highest quality care, education, and research. Half of America's medical schools have achieved departmental recognition, a status which improves faculty and resident integration into all the institutional functions. The broad university support associated with this decision has enhanced the education for students and all other hospital staff, including nurses and doctors in training. Departmental status brings us social, political, and educational parity, which revolutionizes the intellectual environment, dramatically improving the potential for quality patient care.

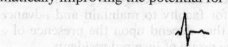

The television show *ER*, the six o'clock news, newspaper headlines, and our roles in people's lives are daily reminders of our importance to New York City. The organization of emergency medicine is well understood as a vital resource permitting our city to maintain its sociopolitical and cultural strength. Just as New Yorkers must prepare for everyday troubles and concerns, they now prepare for terrorism by getting educated about the technical and psychological issues associated with weapons of mass destruction. We have expanded our task as educators so that we can offer risk assessments and analyses that permit survival in a risky environment. We hope to enhance the capacity for individuals to judge the risks we all face with a reduced but appropriate sense of concern.

*Emergency Doctor* describes the life and death experiences in our busy emergency department. We have learned a great deal more. We have added more devoted people to study and care for the human needs of those who come to Bellevue. Many of the best are still here, and they have taken our strategies and skills to new places in need of great, innovative leaders.

The stories in this book are real, representing the devotion of committed people trying to understand problems that have never been previously addressed in an organized manner. The same problems trouble us today in the twenty-first century; reading about our experiences and understanding our efforts remains valuable. Our approach to the care of these people is hopeful. We remain optimistic that we will do a better job today because of what we have learned in the past.

Lewis R. Goldfrank, M.D.
Professor and Chair, Emergency Medicine
New York University School of Medicine
Director, Emergency Medicine Services
Bellevue Hospital/NYU Hospitals/VA Medical Center
Medical Director, New York City Poison Center